Methane and Climate Change

Methane and Climate Change

Edited by
Dave Reay, Pete Smith and André van Amstel

First published by Earthscan in the UK and USA in 2010

For a full list of publications please contact:
Earthscan
2 Park Square, Milton Park, Abingdon, Oxfordshire OX14 4RN
711 Third Avenue, New York, NY 10017

First issued in paperback 2015

Earthscan is an imprint of the Taylor & Francis Group, an informa business

Copyright © Dr David R. Reay, Professor Pete Smith and Dr André van Amstel, 2010
Published by Taylor & Francis.

All rights reserved. No part of this book may be reprinted or reproduced or utilised in any form or by any electronic, mechanical, or other means, now known or hereafter invented, including photocopying and recording, or in any information storage or retrieval system, without permission in writing from the publishers.

Notices

Practitioners and researchers must always rely on their own experience and knowledge in evaluating and using any information, methods, compounds, or experiments described herein. In using such information or methods they should be mindful of their own safety and the safety of others, including parties for whom they have a professional responsibility.

Product or corporate names may be trademarks or registered trademarks, and are used only for identification and explanation without intent to infringe.

ISBN 13: 978-1-138-86693-5 (pbk)
ISBN 13: 978-1-8440-7823-3 (hbk)

Typeset by FiSH Books, Enfield
Cover design by Susanne Harris

A catalogue record for this book is available from the British Library

Library of Congress Cataloging-in-Publication Data

Methane and climate change / edited by Dave Reay, Pete Smith, and André van Amstel.
   p. cm.
  Includes bibliographical references and index.
  ISBN 978-1-84407-823-3 (hardback)
  1. Atmospheric methane—Environmental aspects. 2. Methane—Environmental aspects. 3. Climatic changes. I. Reay, Dave, 1972– II. Smith, Pete, 1965– III. Amstel, André van.
  QC879.85.M48 2010
  551.6—dc22
                        2009052501

Contents

1	Methane Sources and the Global Methane Budget Dave Reay, Pete Smith and André van Amstel	1
2	The Microbiology of Methanogenesis Alfons J. M. Stams and Caroline M. Plugge	14
3	Wetlands Torben R. Christensen	27
4	Geological Methane Giuseppe Etiope	42
5	Termites David E. Bignell	62
6	Vegetation Andy McLeod and Frank Keppler	74
7	Biomass Burning Joel S. Levine	97
8	Rice Cultivation Franz Conen, Keith A. Smith and Kazuyuki Yagi	115
9	Ruminants Francis M. Kelliher and Harry Clark	136
10	Wastewater and Manure Miriam H. A. van Eekert, Hendrik Jan van Dooren, Marjo Lexmond and Grietje Zeeman	151
11	Landfills Jean E. Bogner and Kurt Spokas	175
12	Fossil Energy and Ventilation Air Methane Richard Mattus and Åke Källstrand	201
13	Options for Methane Control André van Amstel	211
14	Summary André van Amstel, Dave Reay and Pete Smith	242
	Contributors	247
	Acronyms and abbreviations	250
	Index	253

To Glyn and Allan

1
Methane Sources and the Global Methane Budget

Dave Reay, Pete Smith and André van Amstel

Introduction

When the late-18th-century Italian physicist Alessandro Volta first identified methane (CH_4) as being the flammable gas in the bubbles that rose from a waterlogged marsh, he could not have guessed how important this gas would prove to be to human society in the centuries that followed. Today CH_4 is used throughout the world as both an industrial and domestic fuel source. Its exploitation has helped drive sustained economic development and it has long provided a lower-carbon energy alternative to coal and oil. As an energy source it remains highly attractive and much sought after; indeed, energy from CH_4 (in the form of natural gas) has played a major role in the UK meeting its Kyoto Protocol commitments to reduce greenhouse gas emissions by allowing a switch from coal firing of power stations to gas firing. However, CH_4 is now increasingly being considered as a leading greenhouse gas in its own right.

At the time that Volta was collecting his bubbles of marsh-gas, CH_4 concentrations in the atmosphere stood at around 750 parts per billion (ppb) and it was almost a century later that John Tyndall demonstrated its powerful infrared absorption properties and its role as a greenhouse gas. Like the other two main anthropogenic greenhouse gases – carbon dioxide (CO_2) and nitrous oxide (N_2O) – the concentration of CH_4 in the atmosphere has increased rapidly since pre-industrial times. Ice core records, and more recently atmospheric samples, show that since the beginning of the industrial era concentrations have more than doubled to their current high of over 1750 ppb (Figure 1.1). This concentration far exceeds the maximum concentration of CH_4 in the atmosphere at any time in the preceding 650,000 years, and is estimated to be resulting in an additional radiative forcing of about 0.5 watts per metre squared compared to the level in 1750.

2 | METHANE AND CLIMATE CHANGE

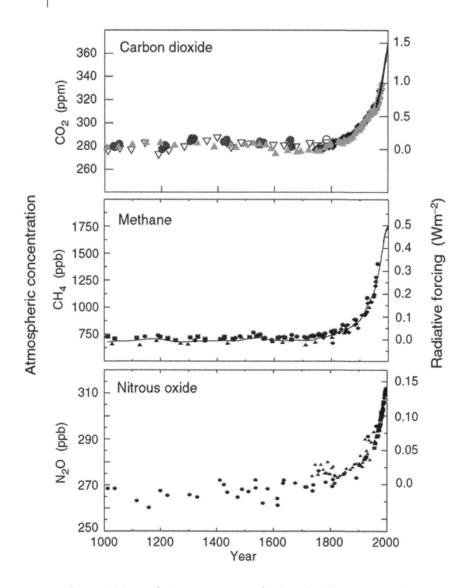

Figure 1.1 Atmospheric concentrations of carbon dioxide, methane and nitrous oxide over the last 1000 years

Source: Reproduced with permission from the Intergovernmental Panel on Climate Change.

Though these concentrations are much lower than those of CO_2 (currently ~386 parts per million – ppm), CH_4 is more effective at absorbing and re-emitting infrared radiation (heat). Indeed, the global warming potential (GWP) of CH_4 on a mass basis is 25 times that of CO_2 over a 100-year time horizon (see Box 1.1).

> **Box 1.1 Global warming potential**
>
> GWP compares the direct climate forcing of different greenhouse gases relative to that of CO_2. The GWP combines the capacity of a gas to absorb infrared radiation, its lifetime in the atmosphere, and the length of time over which its effects on the earth's climate need to be quantified (the time horizon). In the case of CH_4 it is also adjusted to take account of indirect effects via the enhancement of tropospheric ozone, stratospheric water vapour and production of CO_2 resulting from its destruction in the atmosphere. So, as CH_4 has an effective climate-forcing lifetime in the atmosphere of only 12 years, CH_4 has a GWP of 72 over a 20-year time horizon, but a GWP of 25 over a 100-year time horizon and 7.6 over a 500-year time horizon.

The GWP figures for CH_4 provided in Box 1.1 represent the latest provided by the Intergovernmental Panel on Climate Change (IPCC) in its Fourth Assessment Report (IPCC, 2007). However, these values have varied between the different reports based on improved understanding of atmospheric lifetimes and indirect effects. As such, slightly lower values of GWP for CH_4 are often to be found in the literature, with the GWP of 21 for CH_4 over a 100-year time horizon provided in the Second Assessment Report (IPCC, 1995) being very widely used and forming the basis for most national greenhouse gas budget reporting and trading. Throughout this book we therefore default to the 100-year GWP of 21 provided in the Second Assessment Report, unless otherwise stated.

> **Box 1.2 Carbon dioxide equivalents**
>
> When attempting to assess the relative importance of CH_4 fluxes and mitigation strategies the concept of carbon dioxide equivalents (CO_2-eq) is often employed to convert CH_4 fluxes into units directly comparable with CO_2. This is done simply by multiplying the mass of CH_4 by its GWP to give the mass in CO_2-eq. Usually the 100-year time horizon GWP value (for example 21 or 25) is used, so a reduction of 1 tonne of CH_4 would be a reduction of 21 or 25 tonnes CO_2-eq respectively. However, where a shorter time horizon is considered, the GWP increases substantially, a reduction of 1 tonne of CH_4 using a 20-year GWP time horizon yielding a cut of 72 tonnes of CO_2-eq. A 100-year time horizon has become the commonly used benchmark for national greenhouse gas emissions budgets and trading.

During the 1990s and the first few years of the 21st century the growth rate of CH_4 concentrations in the atmosphere slowed to almost zero, but during 2007 and 2008 concentrations increased once again. Recent studies have attributed this new increase to enhanced emissions of CH_4 in the Arctic as a result of high temperatures in 2007, and to greater rainfall in the tropics in 2008 (Dlugokencky et al, 2009). The former response represents a snapshot of a potentially very large positive climate change feedback, with the higher temperatures projected at high latitudes for the 21st century increasing CH_4 emissions from wetlands, permafrosts and CH_4 hydrates. It is to this and the myriad other natural and anthropogenic determinants of CH_4 flux to the atmosphere that this book is directed.

Though CO_2 emissions and their mitigation still dominate much of climate change research and policy, recent years have seen increasing recognition that reducing CH_4 emissions may often provide a more efficient and cost-effective means to mitigate anthropogenic climate change. In addition, current projections of greenhouse gas concentrations and resultant climate forcing in the 21st century require an improved understanding of how natural CH_4 sources will respond to changes in climate. Our aim, therefore, is to provide a synthesis of the current scientific understanding of the major sources of CH_4 around the planet and, where appropriate, to consider how these emissions may change in response to projected climate change. We then focus on the range of CH_4 emission mitigation strategies currently at our disposal and examine the extent to which these can be employed in future decades as part of national and international efforts to address anthropogenic climate change.

The global methane budget

The global CH_4 budget is composed of a wide range of sources (see Table 1.1 and also Figure 4.5 in Chapter 4) balanced by a much smaller number of sinks, any imbalance in these sources and sinks resulting in a change in the atmospheric concentration. There are three main sinks for CH_4 emitted into the atmosphere, with the destruction of CH_4 by hydroxyl (OH) radicals in the troposphere being the dominant one. This process also contributes to the production of peroxy radicals, and it is this that can subsequently lead to the formation of ozone and so induce a further indirect climate-forcing effect of CH_4 in the atmosphere. In addition this reaction with OH radicals reduces the overall oxidizing capacity of the atmosphere – extending the atmospheric lifetime of other CH_4 molecules – and produces CO_2 and water vapour. Each year an estimated 429–507Tg (teragram; 1Tg = 1 million tonnes) of CH_4 are removed from the atmosphere in this way.

The other sinks are much smaller, with ~40Tg CH_4 removed each year by reaction with OH radicals in the stratosphere, and ~30Tg CH_4 removed by CH_4-oxidizing bacteria (methanotrophs) in soils that use the CH_4 as a source of carbon and energy. A relatively small amount of CH_4 is also removed from the atmosphere through chemical oxidation by chlorine in the air and in the surface waters of our seas. Though several chapters within this book refer to the global CH_4 sinks in terms of their impact on net CH_4 emissions – in particular the soil CH_4 sink – the focus of this book is on the sources of

Table 1.1 Global estimates of methane sources and sinks

Natural sources	Methane flux (Tg CH_4 yr^{-1})[a]	Range[b]
Wetlands	174	100–231
Termites	22	20–29
Oceans	10	4–15
Hydrates	5	4–5
Geological	9	4–14
Wild animals	15	15
Wild fires	3	2–5
Total (natural)	238	149–319
Anthropogenic sources		
Coal mining	36	30–46
Gas, oil, industry	61	52–68
Landfills and waste	54	35–69
Ruminants	84	76–92
Rice agriculture	54	31–83
Biomass burning	47	14–88
Total, anthropogenic	336	238–446
Total, all sources (AR4)[c]	574 (582)	387–765
Sinks		
Soils	−30	26–34
Tropospheric OH	−467	428–507
Stratospheric loss	−39	30–45
Total sinks (AR4)	−536 (581)	484–586
Imbalance (AR4)	38 (1)	−199–281

Note: [a] Values represent the mean of those provided in Denman et al (2007, Table 7.6) rounded to the nearest whole number. They draw on eight separate studies, with base years spanning the period 1983–2001. [b] Range is derived from values given in Denman et al (2007, Table 7.6). Values from Chen and Prinn (2006) for anthropogenic sources are not included due to overlaps between source sectors. [c] Values in parentheses denote those provided in the IPCC Fourth Assessment Report (AR4) as the 'best estimates' for the period 2000–2004.
Source: Values derived from Denman et al (2007)

methane, their determinants and their mitigation. Detailed reviews of the key CH_4 sinks and their role globally can be found in Cicerone and Oremland (1988), Crutzen (1991) and Reay et al (2007).

Of the many significant sources of CH_4 on a global scale, both natural and anthropogenic, the bulk have a common basis – that of microbial methanogenesis. Though CH_4 from biomass burning, vegetation and geological or fossil fuel sources may be largely non-microbial in nature, understanding the processes that underpin microbially mediated CH_4 fluxes is central to quantifying and, potentially, reducing emissions from all other major sources. In Chapter 2 'The Microbiology of Methanogenesis', Fons Stams and Caroline Plugge review our current understanding of microbial methanogenesis and the interactions between different microbial communities that result in the bulk of CH_4 emissions to the global atmosphere.

Natural sources

Major natural sources include wetlands, termites and release from onshore and offshore geological sources. Recently, living vegetation has also been suggested as an important natural source of CH_4. Of the globally significant sources of CH_4 to the atmosphere, natural sources are currently outweighed by anthropogenic sources. Together they emit some 582Tg CH_4 each year, with ~200Tg arising from natural sources (Denman et al, 2007). Given the estimated global CH_4 sink of 581Tg per year, the current increase in atmospheric CH_4 concentrations should, therefore, be only 1Tg CH_4 per year. But even with ongoing efforts to reduce anthropogenic emissions, and so arrest the trend of increasing CH_4 concentrations in the atmosphere, future enhancement of natural CH_4 emissions due to climate change threatens to negate some, or all, of these attempts at mitigation.

Wetlands

Wetland CH_4 emissions (excluding rice cultivation) are estimated to total between 100 and 231Tg per year (Denman et al, 2007) – equivalent to around one quarter of global CH_4 emissions. The large range of these estimates reflects the uncertainty as to the underlying determinants of net CH_4 flux in wetland ecosystems, this uncertainty being further compounded by the effects of enhanced global warming. We know that the three key determinants of CH_4 emission from wetlands are temperature (Christensen et al, 2003), water table depth (MacDonald et al, 1998) and substrate availability (Christensen et al, 2003), but the degree of sensitivity of emissions to changes in these determinants remains poorly resolved. Of the three, temperature is most often found to be the dominant factor. For example, over a number of northern wetland sites, soil temperature variations accounted for 84 per cent of the observed variance in CH_4 emissions, with emissions showing a strong positive response to increased temperature (Christensen et al, 2003). The impacts of climate change in the 21st century on these emissions could therefore be substantial. A doubling in CO_2 concentrations (3.4°C warming) is predicted to result in a 78 per cent increase in wetland CH_4 emissions (Shindell et al, 2004). Gedney et al (2004) estimated that this climate feedback mechanism would amplify total anthropogenic radiative forcing by between 3.5 and 5 per cent by 2100. Wetlands then, are critical to the current and future global CH_4 budget.

If we are to successfully mitigate and adapt to climate change in the 21st century it is vital that we improve our understanding of this feedback mechanism. In Chapter 3 'Wetlands', Torben Christensen reviews the scientific basis of wetland CH_4 fluxes, emissions estimates and projected responses to climate change. He concludes that new generations of ecosystem models will allow the incorporation of such feedbacks into climate projections, but that significant gaps remain in our understanding of how tropical wetland CH_4 emissions will respond to changes in precipitation and high-latitude wetland emissions to changes in temperature.

Geological methane

The natural emission of CH_4 from so-called 'geological' sources has often focused on CH_4 hydrates (also called clathrates) – ice-like mixtures of CH_4 and water found in ocean sediments – that are thought to be responsible for between 4 and 5Tg of CH_4 emission to the atmosphere each year. These CH_4 hydrates and the potential of climatic warming to destabilize them has received significant attention in recent years (for example Westbrook et al, 2009). However, in Chapter 4 'Geological Methane', Giuseppe Etiope argues that estimates of emissions from hydrates remain highly speculative and that the overall geological source of CH_4 to the atmosphere is much bigger and more diverse than is commonly reported. He highlights the large losses of CH_4 from seeps, mud volcanoes and geothermal/volcanic areas that cumulatively could be responsible for between 40 and 60Tg CH_4 each year, and on a par with the largest of the anthropogenic CH_4 sources and second only to wetlands as a natural CH_4 source. Etiope reviews the evidence for significant CH_4 losses from onshore and offshore seeps, differentiates between 'natural' emissions of CH_4 associated with coal and oil deposits and those that result from fossil fuel extraction by humans, and assesses how geological CH_4 is classified. Commonly, it is categorized as 'fossil methane' if it is more than 50,000 years old and so radiocarbon free. Finally, he assesses the determinants of these geological CH_4 sources and their dependence on seismic activity, tectonics and magmatism, concluding that the atmospheric greenhouse gas budget of the planet is far from independent of the earth's geophysical processes.

Termites

Though some termite species produce no CH_4 at all and those that do rarely exceed more than half a microgram per termite day, the shear mass of termites globally has given rise to some very large estimates (as much as 310Tg per year) of their contribution to global CH_4 emissions. In Chapter 5 'Termites', David Bignell examines the evidence base and the trend towards smaller global estimates of CH_4 from termites as understanding and measurements have improved. He reviews the differences in CH_4 production rates between species and the reasons for these, assessing the methodologies used for these measurements and highlighting the importance of soil-mediated CH_4 oxidation in determining the net flux of CH_4 from termite colonies. Bignell also examines the issues surrounding the upscaling of CH_4 fluxes and the importance of changes in land use, whether in response to human activity or climate, in determining termite CH_4 emissions. In conclusion he suggests that the importance of termites as a global CH_4 source has probably been overstated in the past, with a more accurate estimate of annual emissions from this source being well below 10Tg and so placing the termite CH_4 source as a relatively minor component of the global CH_4 budget.

The substantial lowering of this estimate suggests that the strength of other CH_4 sources is actually greater than previously thought. As we saw for the geological CH_4 source, much of this 'missing' source can be accounted for by onshore and offshore seeps. However, a novel CH_4 source discovered in 2006 may also help to bridge any global CH_4 budget gap and it is to this source – that of vegetation – that we now turn.

Vegetation

As described in more detail in Chapter 2, the bulk of non-fossil CH_4 emitted to the atmosphere each year is microbially mediated. Methane production (methanogenesis) in wetland soils, for example, involves the microbial mineralization of organic carbon under the anaerobic conditions common to waterlogged soils. In the absence of oxygen, the organic carbon (usually simple carbon compounds such as acetate or CO_2) is used as an alternative terminal electron acceptor and so provides a source of energy for the methanogens. The atmospheric signal of such microbial methanogenesis is such that enhanced CH_4 emission in the tropics during and after periods of heavy rain and waterlogging of soils can be clearly discerned by satellite. An anomaly in this relationship has been observed over some areas of the planet, in particular over the Amazon Basin, where CH_4 concentrations in the atmospheric column appear to be much higher than would be expected given the prevailing conditions in the soil below. Frank Keppler and his team were the first to suggest that such anomalies may be a result of the above-ground vegetation itself producing CH_4 under aerobic conditions, and so adding to the overall concentration of CH_4 in the atmosphere. They provided an initial estimate of the strength of this CH_4 source being between 10 and 40 per cent of global emissions. In Chapter 6 'Vegetation', Andy McLeod and Frank Keppler review the evidence for this novel CH_4 source and the developing postulations as to its mechanism. In particular, they highlight the potential role of UV radiation and reactive oxygen species in determining CH_4 emissions from vegetation. They examine the very limited number of estimates for the global magnitude of this CH_4 source from their own groups and others, and suggest that, even with the large uncertainty that exists in these estimates, the net climate-forcing benefits of the establishment of new forests and enhanced CO_2 sequestration would far exceed any negative effects due to additional CH_4 emissions from the trees.

Biomass burning

Biomass burning accounts for between 14 and 88Tg of CH_4 each year. Methane emissions arising from biomass burning are a result of incomplete combustion and encompass a wide range of sources, including woodlands, peatlands, savanna and agricultural waste. Burning of peat and agricultural waste may produce especially high CH_4 emissions due to the generally high water content and low oxygen availability common to the combustion of these fuel sources. Differentiating between 'natural' and 'anthropogenic' biomass burning is inherently difficult given the coincidence in time and space of many of these events and the difficulty in separating their atmospheric signals. As such, Chapter 7 'Biomass Burning' by Joel Levine, addresses both causes and here is taken as a source that spans both natural and anthropogenic portions of global CH_4 fluxes. He reviews the regional patterns and sources of biomass burning, and the methods used to estimate emissions, suggesting that the bulk of biomass burning and resulting CH_4 emissions globally are anthropogenic in origin. Levine also points to the importance of biomass burning outside of the tropics, highlighting the interaction between reduced precipitation rates due to climate change and enhanced biomass burning in boreal forests. Finally he discusses how changes in climate and land use in the future may alter biomass burning and CH_4 emissions from this source globally. With the changes in climate projected for the 21st century, Levine warns that CH_4 (and CO_2) emissions from biomass burning are likely to increase globally, providing a potentially very important positive feedback mechanism.

Rice cultivation
The frequently waterlogged soils common to many rice fields can provide the anoxic, carbon-rich conditions required for high rates of microbial methanogenesis (see Chapter 2). Most rice paddies are submerged for around a third of the time, though practices vary widely around the world based on rice variety, culture and water availability. As with termites, the estimate of CH_4 from rice cultivation has seen a trend of downward revision in recent years as understanding of its determinants, field measurement and modelling have improved estimates. Nevertheless, with a projected 9 billion people to feed globally by 2050, rice cultivation is likely to comprise a significant proportion of the world's agricultural land and, without intervention, to remain as an important source of CH_4 globally.

In Chapter 8 'Rice Cultivation', Franz Conen, Keith Smith and Kazuyuki Yagi review the estimates of CH_4 emissions from this source, with recent estimates generally being between 25 and 50Tg CH_4 per year. They underline the importance of increasing demand on future emissions and provide an overview of the microbially mediated production and oxidation of CH_4 in rice paddy soils. Various cultivation strategies and locations are then examined and their relative importance in terms of CH_4 emissions assessed. Continuously flooded/irrigated rice emerges as the strongest CH_4 source per unit area, with drought-prone, rain-fed rice having much lower or sometimes zero CH_4 emissions per unit area. Conen et al then examine the ways in which CH_4 emissions per unit yield can be altered through changes in land and water management, rice variety and application of fertilizers and residues. Finally, they review the global assessments of CH_4 emissions from rice cultivation, the methodologies employed and the potential for reducing emissions from this source in the future.

Ruminants

Ruminant livestock, such as cattle, sheep, goats and deer, primarily produce CH_4 as a by-product of feed fermentation in their rumens. The bulk (>90 per cent) of the CH_4 is then emitted through belching – some dairy cattle emitting several hundred litres of CH_4 in this way each day. In 2005, CH_4 emissions from ruminant livestock were estimated to be around 72Tg per year. As with rice agriculture, CH_4 emissions from ruminant livestock are highly dependent on demand pressures and with a global trend of increasing consumption of both meat and dairy products emissions are expected to rise to around 100Tg CH_4 per year by 2010. In Chapter 9 'Ruminants', Frank Kelliher and Harry Clark review the estimates of global and national CH_4 emissions from this source, the ways in which they are calculated and the uncertainties inherent in such estimates. They then examine the role of feedstock type and quality in determining ruminant CH_4 emissions and go on to describe the various strategies available to reduce these emissions in the short, medium and longer term. Such strategies include the reduction in demand for ruminant meat and dairy products, changes in livestock diet and production efficiency, and the use of vaccines.

Manure and wastewater

Microbial methanogenesis in livestock manure and wastewater can produce significant amounts of CH_4 due to the high availability of substrates (acetate, CO_2 and H_2) and the anoxic conditions that tend to prevail. Globally, agricultural waste and wastewater are together responsible for the emissions of between 14 and 25Tg CH_4 each year. As with direct CH_4 emission from ruminants, manure-derived CH_4 emissions are coupled to livestock demand pressures, with increases in demand for meat and dairy products tending to increase manure production and related CH_4 emissions. Similarly, the rapidly increasing human population is itself increasing pressure on sewage and wastewater treatment capacity around the world and has the potential to greatly enhance CH_4 emissions from this source. Emissions from livestock manure are often included in total livestock CH_4 emissions source estimates, but in considering mitigation is it useful to separate these sources. In Chapter 10 'Manure and Wastewater', Miriam van Eekert, Hendrik Jan van Dooren, Marjo Lexmond and Grietje Zeeman review the key processes responsible for manure and wastewater CH_4 emissions and the methods used to estimate them. They then focus on a range of established and putative mitigation options, including anaerobic digestion, manure and sludge handling, and livestock diet manipulations. For both manure and wastewater, anaerobic digestion is shown to have great potential through the effective interception of CH_4 and its use as an alternative energy source with which to replace conventional fossil fuel-derived energy sources.

Landfills

Landfill sites can provide ideal conditions for methanogenesis, with the plentiful supply of substrate held under anoxic conditions making some landfill sites very powerful point sources of CH_4 production and, if uncontrolled, emission. As sewage sludge and agricultural waste may also be incorporated into landfills, the CH_4 source strength for these two categories may overlap somewhat. However, for much of the world, it is CH_4 derived from anaerobic decomposition of municipal rather than agricultural waste that dominates. Early estimates of global CH_4 emissions from landfill were of the order of 70Tg per year, but successful implementation of mitigation strategies has seen a reduction in emissions from this source in many developed nations. In Chapter 11 'Landfills', Jean Bogner and Kurt Spokas review the landfill CH_4 source, its determinants and its measurement. They examine and update progress on mitigating landfill CH_4 emissions using CH_4 collection and the enhancement of CH_4 oxidation rates in landfill cover soils. They conclude that, although landfill CH_4 emissions constitute only a small part (~1.3 per cent) of total anthropogenic GHG emissions globally, improved CH_4 recovery and cover soil oxidation has the potential to further reduce emissions from this source, with the former providing a useful energy source with which to offset fossil fuel use.

Fossil energy

Much of the ~75Tg of CH_4 emission attributable to fossil energy use each year is derived from release during fossil fuel extraction, storage, processing and transportation. Some CH_4 is also emitted during incomplete fossil fuel combustion. At 30–46Tg of CH_4 per year, coal mining and extraction constitutes one of the largest individual source activities of anthropogenic CH_4. The CH_4 is formed as part of the geological process of coal formation and large deposits can then remain trapped within or close to the coal seam until released by mining operations. Methane concentrations between 5 and 15 per cent in the air of coal mines represent an explosion hazard and so ventilation is commonly employed to rid deep mines of this CH_4. In Chapter 12 'Fossil Energy and Ventilation Air Methane', Richard Mattus and Åke Källstrand briefly review the sources of fossil energy CH_4 before focusing on strategies to reduce CH_4 emissions from coal mine ventilation air.

Finally, in Chapter 13 'Options for Methane Control', André van Amstel identifies and reviews a suite of 27 different CH_4 emission mitigation strategies that are proven and that can be deployed immediately. He examines their relative costs and effectiveness, both regionally and globally, between 1990 and 2100 and concludes that many of these strategies can be successfully implemented at little or no net cost in the coming decades.

Conclusion

In introducing the chapters that follow, we have given a brief insight into the complex array of processes that they represent and the latest thinking on how to estimate emissions. We have also provided an indication of the ways in which CH_4 emissions may respond to a changing climate and expanding human population in the 21st century and, most importantly, how emissions from some important sources could be radically reduced through established and emerging mitigation strategies. The potential role of CH_4 mitigation in tackling anthropogenic climate change in the coming decades is immense. For many sectors it represents the so-called 'low-hanging fruit' for mitigation policy, especially in the short and medium term. If we are to avoid 'dangerous climate change' then understanding CH_4 emissions and radically reducing them must be part of any global response.

Acknowledgements

Pete Smith is a Royal Society-Wolfson Research Merit Award holder. Dave Reay's research on methane fluxes is supported by the Natural Environment Research Council, UK.

References

Chen, Y.-H., and Prinn, R. G. (2006) 'Estimation of atmospheric methane emissions between 1996 and 2001 using a three-dimensional global chemical transport model', *Journal of Geophysical Research*, vol 111, D10307, doi:10.1029/2005JD006058

Christensen, T. R., Ekberg, A., Ström, L., Mastepanov, M., Panikov, N., Öquist, M., Svensson, B. H., Nykänen, H., Martikainen, P. J. and Oskarsson, H. (2003) 'Factors controlling large scale variations in methane emissions from wetlands', *Geophysical Research Letters*, vol 30, pp1414, doi:10.1029/2002GL016848

Cicerone, R. J. and Oremland, R. S. (1988) 'Biogeochemical aspects of atmospheric methane', *Global Biogeochemical Cycles*, vol 2, pp299–327

Crutzen, P. (1991) 'Methane's sinks and sources', *Nature*, vol 350, pp380–381

Denman, K. L., Chidthaisong, A., Ciais, P., Cox, P. M., Dickinson, R. E., Hauglustaine, D., Heinze, C., Holland, E., Jacob, D., Lohmann, U., Ramachandran, S., da Silvas Dias, P. L., Wofsy, S. C. and Zhang, X. (2007) 'Couplings between changes in the climate system and biochemistry', in S. Solomon, D. Qin, M. Manning, Z. Chen, M. Marquis, K. B. Averyt, M. Tignor and H. L. Miller (eds) *Climate Change 2007: The Physical Science Basis*, Cambridge University Press, Cambridge, pp499–587

Dlugokencky, E. J., Bruhwiler, L., White, J. W. C., Emmons, L. K., Novelli, P. C., Montzka, S. A., Masarie, K. A., Lang, P. M., Crotwell, A. M., Miller, J. B. and Gatti, L. V. (2009) 'Observational constraints on recent increases in the atmospheric CH_4 burden', *Geophysical Research Letters*, vol 36, L18803, doi:10.1029/2009GL039780

Gedney N., Cox, P. M. and Huntingford, C. (2004) 'Climate feedback from wetland methane emissions', *Geophysical Research Letters*, vol 31, L20503

IPCC (Intergovernmental Panel on Climate Change) (1995) *Contribution of Working Group I to the Second Assessment of the Intergovernmental Panel on Climate Change*, J. T. Houghton, L. G. Meira Filho, B. A. Callender, N. Harris, A. Kattenberg and K. Maskell (eds), Cambridge University Press, Cambridge, UK

IPCC (2007) *Climate Change 2007: The Physical Science Basis, Contribution of Working Group I to the Fourth Assessment Report of the Intergovernmental Panel on Climate Change*, S. Solomon, D. Qin, M. Manning, Z. Chen, M. Marquis, K. B. Averyt, M. Tignor and H. L. Miller (eds), Cambridge University Press, Cambridge, UK and New York, NY

MacDonald, J. A., Fowler, D., Hargreaves, K. J., Skiba, U., Leith, I. D. and Murray, M. B. (1998) 'Methane emission rates from a northern wetland; response to temperature, water table and transport', *Atmospheric Environment*, vol 32, pp3219–3227

Reay, D., Hewitt, C. N., Smith, K. and Grace, J. (2007) *Greenhouse Gas Sinks*, CABI, Wallingford, UK

Shindell, D. T., Walter, B. P. and Faluvegi, G. (2004) 'Impacts of climate change on methane emissions from wetlands', *Geophysical Research Letters*, vol 31, L21202

Westbrook, G. K. Thatcher, K. E., Rohling, E. J., Piotrowski, A. M., Pälike, H., Osborne, A. H., Nisbet, E. G., Minshull, T. A., Lanoisellé, M., James, R. H., Huhnerbach, V., Green, D., Fisher, R. E., Crocker, A. J., Chabert, A., Bolton, C., Beszczynska-Möller, A., Berndt, C. and Aquilina, A. (2009) 'Escape of methane gas from the seabed along the West Spitsbergen continental margin', *Geophysical Research Letters*, vol 36, L15608, doi:10.1029/2009GL039191

2
The Microbiology of Methanogenesis

Alfons J. M. Stams and Caroline M. Plugge

Introduction

Anaerobic decomposition of organic matter to CH_4 and CO_2 is a complex microbial process that requires the syntrophic cooperation of anaerobic bacteria and methanogenic archaea. In short, biopolymers are hydrolysed and fermented, and the products formed are funnelled to compounds that are used by methanogens (Figure 2.1). Polysaccharides yield sugars, while proteins are converted to mixtures of amino acids and small peptides. Lipids are degraded to glycerol and long chain fatty acids. The general pattern of anaerobic mineralization of organic matter is that fermentative bacteria degrade easily degradable compounds such as sugars, amino acids, purines, pyrimidines and glycerol to a variety of fatty acids, CO_2, formate and hydrogen. Then, acetogenic bacteria degrade (higher) fatty acids to acetate, CO_2 and hydrogen, and formate. These compounds are then the substrates for methanogens (Schink and Stams, 2006; Stams and Plugge, 2009). These processes take place simultaneously, but because of the different growth rates and activities of the microorganisms involved, the different processes are partially uncoupled, resulting in the accumulation of organic acids. Methanogenesis is a dynamic process in the sense that methanogens strongly influence the metabolism of fermentative and acetogenic bacteria by means of interspecies hydrogen transfer (Schink and Stams, 2006; Stams and Plugge, 2009). However, most importantly, decomposition is always directed to CH_4 and CO_2, ammonium and minor amounts of hydrogen sulphide, provided that all functional groups of microorganisms are present.

Methanogenic archaea

Methanogens are microorganisms that produce CH_4. They are strictly anaerobic and belong to the archaea. They are a phylogenetically diverse group, classified into five established orders: *Methanobacteriales*,

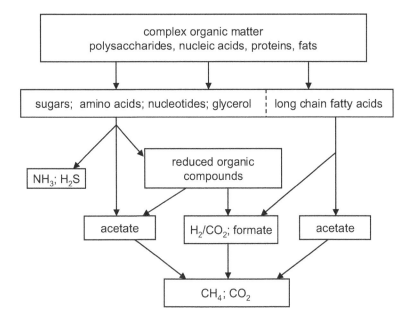

Figure 2.1 General scheme of the anaerobic digestion process

Methanococcales, *Methanomicrobiales*, *Methanosarcinales* and *Methanopyrales*, and further divided into 10 families and 31 genera (Liu and Whitman, 2008). Methanogens have been isolated from a wide variety of anaerobic environments, including marine and freshwater sediments, human and animal gastrointestinal tracts, anaerobic digestors and landfills and geothermal and polar systems. The habitats of methanogens differ largely in temperature, salinity and pH. Although methanogens are very diverse phylogenetically, they are physiologically very restricted. They can grow on a number of simple organic molecules and hydrogen (Table 2.1). Methanogenic substrates can be divided into three major types (Liu and Whitman, 2008; Thauer et al, 2008):

1. H_2 (hydrogen)/CO_2, formate and carbon monoxide (CO);
2. methanol and methylated compounds;
3. acetate.

The general pathway of methanogenesis is presented in Figure 2.2. More complex organic substances are not degraded by methanogens, though a few species can use ethanol and pyruvate.

Most methanogens can reduce CO_2 to CH_4 with H_2 as electron donor. Many of these hydrogenotrophic methanogens can also use formate or CO as electron donor. In hydrogenotrophic methanogenesis CO_2 is reduced successively to CH_4 through formyl, methylene and methyl levels. The C_1 moiety is carried by special coenzymes, methanofuran (MFR),

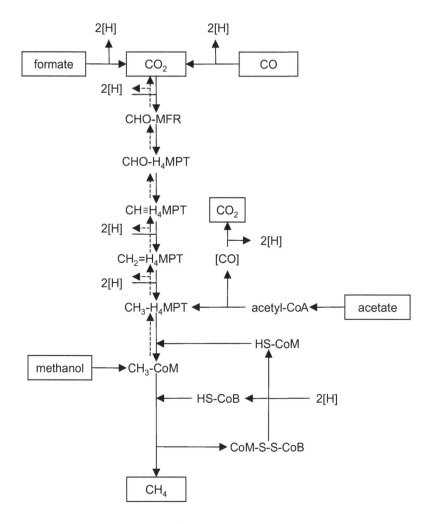

Figure 2.2 Combined pathways of methanogenesis from H_2/CO_2 (CO, formate), methanol and acetate

Note: MFR = methanofuran; H_4MPT = tetrahydromethanopterin; HS-CoM = coenzyme M; HS-CoB = coenzyme B.

tetrahydromethanopterin (H_4MPT) and coenzyme M (HS-CoM). In the first step, CO_2 binds to MFR and is reduced to the formyl level. In this reduction step, ferredoxin (Fd), which is reduced with H_2, is the electron donor. The formation of formyl-MFR is an endergonic conversion, which is driven by an ion gradient. The formyl group is then transferred to H_4MPT, forming formyl-H_4MPT. The formyl group is dehydrated to methenyl group, which is subsequently reduced to methylene-H_4MPT and then to methyl-H_4MPT. In these two reduction steps, reduced factor F_{420} ($F_{420}H_2$) is the electron donor. The methyl group is transferred to CoM, forming methyl-CoM. In the final

Table 2.1 Energy conserving reactions of methanogenic archaea

		$\Delta G^{\circ\prime}$ [kJ/CH$_4$]
$4H_2 + CO_2$	$\rightarrow CH_4 + 2H_2O$	−131
4 formate⁻ + 4H⁺	$\rightarrow CH_4 + 3CO_2 + 2H_2O$	−145
$4CO + 2H_2O$	$\rightarrow CH_4 + 3CO_2$	−211
Acetate⁻ + H⁺	$\rightarrow CH_4 + CO_2$	−36
4 methanol	$\rightarrow 3CH_4 + CO_2 + 2H_2O$	−106
H_2 + methanol	$\rightarrow CH_4 + H_2O$	−113

Source: Gibbs free energy changes from Thauer et al (1977)

reduction methyl-CoM is reduced to CH$_4$ by methyl coenzyme M reductase. Methyl-CoM reductase is the key enzyme in methanogenesis. Coenzyme B (HS-CoB) is the electron donor in this reduction, after oxidation a heterodisulphide is formed with HS-CoM (CoM-S-S-CoB). The heterodisulphide is reduced to HS-CoB and HS-CoM. The methyl transfer from H$_4$MPT to HS-CoM and the reduction of CoM-S-S-CoB are the steps in which energy conservation takes place (Liu and Whitman 2008; Thauer et al, 2008).

The second substrate type is methyl-containing compounds, including methanol, methylated amines and methylated sulphides. Methanogens of the order *Methanosarcinales* and *Methanosphaera* convert methylated compounds. The methyl group is first transferred to a corrinoid protein and then to HS-CoM, involving methyltransferases. Methyl-CoM enters the methanogenesis pathway and is reduced to CH$_4$. The electrons required for this reduction are obtained from the oxidation of methyl-CoM to CO$_2$, which proceeds via a reverse of the described hydrogenotrophic methanogenesis pathway. Methylotrophic growth of some methanogens (*Methanomicrococcus blatticola* and *Methanosphaera* spp) is H$_2$-dependent (Sprenger et al, 2005; Liu and Whitman 2008).

Acetate is the major intermediate in the anaerobic food chain; about two thirds of biologically generated CH$_4$ is derived from acetate. Surprisingly, only two genera are known to use acetate for methanogenesis: *Methanosarcina* and *Methanosaeta* (Jetten et al, 1992). Acetate is split into CH$_4$ and CO$_2$ after activation to acetyl-CoA and then split into methyl-CoM and CO. Methyl-CoM is reduced to CH$_4$, while CO is oxidized to CO$_2$. *Methanosarcina* is a versatile methanogen. It shows fast growth on methanol and methylamine, but growth on acetate is slower. Many species also utilize H$_2$/CO$_2$ but not formate. *Methanosaeta* is a specialist that uses only acetate. *Methanosaeta* can use acetate at concentrations as low as 5–20 micromolar (µM), while *Methanosarcina* requires a minimum concentration of about 1 millimolar (mM). The difference in acetate affinity is due to the different acetate activation mechanism. *Methanosarcina* uses the low-affinity acetate kinase/phosphotransacetylase system to form acetyl-CoA, while *Methanosaeta* uses the high-affinity acetyl-CoA synthase, which requires a higher energy investment than acetate kinase (Jetten et al, 1992).

Fermentation of sugars

Sugars can be fermented by a variety of different microorganisms via different pathways leading to typical end products (Gottschalk, 1985). Generally, C6 sugars are degraded by glycolysis or the Entner-Doudoroff pathway to pyruvate, while C5 sugars are converted via a combined pentose pathway and the glycolytic or the Entner-Doudoroff pathway to pyruvate. Conversion of sugars to pyruvate results in the reduction of nicotinamide adenine dinucleotide (NAD^+) to form NADH. The further metabolism of pyruvate depends on the biochemical mechanism by which sugar-fermenting microorganisms dispose reducing equivalents. Facultative aerobic microorganisms perform a mixed acid fermentation, resulting in the formation of ethanol, lactate, succinate, formate and butanediol. These bacteria produce formate by pyruvate:formate lyase. Formate is split to H_2 and CO_2 by formate:hydrogen lyase. Alcoholic fermentation, lactic acid fermentation, homoacetogenic fermentation, propionic acid and butyric acid fermentation are examples of specific fermentations carried out by anaerobic microorganisms. The combined occurrence of all these fermentations by mixed microbial communities will yield a variety of products. Except for propionate, butyrate and long chain fatty acids, the reduced compounds are fermented further by specific microorganisms.

The utilization of hydrogen by methanogens affects the metabolism of fermentative microorganisms that have the ability to use protons as electron sinks. A typical example is the fermentation of glucose by *Ruminococcus albus* (Ianotti et al, 1973). In pure culture it forms acetate, CO_2, hydrogen and ethanol, while in the coculture ethanol is not formed (Figure 2.3).

R. albus degrades glucose via a glycolytic pathway, leading to the formation of NADH and reduced Fd. The oxidation of reduced Fd is energetically easy to couple to hydrogen formation, while H_2 formation from NADH is only possible at a low hydrogen partial pressure:

$$2Fd_{(red)} + 2H^+ \rightarrow 2Fd_{(ox)} + H_2 \qquad \Delta G^{o\prime} = +3.1 kJ/mol$$

$$NADH + H^+ \rightarrow NAD^+ + H_2 \qquad \Delta G^{o\prime} = +18.1 kJ/mol$$

At a partial pressure of hydrogen of 1 pascal (Pa), created by methanogens, the $\Delta G^{o\prime}$ of the two conversions is around −26 and −11 kilojoule per mol (kJ/mol), respectively. During sugar fermentation by *R. albus* in pure culture, hydrogen accumulates, and due to this, NADH oxidation to proton reduction is no longer possible. As an alternative, acetyl-CoA or acetaldehyde is used as electron sink to form ethanol. In coculture with a hydrogen scavenger,

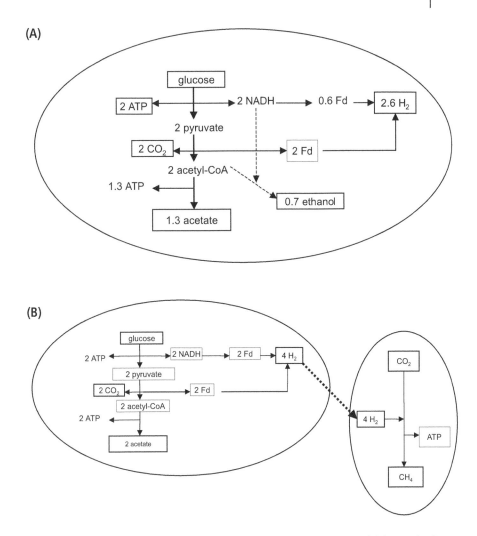

Figure 2.3 Sugar fermentation by *Ruminococcus albus* in (A) pure and (B) mixed culture

Note: ATP = adenosine triphosphate

hydrogen is removed efficiently, and ethanol is not produced. Similar effects have also been observed with other sugar-fermenting microorganisms forming ethanol, lactate, succinate, butyrate or propionate (Stams, 1994; Schink and Stams, 2006). Some sugar-fermenting bacteria can only convert sugars to acetate, H_2, CO_2 and formate. These bacteria strictly depend for growth on hydrogen removal by methanogens (Krumholz and Bryant, 1986; Müller et al, 2008).

Mineralization of amino acids

Proteins are composed of about 20 structurally different amino acids, which require different biochemical pathways for degradation. Anaerobic degradation of amino acids by mixed methanogenic communities is more complex than described above for sugars. The degradation involves oxidation and reduction reactions of one or more amino acids, termed the Stickland reaction. This is a well-known mechanism by which *Clostridia* degrade amino acids. In the Stickland reaction, an oxidative reaction with one amino acid is coupled to the reductive degradation of another. A classical Stickland mixture is alanine plus glycine, in which the selenium-dependent glycine reductase plays a crucial role (Andreesen, 1994, 2004), but many other couples have been described (Barker, 1981). Amino acid degradation is affected by methanogens. Nagase and Matsuo (1982) observed that in mixed methanogenic communities the degradation of alanine, valine and leucine was repressed by inhibition of methanogens, while Nanninga and Gottschal (1985) could stimulate the degradation of these amino acids by the addition of hydrogen-utilizing anaerobes. Several anaerobic bacteria grow syntrophically on amino acids in coculture with methanogens (McInerney, 1988; Stams, 1994). The initial step in the degradation of alanine, valine, leucine and isoleucine is an NAD^+- or nicotinamide adenine dinucleotide phosphate ($NADP^+$)-dependent deamination to the corresponding keto acid. The $\Delta G^{0\prime}$ of this deamination when coupled to hydrogen formation is about $+60 kJ/mol$. Thus, methanogens are needed to pull this reaction. The keto acid is converted further to a fatty acid, a reaction that energetically is much more favourable ($\Delta G^{0\prime} \sim -50 kJ/mol$).

Mineralization of nucleic acids

The hydrolysis of ribonucleic acid (RNA) and deoxyribonucleic acid (DNA) results in the formation of a C5 sugar, ribose and deoxyribose, respectively, and purines and pyrimidines. Purines and pyrimidines are easily fermented under anaerobic conditions (Gottschalk, 1986). Guanine, adenine and a number of other heterocyclic compounds such as hypoxanthine, ureate and xanthine are fermented by *Clostridium* strains (Berry et al, 1987). Uracil is converted to β-alanine, CO_2 and ammonium by *C. glycolicum* and *C. uracillum*, while *C. sporogenes* can transform cytosine and thymine (Hilton et al, 1975). Up to now, no research has been performed on the effect of methanogens on the degradation of nucleic acids.

Mineralization of fats

Fats are cleaved into glycerol and long chain fatty acids. Glycerol is easily fermented. In a few enzymatic steps glycerol is converted to intermediates of a glycolytic pathway. Long chain fatty acids are degraded via the so-called ß-cleavage (McInerney et al, 2008; Sousa et al, 2009). In a cascade of reactions acetyl groups are cleaved off, yielding acetate and hydrogen. Long chain fatty acids are first activated to a HS-CoA derivative and subsequently acetyl-CoA units are cleaved off. Oxidation of long chain fatty acids to acetate and hydrogen is energetically difficult and can only occur by syntrophic communities. Bacteria belonging to the genera *Syntrophomonas* and *Syntrophus* are known for their ability to grow on long chain fatty acids in syntrophy with methanogens. Bacteria that degrade long chain fatty acids are also able to grow with butyrate.

Syntrophic degradation of propionate and butyrate

Propionate and butyrate are important products of polysaccharide and protein fermentation. These fatty acids are degraded by obligate syntrophic consortia of bacteria and methanogens. Methanogens are needed to remove the products acetate and hydrogen. Only at a low hydrogen partial pressure is the degradation of these compounds energetically feasible (Table 2.2).

Table 2.2 Reactions involved in syntrophic propionate and butyrate degradation

	$\Delta G^{\circ\prime}$ [kJ/mol]	$\Delta G'$ [kJ/mol]
Acetogenic reaction		
Propionate$^-$ + 2H$_2$O → Acetate$^-$ + CO$_2$ + 3H$_2$	+72	−21
Butyrate$^-$ + 2H$_2$O → 2 Acetate$^-$ + H$^+$ + 2H$_2$	+48	−22
Methanogenic reaction		
Acetate$^-$ + H$^+$ + H$_2$O → CO$_2$ + CH$_4$	−36	−36
4H$_2$ + CO$_2$ → CH$_4$ + 2H$_2$O	−131	−36
Overall reaction		
Propionate$^-$ + H$^+$ + 0.5H$_2$O → 1.75CH$_4$ + 1.25CO$_2$	−62	−84
Butyrate$^-$ + H$^+$ + H$_2$O → 2.5CH$_4$ + 1.5CO$_2$	−90	−112

Note: $\Delta G'$ values are calculated for P_{H2} = 1Pa, P_{CH4} = P_{CO2} = 10^{4}Pa and 10mM for other compounds.

Boone and Bryant (1980) described *Syntrophobacter wolinii*, a bacterium that grows in syntrophic cooperation with methanogens. Since then, several other bacteria have been described that grow in a similar way. These include Gram-negative bacteria (*Syntrophobacter* and *Smithella*) and Gram-positive bacteria (*Pelotomaculum* and *Desulfotomaculum*). Phylogenetically both groups are related to sulphate-reducing bacteria, and some indeed can grow by sulphate reduction (*Syntrophobacter* and *Desulfotomaculum* spp).

Two pathways for propionate metabolism are known, the methylmalonyl-CoA pathway and a dismutation pathway. In the latter pathway two propionate molecules are converted to acetate and butyrate, the butyrate being degraded to acetate and hydrogen as described below. Thus far, this pathway is only found in *Smithella propionica* (Liu et al, 1999; de Bok et al, 2001). The methylmalonyl-CoA pathway is found in the other syntrophic propionate-oxidizing bacteria (McInerney et al, 2008). In *S. fumaroxidans* the activation of propionate to propionyl-CoA occurs by transfer of a HS-CoA group from acetyl-CoA, and the synthesis of methylmalonyl-CoA by transfer of a carboxyl group from oxaloacetate by a transcarboxylase. Methylmalonyl-CoA is rearranged to form succinyl-CoA, which is converted to succinate, generating adenosine triphosphate (ATP) from adenosine diphosphate (ADP). Succinate is oxidized to fumarate, which is then hydrated to malate and oxidized to oxaloacetate. By decarboxylation pyruvate is formed, which is oxidized in a HS-CoA-dependent decarboxylation to acetyl-CoA and finally to acetate. The pathway (Figure 2.4a) predicts that one ATP is made by substrate-level phosphorylation per propionate degraded.

However, the production of hydrogen (or formate) from electrons derived from the oxidation of succinate is energetically difficult. In this oxidation flavin adenine dinucleotide (FAD) is involved as redox mediator. The energetic of $FADH_2$ conversion to FAD and H_2 is highly endergonic:

$$FADH_2 \rightarrow FAD + H_2 \qquad \Delta G^{o\prime} = +37.4 kJ/mol$$

Even at a partial pressure of hydrogen of 1Pa this reaction is not feasible. The Gibbs free energy change is still positive. Molar growth yield studies with *Syntrophobacter fumaroxidans* have indicated that two thirds of an ATP is required to push succinate oxidation to fumarate by means of a reversed electron transfer process. Thus, one third ATP per propionate is available for the bacterium to grow (van Kuyk et al, 1998).

McInerney et al (1981) describe *Syntrophomonas wolfei*, a bacterium that degraded butyrate and some other short chain fatty acids in syntrophy with methanogens. Since then several other butyrate-oxidizing bacteria have been described. Mesophilic bacteria capable of syntrophic butyrate metabolism are all *Syntrophomonas* species (McInerney et al, 2008).

Butyrate is oxidized via β-oxidation (Figure 2.4b). Butyrate is activated to butyryl-CoA by the transfer of the HS-CoA group from acetyl-CoA. Butyryl-CoA is then converted to two acetyl-CoA, involving two oxidation steps, the conversion of butyryl-CoA to crotonyl-CoA (FAD dependent), and oxidation of 3-hydroxybutyryl-CoA to acetoacetyl-CoA (NAD^+ dependent). In a similar fashion as described above reversed electron transfer is required to couple $FADH_2$ oxidation to hydrogen formation.

Conclusions

Methanogenesis is a process in which different types of microorganisms interact to degrade organic matter to CO_2 and CH_4. Methanogens affect the

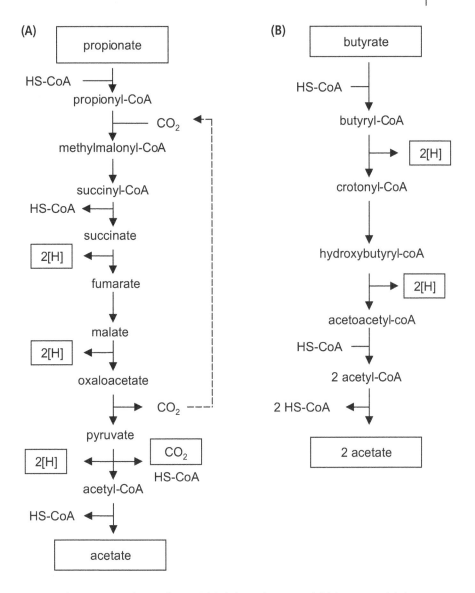

Figure 2.4 Pathway of syntrophic (A) propionate and (B) butyrate oxidation

metabolism of fermentative bacteria and acetogenic bacteria. Some of these obligate and facultative interactions have been described here, but in reality the anaerobic food chain is even more complex. There is a group of acetogenic bacteria that is able to reduce CO_2 with hydrogen to acetate. These acetogens compete with methanogens for the available hydrogen, and they can degrade sugars to solely acetate (Drake et al, 2008). However, in the presence of hydrogen-consuming methanogens they degrade sugars to acetate, hydrogen and CO_2 (Winter and Wolfe, 1980). Some anaerobic bacteria are able to use acetate as a terminal electron acceptor and form propionate or butyrate as a reduced end product (Bornstein and Barker, 1948; Laanbroek et al, 1982). In addition, the methanogenic substrates acetate, methanol and formate can also be degraded by syntrophic communities of bacteria and methanogenic archaea (Schnürer et al, 1996; Dolfing et al, 2008).

The anaerobic food chain changes completely when inorganic electron acceptors like sulphate enter the methanogenic zone. In that case sulphate-reducing bacteria will out-compete methanogenic archaea for hydrogen, formate and acetate, and syntrophic methanogenic communities for substrates like propionate and butyrate (Muyzer and Stams, 2008). Interestingly, sulphate reducers can also grow without sulphate and in that case they grow in syntrophic association with methanogens. Thus, sulphate reducers may compete with methanogens and grow in syntrophy with methanogens, depending on the prevailing environmental conditions (Muyzer and Stams, 2008).

Acknowledgements

Our research was supported by grants of the divisions Chemical Sciences (CW) and Earth and Life Sciences (ALW) and the Technology Foundation (STW) of the Netherlands Science Foundation (NWO) and the Darwin Center for Biogeology.

References

Andreesen, J. R. (1994) 'Glycine metabolism in anaerobes', *Antonie van Leeuwenhoek*, vol 66, pp223–237

Andreesen, J. R. (2004) 'Glycine reductase mechanism', *Current Opinion in Chemical Biology*, vol 8, pp454–461

Barker, H. A. (1981) 'Amino acid degradation by anaerobic bacteria', *Annual Reviews of Biochemistry*, vol 50, pp23–40

Berry, D. F., Francis, A. J. and Bollag, J.-M. (1987) 'Microbial metabolism of homocyclic and heterocyclic aromatic compounds under anaerobic conditions', *Microbiological Reviews*, vol 51, pp43–59

Boone, D. R. and Bryant, M. P. (1980) 'Propionate-degrading bacterium, *Syntrophobacter wolinii* sp. nov. gen. nov., from methanogenic ecosystems', *Applied and Environmental Microbiology*, vol 40, pp626–632

Bornstein, B. T. and Barker, H. A. (1948) 'The energy metabolism of *Clostridium kluyveri* and the synthesis of fatty acids', *Journal of Biological Chemistry*, vol 172, pp659–669

de Bok, F. A. M., Stams, A. J. M., Dijkema, C. and Boone, D. R. (2001) 'Pathway of propionate oxidation by a syntrophic culture of *Smithella propionica* and *Methanospirillum hungatei*', *Applied and Environmental Microbiology*, vol 67, pp1800–1804

Dolfing, J., Jiang, B., Henstra, A. M., Stams, A. J. M. and Plugge, C. M. (2008) 'Syntrophic growth on formate: a new microbial niche in anoxic environments', *Applied and Environmental Microbiology*, vol 74, pp6126–6131

Drake, H. L., Gössner, A. S. and Daniel, S. L. (2008) 'Old acetogens, new light',

Annals of the New York Academy of Sciences, vol 1125, pp100–128

Gottschalk, G. (1985) '*Bacterial Metabolism*', 2nd Edition, Springer Verlag, New York

Hilton, M. G., Mead, G. C. and Elsden, S. R. (1975) 'The metabolism of pyrimidines by proteolytic clostridia', *Archives of Microbiology*, vol 102, pp145–149

Ianotti, E. L., Kafkewitz, D., Wolin, M. J. and Bryant, M. P. (1973) 'Glucose fermentation products by *Ruminococcus albus* grown in continuous culture with *Vibrio succinogenes*: Changes caused by interspecies transfer of H_2', *Journal of Bacteriology*, vol 114, pp1231–1240

Jetten, M. S. M., Stams, A. J. M. and Zehnder, A. J. B. (1992) 'Methanogenesis from acetate: A comparison of the acetate metabolism in *Methanothrix soehngenii* and *Methanosarcina* spp.', *FEMS Microbiological Reviews*, vol 88, pp181–198

Keltjens, J. T. and van der Drift, C. (1986) 'Electron transfer reactions in methanogens', *FEMS Microbiological Reviews*, vol 39, pp259–303

Krumholz, L. R. and Bryant, M. P. (1986) '*Syntrophococcus sucromutans* sp. nov. gen. nov. uses carbohydrates as electron donors and formate, methoxymonobenzoids or *Methanobrevibacter* as electron acceptor systems', *Archives of Microbiology*, vol 143, pp313–318

Laanbroek, H. J., Abee, T. and Voogd, I. L. (1982) 'Alcohol conversions by *Desulfobulbus propionicus* Lindhorst in the presence and absence of sulphate and hydrogen', *Archives of Microbiology*, vol 133, pp178–184

Liu, Y., and Whitman, W. B. (2008) 'Metabolic, phylogenetic, and ecological diversity of the methanogenic archaea', *Annals of the New York Academy of Sciences*, vol 1125, pp171–189

Liu, Y., Balkwill, D. L., Aldrich, H. C., Drake, G. R. and Boone, D. R. (1999) 'Characterization of the anaerobic propionate-degrading syntrophs *Smithella propionica* gen. nov., sp. nov. and *Syntrophobacter wolinii*', *International Journal of Systematic Bacteriology*, vol 49, pp545–556

McInerney, M. J. (1988) 'Anaerobic hydrolysis and fermentation of fats and proteins', in A. J. B. Zehnder (ed) *Biology of anaerobic microorganisms*, John Wiley & sons, New York, pp373–415

McInerney, M. J., Bryant, M. P., Hespell, R. B. and Costerton, J. W. (1981) '*Syntrophomonas wolfei* gen. nov. sp. nov, an anaerobic syntrophic, fatty acid-oxidizing bacterium', *Applied and Environmental Microbiology*, vol 41, pp1029–1039

McInerney, M. J., Struchtemeyer, C. G., Sieber, J., Mouttaki, H., Stams, A. J. M., Schink, B., Rohlin, L. and Gunsalus, R. P. (2008) 'Physiology, ecology, phylogeny, and genomics of microorganisms capable of syntrophic metabolism', *Annals of the New York Academy of Sciences*, vol 1125, pp58–72

Müller, N., Griffin, B. M., Stingl, U. and Schink, B. (2008) 'Dominant sugar utilizers in sediment of Lake Constance depend on syntrophic cooperation with methanogenic partner organisms', *Environmental. Microbiology*, vol 10, pp1501–1511

Muyzer, G. and Stams, A. J. M. (2008) 'The ecology and biotechnology of sulphate-reducing bacteria', *Nature Reviews Microbiology*, vol 6, pp441–454

Nagase, M. and Matsuo, T. (1982) 'Interaction between amino-acid degrading bacteria and methanogenic bacteria in anaerobic digestion', *Biotechnology and*

Bioengineering, vol 24, pp2227–2239

Nanninga, H. J. and Gottschal, J. C. (1985) 'Amino acid fermentation and hydrogen transfer in mixed cultures', *FEMS Microbiology Ecology*, vol 31, pp261–269

Schink, B. and Stams, A. J. M. (2006) 'Syntrophism among prokaryotes', in M. Dworkin, S. Falkow, E. Rosenberg, K. H. Schleifer and E. Stackebrandt (eds) *The Prokaryotes: An Evolving Electronic Resource for the Microbiological Community*, 3rd Edition, vol 2, Springer-Verlag, New York, pp309–335

Schnürer, A., Schink, B. and Svensson, B. H. (1996) '*Clostridium ultunense* sp. nov., a mesophilic bacterium oxidizing acetate in syntrophic association with a hydrogenotrophic methanogenic bacterium', *International Journal of Systematic Bacteriology*, vol 46, pp1145–1152

Sousa, D. Z., Smidt, H., Alves, M. M. and Stams. A. J. M. (2009) 'Ecophysiology of syntrophic communities that degrade saturated and unsaturated long-chain fatty acids', *FEMS Microbiology Ecology* vol 68, pp257–272

Sprenger, W. W., Hackstein, J. H. and Keltjens, J. T. (2005) 'The energy metabolism of *Methanomicrococcus blatticola*: Physiological and biochemical aspects', *Antonie van Leeuwenhoek*, vol 87, pp289–299

Stams, A. J. M. (1994) 'Metabolic interactions between anaerobic bacteria in methanogenic environments', *Antonie Van Leeuwenhoek*, vol 66, pp271–294

Stams, A. J. M. and Plugge, C. M. (2009) 'Electron transfer in syntrophic communities of anaerobic bacteria and archaea', *Nature Review Microbiology*, vol 7, pp568-577

Thauer, R. K., Jungermann, K. and Decker, K. (1977) 'Energy conservation in chemotrophic anaerobic bacteria', *Bacteriological Reviews*, vol 41, pp100–180

Thauer, R. K., Kaster, A. K., Seedorf, H., Buckel, W. and Hedderich, R. (2008) 'Methanogenic archaea: Ecologically relevant differences in energy conservation', *Nature Reviews Microbiology*, vol 6, pp579–591

Van Kuijk, B. L. M., Schlösser, E. and Stams, A. J. M. (1998) 'Investigation of the fumarate metabolism of the syntrophic propionate-oxidizing bacterium strain MPOB', *Archives of Microbiology*, vol 169, pp346–352

Winter, J. and Wolfe, R. S. (1980) 'Methane formation from fructose by syntrophic associations of *Acetobacterium woodii* and different strains of methanogens', *Archives of Microbiology*, vol 124, pp73–79

3
Wetlands

Torben R. Christensen

Introduction and a piece of science history

Wetlands represent a pivotal source of atmospheric CH_4. In a world without anthropogenic enhanced emissions of CH_4, the dynamics of global wetland emissions would be the primary source-driven impact on the atmospheric concentrations. Hence, wetlands attract a lot of attention in palaeo studies of past concentrations such as the analysis of ice cores (Chappellaz et al, 1993; Loulergue et al, 2008). Temperature-driven variations in tropical wetland emissions as well as periglacial development of northern wetlands have been shown to impact strongly on the Holocene development of CH_4 concentrations in the atmosphere (Loulergue et al, 2008), but also on dynamics of emissions of volatile organic compounds and their impact on the atmospheric capacity to break down CH_4 (Harder et al, 2007). The balance between these processes has determined past natural dynamics of CH_4 in the atmosphere.

The presence of a greenhouse effect in the atmosphere determining climate was first proposed in the early part of the 19th century by French authors Fourier and later Pouillet (Handel and Risbey, 1992). Tyndall (1861) was the first to note that changes in atmospheric concentrations of CO_2 might influence climate. In an apparently little-noticed paper by Hunt (1863) it was first suggested that as well as carbon dioxide, other gases including 'marsh gas' (methane) could also be affecting climate (Handel and Risbey, 1992). Arrhenius (1896) provided the first quantitative discussion of the effect of CO_2 on climate and later made the suggestion that man-made emission of this gas could cause changes in climate (Arrhenius, 1908).

Although the emission of 'marsh gas' had been well known for decades before, CH_4 was not actually discovered in the atmosphere before the middle of the 20th century (Migeotte, 1948). In the following decades, various authors gave the first accounts of atmospheric CH_4 (see Wahlen, 1993). Ehhalt (1974) made the first estimation of global emissions including from wetlands, tundra and fresh waters, although only very few real ecosystem–atmosphere flux measurements were available at that time. The first such wetland CH_4 flux measurements were carried out in connection with the International Biological

Program (IBP) in the late 1960s and early 1970s. These studies included the work by Clymo and Reddaway (1971) at Moor House in Britain, and Svensson (1976) who investigated a subarctic mire in Northern Sweden. These studies were carried out as pure biological investigations with no relation to climate change issues. The latter more recent issues form the background for a dramatically increasing number of studies of wetland CH_4 emissions over the past decades, the status of which are briefly reviewed in this chapter.

Processes

Being produced from anaerobic decomposition of organic material in waterlogged anaerobic parts of the soil, wetland environments have for a long time been known to be significant contributors to atmospheric CH_4 (Ehhalt, 1974; Fung et al, 1991; Bartlett and Harriss, 1993). In these wet anaerobic environments, CH_4 is formed through the microbial process of methanogenesis (see also Chapter 2). Methane formation follows from a complex set of ecosystem processes that begins with the primary fermentation of organic macromolecules to acetic acid, other carboxylic acids, alcohols, CO_2 and hydrogen. This is then followed by the secondary fermentation of the alcohols and carboxylic acids to acetate, H_2 and CO_2, which are fully converted to CH_4 by methanogenic bacteria (Cicerone and Oremland, 1988; Conrad, 1996). The controls on this sequence of events span a range of factors, most notably temperature, the persistence of anaerobic conditions, gas transport by vascular plants as well as supply of labile organic substrates (Whalen and Reebugh, 1992; Davidson and Schimel, 1995; Joabsson and Christensen, 2001; Ström et al, 2003). Figure 3.1 shows the variety of controls on CH_4 formation rates at different spatial and temporal scales.

Methane is, however, not only being produced, but also consumed in aerobic parts of the soil. This takes place through the microbial process of methanotrophy, which can even take place in dry soils with the bacteria living off the atmospheric concentration of CH_4 (Whalen and Reebugh, 1992; Moosavi and Crill, 1997; Christensen et al, 1999). Methanotrophy is responsible for the oxidation of an estimated 50 per cent of the CH_4 produced at depth in the soil (Reeburgh et al, 1994) and, as such, is as important a process for net CH_4 emissions as is the methanogenesis. The anaerobic process of methanogenesis is much more responsive to temperature than CH_4 oxidation. The mechanistic basis for this difference is not clear, but the ecosystem consequences are rather straightforward: soil warming in the absence of any other changes will accelerate emission (which is the difference between production and consumption), in spite of the simultaneous stimulation of the two opposing processes (Ridgwell et al, 1999). There may be a buffering effect of temperature changes at greater soil depths, where the methanogenesis mainly takes place. But, in the absence of other changes, warming still favours increasing production and net emission of CH_4.

Controls on methanogenesis

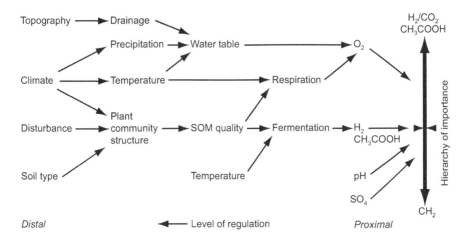

Figure 3.1 Major controls on the pathways to methane formation

Note: Distal and proximal controlling parameters are indicated as well as hierarchy of importance in a complex ecosystem context.
Source: Based on Schimel (2004)

The controls on CH_4 emissions are, hence, a rather complex set of processes working in opposing directions. Early empirical models of wetland CH_4 exchanges suggested sensitivity to climate change (Roulet et al, 1992; Harriss et al, 1993). A simple mechanistic model of tundra CH_4 emissions including the combined effects of temperature, moisture and active layer depth also suggested significant changes in CH_4 emissions as a result of climate change (Christensen and Cox, 1995). Wetland CH_4 emission models have grown in complexity (Panikov, 1995; Christensen et al, 1996; Cao et al, 1996; Walter and Heimann, 2000; Granberg et al, 2001; Wania, 2007) as the mechanistic understanding of the most important processes controlling CH_4 fluxes have improved. Autumn and winter processes have also been found to have a strong influence on net annual emissions of CH_4 (Panikov and Dedysh, 2000; Mastepanov et al, 2008). Variations in CH_4 emissions at the large regional–global scale have been found to be driven largely by temperature (Crill et al, 1992; Harriss et al, 1993), but with important modulating effects of vascular plant species composition superimposed (Christensen et al, 2003a; Ström et al, 2003). From the perspective of empirical studies, then, an initial warming is expected to lead to increased CH_4 emissions, but the scale of this increase depends on associated changes in soil moisture conditions, and the secondary effects of changes in vegetation composition.

The highest emissions are generally associated with stagnant constant high water table levels combined with highly organic soils (often peat). Plant productivity can further amplify the source strength of CH_4 production, and this interaction has been studied at scales ranging from below-ground microbial investigations (Panikov, 1995; Thomas et al, 1996; Joabsson et al, 1999) to large-scale vegetation models linked to CH_4 parameterizations (Cao et al, 1996; Christensen et al, 1996; Walter and Heimann, 2000; Zhuang et al, 2004; Sitch et al, 2007). Various studies have attributed the relationship to different mechanisms such as:

1 stimulation of methanogenesis by increasing C-substrate availability (input of organic substances to soil through root exudation and litter production);
2 build-up of plant-derived peat deposits that retain water and provide an anoxic soil environment;
3 removal of mineral plant nutrients such as nitrate and sulphate, which are competitive inhibitors of methanogenesis (competitive electron acceptors);
4 enhancement of gas transport from methanogenic soil layers to the atmosphere via root aerenchyma acting as gas conduits that bypass zones of potential CH_4 oxidation in the soil.

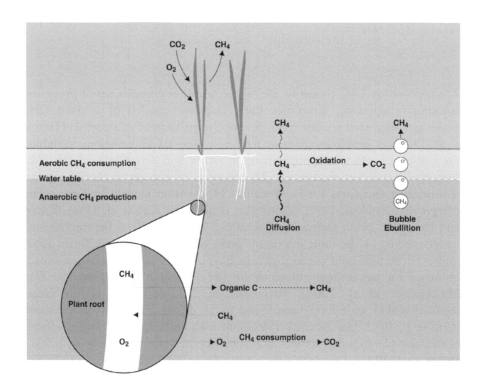

Figure 3.2 Major influences that have impacts on net methane emissions from wetland environments

Note: The processes associated with the vascular plants play a pivotal role as discussed in the text
Source: Figure modified from Joabsson and Christensen (2001)

In addition to these stimulatory effects on net CH_4 emissions, certain plants may also reduce emissions through actively oxidizing the root vicinity (rhizospheric oxidation). Figure 3.2 summarizes the ways in which plants may affect CH_4 emissions from wetlands.

Wetland emission estimates

Global emissions of CH_4 from wetlands range between 100 and 231Tg CH_4 yr^{-1} in six studies reviewed by Denman et al (2007), which is to be compared with a range of global total sources ranging from 503 to 610Tg CH_4 yr^{-1}. Regardless of all uncertainties, and common for all these studies, is that wetlands is placed as the largest single source of atmospheric CH_4, even when considering all anthropogenic emissions. It is interesting to note that the mean estimate and variation between wetland emission estimates presented in the IPCC AR4 (Denman et al, 2007) are very similar to early estimates made of global wetland emissions. Figure 3.3 compares the global total emissions and the proportion estimated for the wetland contribution in the IPCC AR4 and one of the first published attempts at putting together the global CH_4 budget (Ehhalt, 1974). The overall emission estimate has been decreased and the

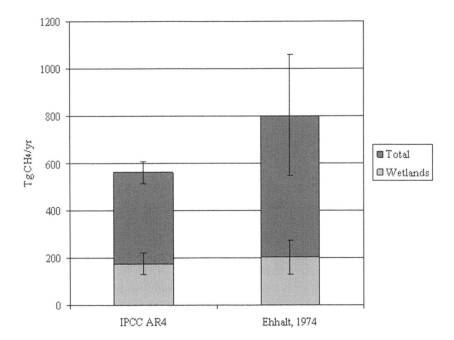

Figure 3.3 The global atmospheric burden of methane as estimated by the latest IPCC AR4 report and the proportion estimated to be originating from global wetlands

Note: The comparison is with the first global methane budget compiled by Ehhalt (1974).
Source: Denman et al (2007)

uncertainty reduced but the average 'consensus' wetland contribution estimate of today is remarkably similar to that from Ehhalt's 1974 study.

Recent inversion studies indicate that northern wetlands including arctic tundra contribute about 30 per cent to the global wetland emissions, and also that there is substantial inter-annual variability of CH_4 sources (Bousquet et al, 2006), with the inter-annual variability in the Arctic associated with climate variability. Bousquet et al (2006) attribute a significant fraction of the observed slowdown in the global growth rate of CH_4 during the early part of 2000s to a reduction of CH_4 emissions from wetlands caused by a drying trend after 1999. Recently (2007–2008) the growth rate has increased again, and it has been suggested that this may also be linked to increasing wetland emissions, in particular at northern latitudes (Dlugokencky et al, 2009, personal communication), although hydroxyl radical (OH) chemistry has also been suggested as a key influence on the recent changes (Rigby et al, 2008).

From a ground-based measurement perspective, extrapolated northern wetland emission estimates have for a long time ranged between 20 and 100Tg CH_4 yr^{-1}. Sebacher et al (1986) estimated 45–106Tg CH_4 yr^{-1} for arctic and boreal wetlands, Crill et al (1988) estimated 72Tg CH_4 yr^{-1} for undrained peatlands north of 40°N. Whalen and Reeburgh (1992) estimated 42±26Tg CH_4 yr^{-1} from measurements in wet meadow and tussock shrub tundra, and from similar measurements in comparable habitats on the North Slope of Alaska, Christensen (1993) estimated 20±5Tg CH_4 yr^{-1}. Reviewing the literature available at the time Bartlett and Harriss (1993) estimated a mean emission from northern wetlands north of 45°N of 38Tg CH_4 yr^{-1}, a value not far from recent estimates of 42–45Tg CH_4 yr^{-1} using inverse modelling for the northern hemisphere to derive a total emission estimate (Chen and Prinn, 2006).

Globally, freshwater lakes have also early been suggested as major sources of atmospheric CH_4. Ehhalt (1974) estimated global lake emissions to be between 1.25 and 25Tg CH_4 yr^{-1}. Subarctic and arctic lake systems in Alaska, as well as in Siberia, have recently seen renewed attention with substantial CH_4 emissions observed (Walter et al, 2006), which may only partially have been captured by earlier attempts to extrapolate CH_4 emissions globally (for example Ehhalt, 1974; Matthews and Fung, 1987). A broad lake survey reported by Bastviken et al (2004), including comparable Swedish lakes, also reported significant emissions from boreal, subarctic and arctic lakes. These emission estimates are similar to major studies of both tropical freshwater (Bartlett et al, 1988) and tundra lake ecosystems (Bartlett et al, 1992). As is also the case with CO_2 emissions, the estimates assigned to lakes should be differentiated between (1) small and large lakes and (2) the presence and absence of permafrost (Bastviken et al, 2004; Walter et al, 2006). Recently, Walter et al (2007) used data from Siberia, Alaska and the literature to estimate that lakes in this northern region (excluding large lakes likely to have low emissions) emit 15–35Tg CH_4 yr^{-1}, most of it through bubbling (ebullition).

A few studies have attempted to scale up tropical lake and flooded ecosystem CH_4 fluxes. Bartlett et al (1988) estimated central Amazonian emissions alone to be in the range of 3–21Tg CH_4 yr^{-1}, an estimate Melack et al (2004) revised upwards, using remote sensing, to approximately 22Tg CH_4 yr^{-1}.

Seasonal dynamics and local-scale emissions

In general, northern wetlands show a distinct seasonal pattern with a growing season peak dominating annual emissions (for example Crill et al, 1988; Whalen and Reeburgh, 1992; Rinne et al, 2007). When emissions are at the growing season peak in high-emitting northern wetlands they average 5–10mg CH_4 m^{-2} hr^{-1}. Recently, high northern permafrost wetlands have been shown to have some interesting additional peak emissions associated with the freeze-in period (Mastepanov et al, 2008). The generality of this feature and frequency of occurrence remains to be documented in more detail.

Tropical emissions are closely related to the seasonal flooding appearing in major parts of the wetland areas. The levels of emission are generally higher than for northern wetlands during the flooded seasons but they drop dramatically during the non-flooded part of the year. Average peak season emission may exceed 15mg CH_4 m^{-2} hr^{-1}. Figure 3.4 shows a generalized diagram comparing seasonal variation in emissions from the high northern tundra over northern wetlands in general to the tropical regions.

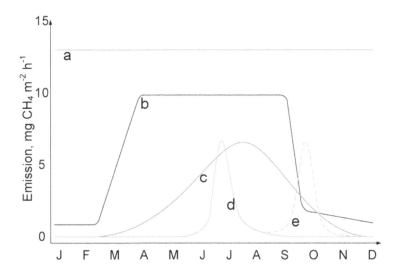

Figure 3.4 A general schematic illustration of differences in seasonal variations of methane emissions in tropical and northern wetland settings

Note: Tropical emissions are primarily influenced by the spatial scale of flooding while northern wetlands has a distinct temperature response. Line 'a' represents a constantly flooded tropical high-emitting wetland. Line 'b' a very common seasonally flooded tropical wetland. Line 'c' is a general representation of a northern seasonally temperature-dependent emission pattern and line 'd' represents the special dynamics associated with permafrost environments and recently discovered freeze-in bursts associated with those (stippled line 'e').
Source: The generalized schematic is based on flux measurements presented in Bartlett et al (1988), Melack et al (2004), Mastepanov et al (2008) and Jackowicz-Korczyński et al (in press)

The emission pathways may also vary seasonally. During the growing season, where there are vegetated surfaces, the percentage of the flux arising from ebullition may be below 50 per cent as the vascular transport mechanisms by the plants dominate. Purely diffusive flux is normally a very minor part of the overall flux (Bartlett et al, 1988; Christensen et al, 2003b). During the non-growing season in northern wetland, ebullition flux may be relatively more dominant as an emission pathway, and physical processes influencing emissions of stored gas are important (Mastepanov et al, 2008).

Changing emissions

Bousquet et al (2006) suggest a substantial inter-annual variability in wetland CH_4 emissions. In the tropics, this variability is most strongly affected by the variability in wetland extent, where northern latitudes seem most strongly affected by climate variability. As climate warming is expected to be strongest at high northern latitudes, and the wetland emission sensitivity to climate is highest here, special attention should be given to potential changes in the high-latitude emissions.

Dramatic changes in the high northern emissions as a consequence of climate warming may take place if permafrost melts at an increasing pace, and critical parts of the surface soil and lake environments become warmer and wetter. At least two different mechanisms behind increased CH_4 emissions may occur: (1) increased anaerobic near-surface breakdown of organic matter and stored peat as the active layer becomes deeper, and if soils in turn become warmer and wetter; and (2) increased breakdown of old organic deposits in expanding thaw lakes.

Recent findings suggest that, in part, both processes are already happening. Studies in Alaska, Canada and northern Scandinavia suggest wetter conditions in areas at the permafrost margin where it is receding (Turetsky et al, 2002; Payette et al, 2004; Malmer et al, 2005). This type of development has been documented to cause increasing CH_4 emissions over the landscape as a whole (Christensen et al, 2004; Johansson et al, 2006). Likewise, thaw lakes are well-known sources of CH_4 (Bartlett et al, 1992) and the increasing number and extent of these in Siberia and Alaska have been documented (Walter et al, 2006), and shown to have large potential implications for the global atmospheric CH_4 budget, at least in the short term (Walter et al, 2007).

Sensitivity and the future of emissions

The dominating tropical wetland emissions are governed less by temperature than the northern wetlands. In the tropical regions, the seasonality and length of the flooded periods will be the main factor determining any major changes in the atmospheric burden of CH_4. On the contrary, the estimated circumpolar CH_4 emissions of 30–60Tg CH_4 yr^{-1} (also reviewed by McGuire et al, 2009) are more directly sensitive to climate warming, and may hold a significant potential for feedback to a changing climate. Large-scale CH_4 flux models are currently not as advanced as general carbon cycling models and few allow for climate change scenario-based projections of changes in the future. Early attempts to assess and model tundra CH_4 emissions driven by climate change all indicated a potential increase in emissions (Roulet et al, 1992; Harriss et al, 1993; Christensen et al, 1995) but more advanced mechanistic models (Walter and Heimann, 2000; Granberg et al, 2001) are now approaching the stage where they, in further developed forms, will be fully coupled with Global Circulation Model (GCM) predictions to assess circumpolar CH_4 emissions in the future (Sitch et al, 2007; Wania, 2007). A critical factor is, as eluded to above, not only the mechanistic responses of soil processes (dominant in the northern wetland response) but also the geographical extent of wetlands (dominant feature for tropical wetlands), and how these may change in the future. Combined predictive hydrological and ecosystem process modelling is needed for an improved certainty in projections of changes in global wetland emissions. There is, however, little doubt that with climate scenarios of warming and wetting of the soils, there will undoubtedly be increases in CH_4 emissions, while with warming and drying there will be few changes, or a decline of emissions relative to the current scale.

The model developed by Walter and Heimann (2000) and applied by Walter et al (2001a, 2001b) has seen widespread uses for predictive purposes. Shindell et al (2004) used this model to examine the potential feedback of wetland CH_4 emissions on climate change. They simulated a 78 per cent increase in CH_4 emissions globally in a scenario with a doubling of the atmospheric CO_2. This increase was significant in the tropical regions but also included a doubling of northern wetland emissions from a modelled 24–48Tg CH_4 yr^{-1}. Gedney et al (2004) estimated substantial increases in wetland emissions, doubling globally by 2100. This led in their model to an overall increase of approximately 5 per cent in radiative forcing compared with the CO_2 feedback effect modelled by Cox et al (2000). But, as noted by Limpens et al (2008), this has to be seen in the light of the latter being a factor of two larger than any other coupled carbon–climate simulation over the 21st century (Friedlingstein et al, 2006). The wetland CH_4 emission feedback effect therefore may well be greater than suggested by the model of Gedney et al (2004). This is exemplified by improved process modelling based on the Walter model presented by Wania (2007), which suggests an increase in emissions by over 250 per cent in vast wetland areas by the end of this century.

Conclusions

This chapter has briefly looked at research into CH_4 emissions from wetlands over recent decades. Although the estimates of the overall scale of emissions from wetlands have not changed much since the early 1970s, significant progress has been made in our understanding of the dynamics and controls on the emissions and the relative contributions from tropical versus northern wetlands. Ecosystem models are developing towards the stage where the feedback mechanism from changed CH_4 emissions from wetlands in a warming climate may be fully integrated with climate change predictions. But significant challenges remain, in particular in our understanding of the seasonality of wetland emissions in the tropical region as driven by changes in hydrology and the climate sensitivity of northern wetland emissions, especially during the shoulder seasons and as affected by thawing permafrost.

References

Arrhenius, S. (1896) 'On the influence of carbonic acid in the air upon the temperature on the ground', *Philosophical Magazine* (Ser. 5), vol 41, pp237–276

Arrhenius, S. (1908) *Worlds in the making*, Harper, New York

Bartlett, K. B. and Harriss, R. C. (1993) 'Review and assessment of methane emissions from wetlands', *Chemosphere*, vol 26, pp261–320

Bartlett, K. B., Crill, P. M., Sebacher, D. I., Harriss, R. C., Wilson, J. O. and Melack, J. M. (1988) 'Methane flux from the central Amazonian floodplain', *Journal of Geophysical Research*, vol 93, pp1571–1582

Bartlett, K. B., Crill, P., Sass, R. C., Harriss, R. C. and Dise, N. B. (1992) 'Methane emissions from tundra environments in the Yukon-Kuskokwim Delta, Alaska', *Journal of Geophysical Research*, vol 97, pp16645–16660

Bastviken, D., Cole, J., Pace, M. and Tranvik, L. (2004) 'Methane emissions from lakes: Dependence of lake characteristics, two regional assessments, and a global estimate', *Global Biogeochemical Cycles*, vol 18, GB4009, doi:10.1029/2004GB002238

Bousquet, P., Ciais, P., Miller, J. B., Dlugokencky, E. J., Hauglustaine, D. A., Prigent, C., van der Werf, G., Peylin, P., Brunke, E., Carouge, C., Langenfelds, R. L., Lathiere, J., Ramonet, P. F. M., Schmidt, M., Steele, L. P., Tyler, S. C. and White, J. W. C. (2006) 'Contribution of anthropogenic and natural sources methane emissions variability', *Nature*, vol 443, pp439–443

Cao, M. K., Marshall, S. and Gregson, K. (1996) 'Global carbon exchange and methane emissions from natural wetlands: Application of a process-based model', *Journal of Geophysical Research – Atmospheres*, vol 101, pp14399–14414

Chappellaz, J. A., Fung, I. Y. and Thompson, A. M. (1993) 'The atmospheric CH_4 increase since the last glacial maximum: 1. Source estimates', *Tellus*, vol 45B, pp228–241

Chen, Y. H. and Prinn, R. G. (2006) 'Estimation of atmospheric methane emissions between 1996 and 2001 using a three-dimensional global chemical transport model', *Journal of Geophysical Research – Atmospheres*, vol 111, D10307

Christensen, T. R. (1993) 'Methane emission from Arctic tundra', *Biogeochemistry*, vol 21, pp117–139

Christensen, T. R. and Cox, P. (1995) 'Response of methane emission from Arctic tundra to climatic change: Results from a model simulation', *Tellus*, vol 47B, pp301–310

Christensen, T. R., Jonasson, S., Callaghan, T.V. and Havström, M. (1995) 'Spatial variation in high latitude methane flux along a transect across Siberian and

Eurasian tundra environments', *Journal of Geophysical Research*, vol 100D, pp21035–21045

Christensen, T. R., Prentice, I. C., Kaplan, J., Haxeltine, A. and Sitch, S. (1996) 'Methane flux from northern wetlands and tundra: An ecosystem source modelling approach', *Tellus*, vol 48B, pp651–660

Christensen, T. R., Michelsen, A. and Jonasson, S. (1999) 'Exchange of CH_4 and N_2O in a subarctic heath soil: Effects of inorganic N and P amino acid addition', *Soil Biology and Biochemistry*, vol 31, pp637–641

Christensen T. R., Joabsson, A., Ström, L., Panikov, N., Mastepanov, M., Öquist, M., Svensson, B. H., Nykänen, H., Martikainen, P. and Oskarsson, H. (2003a) 'Factors controlling large scale variations in methane emissions from wetlands', *Geophysical Research Letters*, vol 30, pp1414

Christensen, T. R., Panikov, N., Mastepanov, M., Joabsson, A., Öquist, M., Sommerkorn, M., Reynaud, S. and Svensson, B. (2003b) 'Biotic controls on CO_2 and CH_4 exchange in wetlands – a closed environment study', *Biogeochemistry*, vol 64, pp337–354

Christensen, T. R., Johansson, T., Malmer, N., Åkerman, J., Friborg, T., Crill, P., Mastepanov, M. and Svensson, B. (2004) 'Thawing sub-arctic permafrost: Effects on vegetation and methane emissions', *Geophysical Research Letters*, vol 31, L04501, doi:10.1029/2003GL018680

Cicerone, R. J. and Oremland, R. S. (1988) 'Biogeochemical aspects of atmospheric methane', *Global Biogeochemical Cycles*, vol 2, pp299–327

Clymo, R. S. and Reddaway, E. J. F., (1971) 'Productivity of *Sphagnum* (bog-moss) and peat accumulation', *Hidrobiologia*, vol 12, pp181–192

Conrad, R. (1996) 'Soil microorganisms as controllers of atmospheric trace gases (H_2, CO, CH_4, OCS, N_2O and NO)', *Microbiological Reviews*, vol 60, pp609–640

Cox, P. M., Betts, R. A., Jones, C. D., Spall, S. A. and Totterdell, I. J. (2000) 'Acceleration of global warming due to carbon-cycle feedbacks in a coupled climate model', *Nature*, vol 408, pp184–187, 750

Crill, P. M., Bartlett, K. B., Harriss, R. C., Gorham, E., Verry, E. S., Sebacher, D. I., Madzar, L. and Sanner, W. (1988) 'Methane flux from Minnesota peatlands', *Global Biogeochemical Cycles*, vol 2, pp371–384

Crill, P., Bartlett, K. and Roulet, N. (1992) 'Methane flux from boreal peatlands', *Suo*, vol 43, pp173–182

Davidson, E. A. and Schimel, J. P. (1995) 'Microbial processes of production and consumption of nitric oxide, nitrous oxide, and methane', in P. Matson and R. Harriss (eds) *Methods in Ecology: Trace Gases*, Blackwell Scientific, Oxford, pp327–357

Denman, K. L., Brasseur, G., Chidthaisong, A., Ciais, P., Cox, P. M., Dickinson, R. E., Hauglustaine, D., Heinze, C., Holland, E., Jacob, D., Lohmann, U., Ramachandran, S., da Silva Dias, P. L., Wofsy, S. C. and Zhang, X. (2007) 'Couplings between changes in the climate system and biogeochemistry', in S. Solomon, D. Qin, M. Manning, Z. Chen, M. Marquis, K. B. Averyt, M. Tignor and H. L. Miller (eds) *Climate Change 2007: The Physical Science Basis. Contribution of Working Group I to the Fourth Assessment Report of the Intergovernmental Panel on Climate Change*, Cambridge University Press, Cambridge and New York, pp499–587

Ehhalt, D. H. (1974) 'The atmospheric cycle of methane', *Tellus*, vol 26, pp58–70

Friedlingstein, P., Cox, P., Betts, R., Bopp, L., Von Bloh, W., Brovkin, V., Cadule, P., Doney, S., Eby, M., Fung, I., Bala, G., John, J., Jones, C., Joos, F., Kato, T., Kawamiya, M., Knorr, W., Lindsay, K., Matthews, H. D., Raddatz, T., Rayner, P., Reick, C., Roeckner, E., Schnitzler, K. G., Schnur, R., Strassmann, K., Weaver, A. J., Yoshikawa, C. and Zeng, N. (2006) 'Climate–carbon cycle feedback analysis: Results from the (CMIP)-M-4 model intercomparison', *Journal of Climate*, vol 19, pp3337–3353

Fung, I., John, J., Lerner, J., Matthews, E., Prather, M., Steele, L. P. and Fraser, P. J. (1991) 'Three-dimensional model synthesis of the global methane cycle', *Journal of Geophysical Research*, vol 96, pp13033–13065

Gedney, N., Cox, P. M. and Huntingford C. (2004) 'Climate feedback from wetland methane emissions', *Geophysical Research Letters*, vol 31, L20503

Granberg, G., Ottosson-Lofvenius, M., Grip, H., Sundh, I. and Nilsson, M. (2001) 'Effect of climatic variability from 1980 to 1997 on simulated methane emission from a boreal mixed mire in northern Sweden', *Global Biogeochemical Cycles*, vol 15, pp977–991

Handel, M. D. and Risbey, J. S. (1992) 'An annotated-bibliography on the greenhouse-effect and climate change', *Climatic Change*, vol 21, pp97–253

Harder, S. L., Shindell, D. T., Schmidt, G. A. and Brook, E. J. (2007) 'A global climate model study of CH_4 emissions during the Holocene and glacial-interglacial transitions constrained by ice core data', *Global Biogeochemical Cycles*, vol 21, GB1011

Harriss, R., Bartlett, K., Frolking, S. and Crill, P. (1993) 'Methane emissions from northern high-latitude wetlands', in R. S. Oremland (ed) *Biogeochemistry of Global Change: Radiatively Active Trace Gases*, Chapman & Hall, New York, pp449–486

Hunt, T. S. (1863) 'On the Earth's climate in palaeozoic times', *Philosophical Magazine*, vol IV, pp323–324

Jackowicz-Korczyński, M., Christensen T. R., Bäckstrand, K., Crill, P., Friborg, T., Mastepanov, M. and Ström, L. 'Annual cycle of methane emission from a subarctic peatland', *Journal of Geophysical Research, Biogeosciences*, in press

Joabsson, A., and Christensen, T. R. (2001) 'Methane emissions from wetlands and their relationship with vascular plants: An Arctic example', *Global Change Biology*, vol 7, pp919–932

Joabsson, A., Christensen, T. R. and Wallén, B. (1999) 'Vascular plant controls on methane emissions from northern peatforming wetlands', *Trends in Ecology and Evolution*, vol 14, pp385–388

Johansson, T., Malmer, N., Crill, P. M., Mastepanov, M. and Christensen, T. R. (2006) 'Decadal vegetation changes in a northern peatland, greenhouse gas fluxes and net radiative forcing', *Global Change Biology*, vol 12, pp2352–2369

Limpens, J., Berendse, F., Blodau, C., Canadell, J. G., Freeman, C., Holden, J., Roulet, N., Rydin, H. and Schaepman-Strub, G. (2008) 'Peatlands and the carbon cycle: From local processes to global implications – a synthesis', *Biogeosciences*, vol 5, pp1475–1491

Loulergue, L., Schilt, A., Spahni, R., Masson-Delmotte, V., Blunier, T., Lemieux, B., Barnola, J. M., Raynaud, D., Stocker, T. F. and Chappellaz, J. (2008) 'Orbital and millennial-scale features of atmospheric CH_4 over the past 800,000 years', *Nature*, vol 453, pp383–386

Malmer, N., Johansson, T., Olsrud, M. and Christensen, T. R. (2005) 'Vegetation, climatic changes and net carbon sequestration', *Global Change Biology*, vol 11, pp1895–1909

Mastepanov, M., Sigsgaard, C., Dlugokencky, E. J., Houweling, S., Strom, L., Tamstorf, M. P. and Christensen, T. R. (2008) 'Large tundra methane burst during onset of freezing', *Nature*, vol 456, pp628–631

Matthews, E. and Fung, I. (1987) 'Methane emission from natural wetlands: Global distribution, area, and environmental characteristics of sources', *Global Biogeochemical Cycles*, vol 1, pp61–86

McGuire, A. D., Anderson, L. G., Christensen, T. R., Dallimore, S., Guo, L., Hayes, D. J., Heimann, M., Lorenson, T. D., Macdonald, R. W. and Roulet, N. (2009) 'Sensitivity of the carbon cycle in the Arctic to climate change', *Ecological Monographs*, vol 79, pp523–555

Melack, J. M., Hess, L. L., Gastil, M., Forsberg, B. R., Hamilton, S. K., Lima, I. B. T. and Novo, E. (2004) 'Regionalization of methane emissions in the Amazon Basin with microwave remote sensing', *Global Change Biology*, vol 10, pp530–544

Migeotte, M. V. (1948) 'Spectroscopic evidence of methane in the Earth's atmosphere', *Physical Review*, vol 73, pp519–520

Moosavi, S. C., and Crill, P. M. (1997) 'Controls on CH_4 and CO_2 emissions along two moisture gradients in the Canadian boreal zone', *Journal of Geophysical Research*, vol 102, pp29261–29277

Panikov, N. S. (1995) *Microbial Growth Kinetics*, Chapman & Hall, London

Panikov, N. S. and Dedysh, S. N. (2000) 'Cold season CH_4 and CO_2 emission from boreal peat bogs (West Siberia): Winter fluxes and thaw activation dynamics', *Global Biogeochemical Cycles*, vol 14, pp1071–1080

Payette, S., Delwaide, A., Caccianiga, M. and Beauchemin, M. (2004) 'Accelerated thawing of sub-arctic peatland permafrost over the last 50 years', *Geophysical Research Letters*, vol 31, L18208

Reeburgh, W. S., Roulet, N. T. and Svensson, B. (1994) 'Terrestrial biosphere-atmosphere exchange in high latitudes', in R. G. Prinn (ed) *Global Atmospheric-Biospheric Chemistry*, Plenum Press, New York, pp165–178

Ridgwell, A. J., Marshall, S. J. and Gregson, K. (1999) 'Consumption of atmospheric methane by soils: A process-based model', *Global Biogeochemical Cycles*, vol 13, pp59–70

Rigby, M., Prinn, R. G., Fraser, P. J., Simmonds, P. G., Langenfelds, R. L., Huang, J., Cunnold, D. M., Steele, L. P., Krummel, P. B., Weiss, R. F., O'Doherty, S., Salameh, P. K., Wang, H. J., Harth, C. M., Muhle, J. and Porter, L. W. (2008) 'Renewed growth of atmospheric methane', *Geophysical Research Letters*, vol 35, L22805

Rinne, J., Riutta, T., Pihlatie, M., Aurela, M., Haapanala, S., Tuovinen, J. P., Tuittila, E. S. and Vesala, T. (2007) 'Annual cycle of methane emission from a boreal fen measured by the eddy covariance technique', *Tellus B*, vol 59, pp449–457

Roulet, N., Moore, T., Bubier, J. and Lafleur, P. (1992) 'Northern fens: Methane flux and climatic change', *Tellus*, vol 44, pp100–105

Sebacher, D. I., Harriss, R. C., Bartlett, K. B., Sebacher, S. M. and Grice, S. S. (1986) 'Atmospheric methane sources: Alaskan tundra bogs, an alpine fen, and a subarctic boreal marsh', *Tellus*, vol 38, pp1–10

Schimel, J. (2004) 'Playing scales in the methane cycle: From microbial ecology to the globe', *Proceedings of the National Academy of Sciences of the United States of America*, vol 101, pp12400–12401

Shindell, D. T., Walter, B. P. and Faluvegi, G. (2004) 'Impacts of climate change on methane emissions from wetlands', *Geophysical Research Letters*, vol 31, L21202

Sitch, S., McGuire, A. D., Kimball, J., Gedney, N., Gamon, J., Emgstrom, R., Wolf, A., Zhuang, Q. and Clein, J. (2007) 'Assessing the circumpolar carbon balance of arctic tundra with remote sensing and process-based modeling approaches', *Ecological Applications*, vol 17, pp213–234

Ström, L., Ekberg, A. and Christensen, T. R. (2003) 'Species-specific effects of vascular plants on carbon turnover and methane emissions from a tundra wetland', *Global Change Biology*, vol 9, pp1185–1192

Svensson, B. H. (1976) 'Methane production in tundra peat', in H. G. Schlegel, G. Gottschalk and N. Pfennig (eds) *Microbial Production and Utilization of Gases (H_2, CH_4, CO)*, E. Goltze, Göttingen, pp135–139

Thomas, K. L., Benstead, J., Davies, K. L. and Lloyd, D. (1996) 'Role of wetland plants in the diurnal control of CH_4 and CO_2 fluxes in peat', *Soil Biology and Biochemistry*, vol 28, pp17–23

Turetsky, M. R., Kelman Wieder, R. and Vitt, D. H. (2002) 'Boreal peatland fluxes under varying permafrost regimes', *Soil Biology and Biochemistry*, vol 34, pp907–912

Tyndall, J. (1861) 'The Bakerian Lecture: On the absorbtion and radiation of heat by gases and vapours, and on the physical connexion of radiation, absorption and conduction', *Proceedings of the Royal Society of London*, vol 11, pp100–104

Wahlen, M. (1993) 'The global methane cycle', *Annual Review of Earth and Planetary Science*, vol 21, pp407–426

Walter, B. P. and Heimann, M. (2000) 'A process-based, climate-sensitive model to derive methane emissions from natural wetlands: Application to five wetland sites, sensitivity to model parameters, and climate', *Global Biogeochemical Cycles*, vol 14, pp745–766

Walter B., Heimann, M. and Matthews, E. (2001a) 'Modelling modern methane emissions from natural wetlands: 1. Model description and results', *Journal of Geophysical Research*, vol 106, D24, doi:10.1029/2001JD900165

Walter, B., Heimann, M. and Matthews, E. (2001b) 'Modelling modern methane emissions from natural wetlands: 2. Interannual variations 1982–1993', *Journal of Geophysical Research*, vol 106, D24, doi:10.1029/2001JD900164

Walter, K. M., Zimov, S. A., Chanton, J. P., Verbyla, D. and Chapin, F. S. (2006) 'Methane bubbling from Siberian thaw lakes as a positive feedback to climate warming', *Nature*, vol 443, pp71–75

Walter, K. M., Smith, L. C. and Chapin, F. S. (2007) 'Methane bubbling from northern lakes: Present and future contributions to the methane budget', *Philosophical Transactions of the Royal Society A*, vol 365, pp1657–1676, doi:10.1098/rsta.2007.2036

Wania, R. (2007) 'Modelling northern peatland land surface processes, vegetation dynamics and methane emissions', PhD thesis, University of Bristol, Bristol

Whalen, S. C. and Reeburgh, W. S. (1992) 'Interannual variations in tundra methane emissions: A four-year time-series at fixed sites', *Global Biogeochemical Cycles*, vol 6, pp139–159

Zhuang, Q., Melillo, J. M., Kicklighter, D. W., Prinn, R. G., McGuire, A. D., Steudler, P. A., Felzer, B. S. and Hu, S. (2004) 'Methane fluxes between terrestrial ecosystems and the atmosphere at northern high latitudes during the past century: A retrospective analysis with a process-based biogeochemistry model', *Global Biogeochemical Cycles*, vol 18, GB3010

4
Geological Methane

Giuseppe Etiope

Introduction

Natural emissions of CH_4 are not only produced by contemporary biochemical sources such as wetlands, termites, oceans, wildfires and wild animals, and fossil CH_4 (that is geologically ancient, radiocarbon-free CH_4) is not emitted only by the fossil fuel industry. Beyond CH_4 from the biosphere and CH_4 from anthropogenic sources, a third CH_4 'breath' exists – earth's degassing.

The term 'degassing', in general, makes one think of volcanoes and geothermal manifestations (eruptions, fumaroles, mofettes, hydrothermal springs, either on land or on the seafloor) that release carbon dioxide, water vapour and sulphur gases, but this is only a partial vision of earth's degassing. Earth also exhales hydrocarbons, especially in geologically 'cold' areas, such as sedimentary basins where large quantities of natural gas migrate from shallow or deep rocks and reservoirs to the surface along faults and fractured rocks. The phenomenon is called 'seepage' and the gas is almost totally CH_4, with low quantities (hundreds of ppmv to a few per cent) of other hydrocarbons (mainly ethane and propane) and non-hydrocarbon gases (CO_2, N_2, H_2S, Ar and He). Gaseous hydrocarbons are produced by geologically ancient microbial activity, in shallow and low-temperature sedimentary rocks, and by thermogenic processes in deeper, warmer rocks. Therefore, seepage is a natural source of fossil CH_4.

Until recently, geological seepage has generally been neglected or considered a 'minor source' for CH_4 in the scientific literature (for example Lelieveld et al, 1998). The Second and Third Assessment Reports of the Intergovernmental Panel on Climate Change (Schimel et al, 1996; Prather et al, 2001) only considered gas hydrates as geological sources of methane. Gas hydrates, or CH_4 clathrates as they are sometimes called, are ice-like mixtures of water and CH_4 trapped in oceanic sediments (for example Kvenvolden, 1988). The majority of this gas escaping from melting deep-sea hydrates is dissolved in the seawater column and does not enter the atmosphere. However, global emissions of CH_4 to the atmosphere from hydrates have been reported to be roughly 3Tg y^{-1} (Kvenvolden, 1988) to 10Tg y^{-1} (Lelieveld et al, 1998), highly speculative values since they result from misquotations not supported by direct measurements.

Studies conducted during the last ten years have made it clear that other geological CH_4 sources, much more important than gas hydrates, exist; and there has been a growing consensus regarding the importance of marine (offshore) seepage, independent from gas hydrates, as a global contributor of CH_4 to the atmosphere (for example Judd et al, 2002; Judd and Hovland, 2007). Experimental flux data, acquired since 2001, have provided more and more evidence for large emissions from continental (onshore) gas manifestations, including macroseeps and diffuse microseepage from soils (Etiope et al, 2008; Etiope, 2009, and references therein). Geothermal emissions are subordinate, but worth considering globally, while volcanoes appear not to be substantial CH_4 contributors (Etiope et al, 2007a).

At present, it is clear and unambiguously understood that geological emissions are a significant global source of CH_4; and today, earth's degassing is considered the second highest natural source for CH_4 emissions after wetlands (for example Etiope, 2004; Kvenvolden and Rogers, 2005; Etiope et al, 2008). A new global estimate for geological sources has finally been acknowledged by the IPCC in its Fourth Assessment Report (Denman et al, 2007). Also, geological seepage has been considered as a new source for natural CH_4 in the Emission Inventory Guidebook of the European Environment Agency (EMEP/EEA, 2009) and in the new US Environmental Protection Agency report on Natural Emissions of Methane (US EPA, 2010).

General classification of geological sources

In this section the types and/or source categories for geological CH_4 are described, with particular emphasis on the use of the correct terminology, which is essential for avoiding confusion and misunderstandings when discussing CH_4 sources.

Two main source categories for geological CH_4 can be distinguished: (1) hydrocarbon-generation processes in sedimentary basins (seepage *in sensu strictu*) and (2) geothermal and volcanic exhalations.

Sedimentary seepage

Gas seepage in sedimentary hydrocarbon-prone (petroliferous) basins includes low-temperature CH_4-dominated (generally around 80–99 per cent v/v) gas manifestations and exhalations related to the following four classes:

1. onshore mud volcanoes;
2. onshore seeps (independent of mud volcanism);
3. onshore microseepage;
4. offshore (submarine) macro-seeps (including seafloor mud volcanoes).

Such gas manifestations have historically been an important indicator of subsurface hydrocarbon accumulations, and still drive the geochemical exploration for petroleum and gas today (for example Schumacher and Abrams, 1996; Abrams, 2005). Methane production in these areas can be due to microbial and/or thermogenic processes. Microbial CH_4 forms through the bacterial breakdown of organic material in sediments, providing a distinctive isotopic carbon composition, a $\delta^{13}C$-CH_4 lighter than −60 parts per mil. At greater depths, thermogenic CH_4 produced through the thermal breakdown of organic matter or heavier hydrocarbons has a $\delta^{13}C$-CH_4 composition ranging from −50 to −25 parts per mil. The migration and accumulation of fossil CH_4 in stratigraphic and structural traps has been extensively described in the literature of petroleum geology (Hunt, 1996). Methane gas in these areas is released naturally into the atmosphere mainly through active and permeable faults and fractured rocks, after long-distance migration driven by pressure or density gradients (Etiope and Martinelli, 2002).

Onshore mud volcanoes
Today a large amount of scientific information can be found regarding mud volcanoes, including information describing their formation mechanisms and distributions (for example Milkov, 2000; Dimitrov, 2002; Kopf, 2002). Mud volcanoes are cone-shaped structures formed by the emission of gas, water and sediments, sometimes with oil and/or rocky breccia, in areas where thick sequences of sedimentary rocks are compressed tectonically and often subjected to buoyancy-driven movements. Mud volcanoes, more than 900 structures on land and more than 300 on the ocean's shelves, are distributed along faults, over oil and gas reservoirs of the Alpine-Himalayan, the Pacific Ocean and the Caribbean geological belts (Figure 4.1). The gas of onshore mud volcanoes is mainly thermogenic CH_4 (Etiope et al, 2009) and can be released through continuous (steady-state) exhalations from craters (Figure 4.2), vents (gryphons, bubbling pools or salses) and surrounding soil, or intermittent blow-outs and eruptions (for example Etiope and Milkov, 2004, and references therein).

Other macroseeps
All gas manifestations that are independent of mud volcanism can be referred to as 'other seeps', and include 'water seeps' and 'dry seeps' (Etiope et al, 2009). Water seeps release an abundant gas phase accompanied by a water discharge (bubbling springs, groundwater or hydrocarbon wells), where the water may have a deep origin and may have interacted with gas during its ascent to the surface. Dry seeps release only a gaseous phase, such as the gas venting from outcropping rocks or through the soil horizon or through river/lake beds. Gas bubbling from groundwater filled wells, or from other shallow water bodies, should be considered as dry seeps, since surface water is only crossed by gas flow. Dry gas flow through rocks and dry soils can produce fascinating flames (Figure 4.3). Many seeps naturally burn in the dry and

GEOLOGICAL METHANE | 45

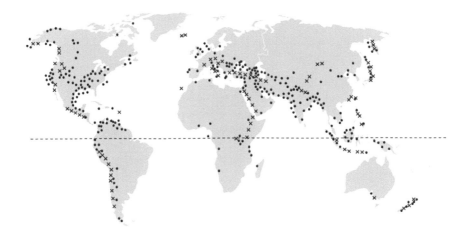

Figure 4.1 Global distribution of geological sources of hydrocarbons
Note: Dots are the main petroleum seepage areas; crosses are the main geothermal and volcanic areas.
Source: Modified from Etiope and Ciccioli (2009)

Figure 4.2 Examples of mud volcano gas exhalations
Note: a) Dashgil, Azerbaijan; b) Regnano, Italy; c) Bakhar satellite, Azerbaijan; d) Paclele Beciu, Romania.
Sources: a) C. Baciu, Babes-Bolyai University; b) G. Etiope, INGV; c) L. Innocenzi, INGV; d) G. Etiope, INGV

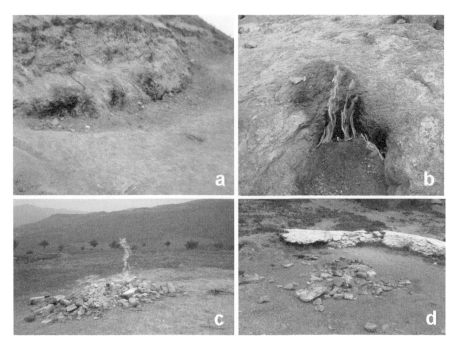

Figure 4.3 Examples of everlasting fire seeps

Note: a) Yanardag, Azerbaijan; b) Chimaera, Turkey; c) Monte Busca, Italy; d) Andreiasu, Romania
Sources: a) L. Innocenzi, INGV; b) H. Hosgormez, University of Istanbul; c) G. Etiope, INGV; d) G. Etiope, INGV

summer seasons, or throughout the year. Many vents can be easily ignited artificially.

Some fires are called 'everlasting' or 'eternal', since the presence of a flame has been continuously reported in historical records. Several continuous seeps are related to ancient religious traditions (such as those related to Zoroastrism in Azerbaijan; see Etiope et al, 2004a) and are still active today in archaeological sites (for example Chimaera seep in Turkey; see Hosgormez et al, 2008). Active seeps occur in almost all of the 112 countries hosting total petroleum systems (TPSs). More than 10,000 seeps are assumed to exist on land (Clarke and Cleverly, 1991) and can be found in all petroliferous areas (Figure 4.1) in correspondence with active tectonic faults.

Microseepage

Microseepage is the slow, invisible, but continuous loss of CH_4 and light alkanes from sedimentary basins. It is the pervasive, diffuse exhalation of CH_4 from the soil and may be responsible for positive fluxes or for a decrease in negative CH_4 flux in dry lands, indicating that methanotrophic consumption in the soil could be lower than the input from underground sources (Etiope and Klusman, 2002; 2010). Positive fluxes are typically a few or tens of mg m^{-2} d^{-1}, and may reach hundreds of mg m^{-2} d^{-1} over wide tectonized and faulted areas. All petroleum basins contain microseepage, as shown by innumerable surveys performed for petroleum exploration (for example Hunt, 1996; Saunders et al, 1999; Wagner et al, 2002; Abrams, 2005, Khan and Jacobson 2008). More than 75 per cent of the world's petroliferous basins contain surface seeps (Clarke and Cleverly, 1991). Klusman et al (1998, 2000) assumed that microseeping areas potentially include all of the sedimentary basins in dry climates, with petroleum and gas generation processes at depth, an area that has been estimated to be ~43.4 × $10^6 km^{-2}$. Flux data available today suggest that microseepage corresponds closely with the spatial distribution of hydrocarbon reservoirs, coal measures and portions of sedimentary basins that are, or that have been, at temperatures >70°C (thermogenesis). Accordingly, Etiope and Klusman (2010) assumed that microseepage may occur within a TPS, a term used in petroleum geology (Magoon and Schmoker, 2000) to describe the whole hydrocarbon-fluid system in the lithosphere including the essential elements and processes needed for oil and gas accumulations, migration and seeps. 42 countries produce 98 per cent of the world's petroleum, 70 countries produce 2 per cent, and 70 countries produce 0 per cent. So a TPS, and consequently the potential for microseepage, occurs in 112 (42 + 70) countries, suggesting that microseepage is potentially a very common phenomenon and widespread on all continents.

The global area of potential microseepage was assessed using an analysis for the distribution of oil/gas fields within all of the 937 petroleum provinces or basins, reported using a GIS data set from the US Geological Survey World Petroleum Assessment 2000 and related maps (Etiope and Klusman, 2010). For each province, a polygon was drawn that enclosed gas/oil field points in interactive maps, and the area was estimated using graphic software. Using this method, it was determined that significant gas/oil field zones occur in at least 120 provinces. The total area of gas/oil field zones was estimated to be between 3.5 and 4.2 million km^2 (Etiope and Klusman, 2010), approximately 7 per cent of global dryland area.

Submarine emissions
Methane is released from the seafloor in sedimentary basins mainly through cold seeps, mud volcanoes and pockmarks (see Judd and Hovland, 2007, for a comprehensive review). Unlike the flow in the subaerial environment, CH_4 seeping into the marine environment meets significant hindrances before entering the atmosphere. Methane passing through seafloor sediments is normally oxidized at the sulphate–methane transition zone (Borowski et al, 1999); if the supply of CH_4 overcomes anaerobic consumption, CH_4 bubbles are able to escape into the water column where they can be partially or completely dissolved and oxidized. The degree of dissolution in seawater depends mainly on the depth of the water, the temperature and the size of the bubbles rising towards the surface. In general, models and field data indicate

that only submarine seeps occurring at depths less than 100–300m have a significant impact on the atmosphere (for example Leifer and Patro, 2002; Schmale et al, 2005).

Geothermal and volcanic emissions

Geothermal volcanic emissions are high-temperature H_2O- or CO_2-dominated gas manifestations where CH_4 is produced by inorganic reactions (abiogenic CH_4) or the thermal breakdown of organic hydrocarbons. Methane concentrations generally range from a few ppmv to a few per cent by volume (for example Taran and Giggenbach, 2003; Fiebig et al, 2004; Etiope et al, 2007a). The distinction between 'volcanic' and 'geothermal' emissions is not always easy, especially in areas where geothermal and volcanic systems are adjacent. In these areas, the CH_4 released from geothermal areas could have originated from fluids related to a close volcanic circulation system (or vice versa). However, a distinction should be made since the two sources produce different emission factors and since the term 'volcanoes' is used in the CH_4 source list of the UNECE/EMEP Atmospheric Emission Inventory Guidebook (EEA, 2004).

Etiope et al (2007a) proposed the following definitions:

1 Volcanic emission – a gas released from active, or historically active volcanoes, either from craters or from flanks. The gas, released from magma and emanating diffusely without ever being dissolved in a subcritical aqueous phase on the way to the surface, would typically have a very high H_2O content and high CO_2/CH_4 ratios.
2 Geothermal emission – a gas released from ipomagmatic (plutonic) or thermometamorphic environments where there is no contemporary magma output at the surface; the gas is generally released from an aqueous hydrothermal solution, by boiling or degassing. Included are gas manifestations in extinct volcanoes, palaeo-volcanic zones and CO_2-rich 'cold' vents in active tectonic zones, where gas originates from deep thermometamorphic processes and faults. Geothermal or volcanic CH_4 can be released from localized sites (point sources: gas vents, mofettes, fumaroles and crater exhalation) and from diffuse and pervasive leakage throughout large areas (area sources). This diffuse degassing from soils is equivalent to the microseepage in sedimentary basins. The only difference is that the CH_4 originates and is abundant in the ascending gas phase (CO_2 is dominant in geothermal diffuse degassing). However, the term 'microseepage' describes emissions from sedimentary (non-geothermal) sources.

Some questions and clarifications

What about abiogenic non-volcanic methane?

A third, apparently minor, category can be attributed to emissions of abiogenic (inorganic) CH_4 in non-volcanic areas. These emissions may derive from low-temperature (non-hydrothermal) serpentinization processes and mantle degassing through deep faults (Abrajano et al, 1988; Sano et al, 1993; Hosgormez et al, 2008). Although several onshore areas of abiogenic seepage are known (for example in Turkey, the Philippines and Oman), fluxes have been studied in only one case (Chimaera seep; see Hosgormez et al, 2008) and there is no evaluation of a global emission estimate (apart from a rough estimate from serpentinization, suggesting about 1.3Tg CH_4 yr^{-1}, that refers mainly to gas output at the seafloor and includes hydrothermal processes; see Emmanuel and Ague, 2007). So, abiogenic non-volcanic CH_4 is a virtual further source that cannot be quantified, at present, in global budgets. Specific studies are needed.

Are emissions from coal beds a natural source?

Methane seepage from coal beds in sedimentary basins is not generally considered as a natural source, being almost always produced from the dewatering of coal strata induced through mining activities (for example Thielemann et al, 2000). A few cases of natural gas seepage associated with coal-bearing strata have been reported. Thielemann et al (2000) identified some thermogenic gas emissions in the Ruhr basin (Germany) apparently unrelated to mining, and Judd et al (2007) reported extensive CH_4-derived authigenic carbonates associated with coal-bearing carboniferous rocks in the Irish Sea. Therefore, the existence of significant natural seepage related to coal beds cannot be excluded and its actual role as a CH_4 source should be assessed through field measurements in coal basins not perturbed by mining.

Fossil versus modern methane

So far, the term 'geological methane' has been used in reference to 'fossil' CH_4 (Etiope and Klusman, 2002; Kvenvolden and Rogers, 2005; Etiope et al, 2008), which is radiocarbon free (older than approximately 50,000 years) and that can be distinguished from 'modern' gas developed from recent organic material in soils or shallow sediments by radiocarbon (^{14}C-CH_4) analyses. However, the CH_4 produced in late Pleistocene and Holocene sediments in estuaries, deltas and bays or trapped beneath permafrost, could also be formally considered geological even though it is not necessarily fossil. 'Recent' gas has been widely discussed by Judd (2004), and Judd and Hovland (2007). Modern microbial CH_4 produced by recent and contemporary microbial activity should be considered in the literature and source categories for peatlands, wetlands and oceans.

Fluxes and emission factors

Microseepage

Emission calculations for microseepage are currently based on averaging field contributions from identifiable homogeneous areas, with calculations of the type E = A × <F>, where A is the area in km^2 or m^2, and <F> is the average flux value (t km^{-2} y^{-1} or kg m^{-2} day^{-1}). Following a statistical analysis of a database of 563 measurements performed by closed-chamber systems in dry soils in different petroliferous basins in the US and Europe (Etiope and Klusman, 2010), microseepage emission factors could be divided into three main classes:

1. level 1: high microseepage (>50mg m^{-2} d^{-1});
2. level 2: medium microseepage (5–50mg m^{-2} d^{-1});
3. level 3: low microseepage (0–5mg m^{-2} d^{-1}).

In general, levels 1 and 2 mainly occurred in sectors hosting macroseepage sites and in sedimentary basins during the winter. Of the 563 flux data points, 276 were positive fluxes (49 per cent) and 3 per cent were in the level 1 range (mean of 210mg m^{-2} d^{-1}). Level 2 represented approximately 12 per cent of the surveyed areas (mean of 14.5mg m^{-2} d^{-1}) and level 3 was common in winter, far from macroseepage zones, accounting for roughly 34 per cent of the surveyed sedimentary zones (mean of 1.4mg m^{-2} d^{-1}). These percentages should be considered as a first spatial disaggregation of emission factors. Many more data points are necessary to get a more reliable and globally representative disaggregation. Actually, microseepage flux depends upon two main geological factors: the amount and pressure of the reservoir gas, and the permeability of the rocks and faults that, in turn, are controlled by tectonic activity and brittle lithologic response. Therefore, it is expected that emission factors are higher in regions characterized by active tectonics, neotectonics and seismicity.

Onshore macroseeps

Onshore macroseep flux measurements are available mainly from manifestations in Europe and Azerbaijan (Etiope, 2009, and references therein), Asia (Yang et al, 2004) and the US (for example Duffy et al, 2007). Macroseep gas fluxes can cover a wide range of values. Single vents or craters of small mud volcanoes (1–5m high) can release up to tens of tonnes of CH$_4$ per year. A whole mud volcano (hosting tens or hundreds of vents) can continuously emit hundreds of tonnes of CH$_4$ per year, and eruptions from mud volcanoes can release thousands of tonnes of CH$_4$ in a few hours. From 1810 until present, more than 250 eruptions from 60 mud volcanoes were observed in Azerbaijan. Dry seeps (non-mud volcanoes) often have higher fluxes, reaching thousands of tonnes per year (Etiope, 2009).

Also important to consider is that around a seep there is always a wide microseepage zone, a halo around a vent zone, and it should be included in calculation for total gas seep emissions. Then, emission calculations from macroseepage zones are determined by summing the macroseep flux component ($E_{macro} = \Sigma F_{vent}$, sum of all of the vent fluxes, measured or estimated) to the microseepage component ($E_{micro} = A \times <F>$).

In all mud volcano areas measured thus far (Italy, Romania, Azerbaijan and Taiwan), the specific flux, including microseepage and macroseeps (excluding the eruptions), was between 100 and 1000 tonnes km^{-2} yr^{-1} (Etiope et al, 2002, 2004a, 2004b, 2007b; Hong and Yang, 2007).

The flux from vents can be determined using several techniques, including closed-chamber systems, inverted funnel systems, flux meters (associated with portable gas chromatographs), semiconductors or infrared laser sensors. These techniques have been described previously by Etiope et al (2002, 2004a). In some cases, emissions from eruptions were estimated using only visual or indirect methods.

Submarine fluxes

Offshore emission data are available mainly from the US (offshore of California and the Gulf of Mexico), the North Sea, the Black Sea, Spain, Denmark, Taiwan and Japan. However, in many cases the data only refer to gas outputs from the seafloor to the water column, and not to the fraction entering the atmosphere. Submarine seeps can release 10^3–10^6 tonnes yr^{-1} of gas over a 10^5km^2 area (Judd et al, 1997). The flux of individual seepage or groups of bubble streams may reach several tonnes per year (Hornafius et al, 1999; Judd, 2004). One of the main problems in extrapolating regionally and globally is the uncertainty associated with the actual area of active seepage.

Submarine gas fluxes are generally estimated on the basis of geophysical images (echo-sounders, seismic, sub-bottom profilers and side-scan sonar records) and bubble parameterization (the size of the bubble plume and single bubbles), sometimes associated with geochemical seawater analysis (for example Judd et al, 1997). Recent studies have proposed using remote sensing techniques based on airborne visible/infrared imaging spectrometry (Leifer et al, 2006).

Geothermal and volcanic fluxes

Volcanoes are not a significant CH_4 source. Methane concentrations in volcanic gases are generally of the order of a few ppmv, with emissions derived by CO_2/CH_4 or H_2O/CH_4 ratios and CO_2 or H_2O fluxes, ranging from a few to tens of tonnes/year (Ryan et al, 2006; Etiope et al, 2007a). Methane emissions from geothermal fluids (where inorganic synthesis, thermo-metamorphism and thermal breakdown of organic matter are substantial), are not negligible, globally.

The gas composition of geothermal vents, mofettes and bubbling springs is generally more than 90 per cent CO_2. The fraction of CH_4 is low, typically 0.01 to 1 per cent, but the amount of the total gas released is of the order of 10^3–10^5t yr^{-1}, and may result in significant emissions of CH_4 to the atmosphere (10^1–10^2t yr^{-1} from individual vents). The specific flux of soil degassing is generally on the order of 1–10t km^{-2} yr^{-1} (Etiope et al, 2007a). In the absence of direct CH_4 measurements, emissions of CH_4 can be estimated from knowing the CO_2 flux and the CO_2/CH_4 concentration ratio or the steam flux and the steam/CH_4 concentration ratio (Etiope et al, 2007a).

Global emission estimates

Global emission estimates for natural geological CH_4 sources are listed in Table 4.1; the latest estimates are also summarized in Figure 4.4.

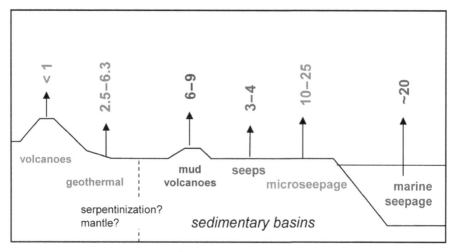

Figure 4.4 Estimates of methane emissions from geological sources

Source: Based on data from Etiope et al (2008)

For global submarine emission estimates, a dual approach was utilized based on the seep flux and on the amount of geological CH_4 produced and available to seep (Kvenvolden et al, 2001). The two approaches produced comparable results, 30 and 10Tg yr^{-1}, respectively, with the average, ~20Tg yr^{-1}, still a consensus value (Judd, 2004) awaiting refinement.

Unlike seabed seeps, onshore global emission estimates were derived on the basis of hundreds of flux measurements performed since 2001; some estimates (mud volcanoes, microseepage and geothermal sources) follow upscaling methods recommended by the EMEP/CORINAIR Guidelines, which are based on the concepts of an 'emission factor' and 'area' or 'point' sources (EEA, 2004; Etiope et al, 2007a).

Table 4.1 Global emissions of methane from geological sources

	Emission (Tg yr^{-1})	Reference
Marine seepage	18–48	Hornafius et al (1999)
	10–30 (20)	Kvenvolden et al (2001)
Mud volcanoes	5–10	Etiope and Klusman (2002)
	10.3–12.6	Dimitrov (2002)
	6	Milkov et al (2003)
	6–9	Etiope and Milkov (2004)
Other macroseeps	3–4	Etiope et al (2008)
Microseepage	>7	Klusman et al (1998)
	10–25	Etiope and Klusman (2010)
Geothermal/volcanic areas	1.7–9.4	Lacroix (1993)
	2.5–6.3[a]	Etiope and Klusman (2002)
	<1[b]	Etiope et al (2008)
	30–70[c]	Etiope and Klusman (2002)
TOTAL	13–36[d]	Judd (2004)
	35–45[e]	Etiope and Milkov (2004)
	45[c, e]	Kvenvolden and Rogers (2005)
	40–60[c]	Etiope (2004); Etiope and Klusman (2010)
	42–64[c]	Etiope et al (2008) – best estimate
	30–80[c]	Etiope et al (2008) – extended range

Note: [a] Volcanoes not considered; [b] only volcanoes; [c] gas hydrates not considered; [d] microseepage not considered; [e] former microseepage estimate.

The CH$_4$ emission estimates from mud volcanoes differ slightly, although they were derived from different data sets and approaches. The latest estimate, by Etiope and Milkov (2004), was based on direct measurements of flux, and is probably the only one that includes both focused venting and diffuse microseepage around craters and vents. The estimate was also based on a classification of mud volcano sizes in terms of area, following a compilation of data from 120 mud volcanoes. Global emissions from other seeps were based on a database of fluxes measured directly or visually estimated from 66 gas seeps in 12 countries, with the assumption that their flux and size distributions were representative of the global macroseep population, at least 12,500 seeps (Etiope et al, 2008).

The most recent global microseepage emission value was derived on the basis of an accurate estimate of the global area of oil and gas fields, and on TPS, the average flux from each of the three microseepage levels recognized in the global data set (563 measurements), assuming that the percentage of occurrence of the three levels (3 per cent, 12 per cent and 34 per cent) was valid at the global scale (Etiope and Klusman, 2010). Since measurements were made in all seasons, seasonal variations were incorporated into the data set.

Upscaling the measurement to all gas/oil field areas would give a total microseepage of the order of 11–13Tg yr^{-1}. Extrapolating to the global potential microseepage area (TPS: ~8 million km^2) would result in an emission in the order of 25Tg yr^{-1}. These estimates are coherent with the lower limit of 7Tg yr^{-1} initially suggested by Klusman et al (1998) and Etiope and Klusman (2002). However, more measurements in various areas and for different seasons are needed to refine the three-level classification and the actual area of seepage. Finally, global geothermal CH$_4$ flux estimates of 0.9–3.2Tg yr^{-1} were preliminarily proposed by Lacroix (1993). A wide data set was then reported by Etiope and Klusman (2002), who conservatively derived a global geothermal flux between 2.5 and 6.3Tg yr^{-1}. Lacroix (1993) also suggested a global volcanic CH$_4$ flux of 0.8–6.2Tg yr^{-1}. More recently, volcanic emissions have been considered as not exceeding 1Tg yr^{-1} (Etiope et al, 2008).

Thus, global geo-CH$_4$ emission estimates, stemming from mud volcanoes plus other seeps, plus microseeps, plus submarine emissions, plus geothermal and volcanic emissions range from 42 to 64Tg yr^{-1} (mean of 53Tg yr^{-1}), almost 10 per cent of total CH$_4$ emissions, representing the second most important natural CH$_4$ source after wetlands. Geo-CH$_4$ sources would then also represent the missing source of fossil CH$_4$ as recognized in the recent re-evaluation of the fossil CH$_4$ budget for the atmosphere (~30 per cent) (see Lassey et al, 2007; Etiope et al, 2008), which implies a total fossil CH$_4$ emission much higher than that due to the fossil fuel industry. Global geo-CH$_4$ emission estimates are on the same level or higher than other sources (such as biomass burning, termites, wild animals, oceans and wildfires) considered by the IPCC (Denman et al, 2007) (Figure 4.5).

Recent studies indicate that earth's degassing also accounts for at least 17 per cent and 10 per cent of global emissions of ethane and propane, respectively (Etiope and Ciccioli, 2009), hydrocarbons that contribute to photochemical pollution and ozone production in the atmosphere.

Uncertainties

Emission factors for microseepage and macroseeps and related ranges are well known. Uncertainties in global emission estimates are mainly due to a poor knowledge of the actual area of shallow submarine macroseeps and, even more importantly, the dryland area of invisible microseepage. It is evident that all seeps and microseepage zones occur within hydrocarbon provinces, in particular within TPSs (Etiope and Klusman, 2010), but actual microseepage areas are not known. Accordingly, three main levels of spatial disaggregation can be defined, with increasing estimates for uncertainty: an area including (encompassing) sites of verified microseepage flux; an area encompassing macroseeps (where microseepage very likely occurs); and an area encompassing oil-gas fields (where microseepage likely occurs).

The classification can be used for upscaling procedures but, presently, there are no detailed maps of verified microseepage. The definition of the area used for emission calculations depends on the recognition of homogeneous

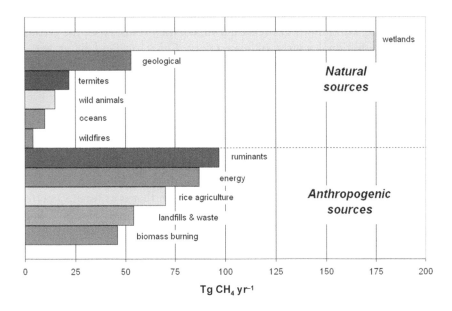

Figure 4.5 Global methane sources revisited

Source: Natural and anthropogenic sources are from IPCC AR4 (Denman et al, 2007; average of values in Table 7.6); geological source is from Etiope et al (2008).

identifiable areas and the spatial variability of the flux measured. Currently, estimations are performed based on the distribution of oil fields, assuming that approximately 50 per cent of oil field areas have positive fluxes of CH_4 from soils, as suggested by preliminary data sets (Etiope and Klusman, 2010). The area identified in oil field maps is then transformed into polygons that are later used in calculations. The polygons drawn are used as a rough method for estimating the emitting area. In addition, the use of polygons most likely results in an over and/or under estimation of emitting areas. Somehow in the overall scenario emission values may be closer in reality than one may think since errors in area estimations are balanced.

Qualitatively, it is known that microseepage is higher in winter and lower in summer, due to differences in methanotrophic activity between the two seasons, removing CH_4 before it can reach the atmosphere. Other short-term or seasonal variability is due to meteorology and soil conditions. Longer-time variability (decades, centuries and millennia) can be induced by endogenic factors (changes of pressure gradients in rocks, tectonic stress, etc.).

For macroseepage, the main source of uncertainty is temporal variations in emissions. The largest fraction of emissions occurs during 'individual' events/eruptions that are difficult to simulate. The resultant emissions are also not easily quantifiable. Thus, calculations are normally done using assumptions of continuous gas release from counted vents. Also, the census of vents is an additional source of uncertainty. Either on land or at sea, the majority of large macroseeps have been identified and studied, but most of the small ones have not been determined, surveyed or characterized.

Conclusions

Geological emissions of gaseous hydrocarbons represent the second natural source of CH_4 after wetlands and are comparable with or higher than other man-made sources. The atmospheric fossil fraction of CH_4, estimated at ~30 per cent, would support a geo-CH_4 source strength that is at least 50 per cent of anthropogenic fossil emissions, ~90–100Tg yr^{-1} (Lassey et al, 2007; Etiope et al, 2008).

Emission factors from geological sources are fairly well known. However, uncertainties arise due to incomplete knowledge of the actual area of microseepage from soils and from submarine emissions. However, the uncertainties of the global emission estimates are lower than or comparable with those of other sources traditionally considered in the literature and in IPCC reports. Further studies, based on direct field measurements (especially for diffuse microseepage and underwater sources) are needed in order to reduce the uncertainties.

Neotectonics, seismicity and magmatism may have a profound influence on planetary degassing because they can control the accumulation, the migration and the discharge of subsurface gases (Morner and Etiope, 2002). The activity of mud volcanoes is also related to seismic events (Mellors et al, 2007). Neotectonic faults and fractures cutting Pleistocene sediments have been recognized as migration pathways for fluids rising from diapirs and hydrocarbon accumulations (for example Revil, 2002). Salt tectonics itself is a powerful factor capable of creating crustal weakness zones that are very effective as gas migration routes (Etiope et al, 2006). Basically, the geological CH_4 source follows the physical laws of gas migration in rocks and sediments (i.e. the transport equations for gas in fractured media; for example Etiope and Martinelli, 2002). Today it is clear that the atmospheric greenhouse gas budget is not independent of the geophysical processes of the solid earth that lead to lithospheric degassing, the third 'breath' of our planet.

References

Abrajano, T. A., Sturchio, N. C., Bohlke, J. K., Lyon, G. L., Poreda, R. J. and Stevens, C. M. (1988) 'Methane-hydrogen gas seeps, Zambales Ophiolite, Philippines: Deep or shallow origin?', *Chem. Geol.*, vol 71, pp211–222

Abrams, M. A. (2005) 'Significance of hydrocarbon seepage relative to petroleum generation and entrapment', *Mar. Petrol. Geol.*, vol 22, pp457–477

Borowski, W. S., Paull, C. K. and Ussler, W. (1999) 'Global and local variations of interstitial sulfate gradients in deep-water, continental margin sediments: Sensitivity to underlying CH_4 and gas hydrates', *Mar. Geol.*, vol 159, pp131–154

Clarke, R. H. and Cleverly, R. W. (1991) 'Petroleum seepage and post-accumulation migration', in W. A. England and A. J. Fleet (eds) *Petroleum Migration*, Geological Society Special Publication N. 59, Geological Society of London, Bath, pp265–271

Denman, K. L., Brasseur, G., Chidthaisong, A., Ciais, P., Cox, P. M., Dickinson, R. E., Hauglustaine, D., Heinze, C., Holland, E., Jacob, D., Lohmann, U., Ramachandran, S., da Silva Dias, P. L., Wofsy S. C. and Zhang, X. (2007) 'Couplings between changes in the climate system and biogeochemistry' in S. Solomon, D. Qin, M. Manning, Z. Chen, M. Marquis, K. B. Averyt, M. Tignor and H. L. Miller (eds) *Climate Change 2007: The Physical Science Basis*, Cambridge University Press, Cambridge and New York, pp499–587

Dimitrov, L. (2002) 'Mud volcanoes – the most important pathway for degassing deeply buried sediments', *Earth-Sci. Rev.*, vol 59, pp49–76

Duffy, M., Kinnaman, F. S., Valentine, D. L., Keller, E. A. and Clark J. F. (2007) 'Gaseous emission rates from natural petroleum seeps in the Upper Ojai Valley', California *Environ. Geosciences.*, vol 14, pp197–207

EEA (European Environment Agency) (2004) *Joint EMEP/CORINAIR Atmospheric Emission Inventory Guidebook*, 4th Edition, European Environment Agency, Copenhagen, http://reports.eea.eu.int/EMEPCORINAIR4/en

EMEP/EEA (European Monitoring and Evaluation Programme/EEA)(2009) *EMEP/EEA Air Pollutant Emission Inventory Guidebook – 2009. Technical Guidance to Prepare National Emission Inventories*, EEA Technical report No 6/2009, European Environment Agency, Copenhagen, doi:10.2800/23924

Emmanuel, S. and Ague, J. J. (2007) 'Implications of present-day abiogenic methane fluxes for the early Archean atmosphere', *Geophys. Res. Lett.*, vol 34, L15810

Etiope, G. (2004) 'GEM – Geologic Emissions of Methane, the missing source in the atmospheric methane budget', *Atm. Environ.*, vol 38, pp3099–3100

Etiope, G. (2009) 'Natural emissions of methane from geological seepage in Europe', *Atm. Environ*, vol 43, pp1430–1443

Etiope, G. and Ciccioli, P. (2009) 'Earth's degassing – A missing ethane and propane source', *Science*, vol 323, no 5913, p478

Etiope, G. and Klusman, R. W. (2002) 'Geologic emissions of methane to the atmosphere', *Chemosphere*, vol 49, pp777–789

Etiope, G. and Klusman, R. W. (2010) 'Microseepage in drylands: Flux and implications in the global atmospheric source/sink budget of methane', *Global Planet. Change*, doi:10.1016/j.gloplacha.2010.01.002, in press

Etiope, G. and Martinelli, G. (2002) 'Migration of carrier and trace gases in the geosphere: An overview', *Phys. Earth Planet. Int.*, vol 129, no 3–4, pp185–204

Etiope, G. and Milkov, A. V. (2004) 'A new estimate of global methane flux from onshore and shallow submarine mud volcanoes to the atmosphere', *Env. Geology*, vol 46, pp997–1002

Etiope, G., Caracausi, A., Favara, R., Italiano, F. and Baciu, C. (2002) 'Methane emission from the mud volcanoes of Sicily (Italy)', *Geophys. Res. Lett.*, vol 29, no 8, doi:10.1029/2001GL014340

Etiope, G., Feyzullaiev, A., Baciu, C. L. and Milkov, A. V. (2004a) 'Methane emission from mud volcanoes in eastern Azerbaijan', *Geology*, vol 32, no 6, pp465–468

Etiope, G., Baciu, C., Caracausi, A., Italiano, F. and Cosma, C. (2004b) 'Gas flux to the atmosphere from mud volcanoes in eastern Romania', *Terra Nova*, vol 16, pp179–184

Etiope, G., Papatheodorou, G., Christodoulou, D., Ferentinos, G., Sokos, E. and Favali, P. (2006) 'Methane and hydrogen sulfide seepage in the NW Peloponnesus petroliferous basin (Greece): Origin and geohazard', *Amer. Assoc. Petrol. Geol. Bulletin*, vol 90, no 5, pp701–713

Etiope, G., Fridriksson, T., Italiano, F., Winiwarter, W. and Theloke, J. (2007a) 'Natural emissions of methane from geothermal and volcanic sources in Europe', *J. Volc. Geoth. Res.*, vol 165, pp76–86

Etiope, G., Martinelli, G., Caracausi, A. and Italiano, F. (2007b) 'Methane seeps and mud volcanoes in Italy: Gas origin, fractionation and emission to the atmosphere', *Geophys. Res. Lett.*, vol 34, doi:10.1029/2007GL030341

Etiope, G., Lassey, K. R., Klusman, R. W. and Boschi, E. (2008) 'Reappraisal of the fossil methane budget and related emission from geologic sources', *Geophys. Res. Lett.*, vol 35, no L09307, doi:10.1029/2008GL033623

Etiope, G., Feyzullayev, A. and Baciu, C. L. (2009) 'Terrestrial methane seeps and mud volcanoes: A global perspective of gas origin', *Mar. Petroleum Geology*, vol 26, pp333–344

Fiebig, J., Chiodini, G., Caliro, S., Rizzo, A., Spangenberg, J. and Hunziker, J. C. (2004) 'Chemical and isotopic equilibrium between CO_2 and CH_4 in fumarolic gas discharges: Generation of CH_4 in arc magmatic-hydrothermal systems', *Geochim Cosmochim Acta*, vol 68, pp2321–2334

Hong, W. L. and Yang, T. F. (2007) 'Methane flux from accretionary prism through mud volcano area in Taiwan – from present to the past', *Proceed. 9th International Conference on Gas Geochemistry*, 1–8 October, 2007, National Taiwan University, pp80–81

Hornafius, J. S., Quigley, D. and Luyendyk, B. P. (1999) 'The world's most spectacular marine hydrocarbon seeps (Coal Oil Point, Santa Barbara Channel, California): Quantification of emissions', *J. Geoph. Res.*, vol 104, pp20703–20711

Hosgormez, H., Etiope, G. and Yalçın, M. N. (2008) 'New evidence for a mixed inorganic and organic origin of the Olympic Chimaera fire (Turkey): A large onshore seepage of abiogenic gas', *Geofluids*, vol 8, pp263–275

Hunt, J. M. (1996) *Petroleum geochemistry and geology*, W. H. Freeman and Co., New York

Judd, A. G. (2004) 'Natural seabed seeps as sources of atmospheric methane', *Env. Geology*, vol 46, pp988–996

Judd, A. G. and Hovland, M. (2007) *Seabed Fluid Flow: Impact on Geology, Biology and the Marine Environment*, Cambridge University Press, Cambridge, UK, web material: www.cambridge.org/catalogue/catalogue.asp?isbn=9780521819503&ss=res

Judd, A. G., Davies, J., Wilson, J., Holmes, R., Baron, G. and Bryden, I. (1997) 'Contributions to atmospheric methane by natural seepages on the UK continental shelf', *Mar. Geology*, vol 137, pp165–189

Judd, A. G., Hovland, M., Dimitrov, L. I., Garcia Gil, S. and Jukes, V. (2002) 'The geological methane budget at Continental Margins and its influence on climate change', *Geofluids*, vol 2, pp109–126

Judd, A. G., Croker P., Tizzard, L. and Voisey, C. (2007) 'Extensive methane-derived authigenic carbonates in the Irish Sea', *Geo-Marine Lett.*, vol 27, pp259–268

Khan, S. D. and Jacobson, S. (2008) 'Remote sensing and geochemistry for detecting hydrocarbon microseepages', *GSA Bulletin*, vol 120, no 1–2, pp96–105

Klusman, R. W., Jakel, M. E. and LeRoy, M. P. (1998) 'Does microseepage of methane and light hydrocarbons contribute to the atmospheric budget of methane and to global climate change?', *Assoc. Petrol. Geochem. Explor. Bull.*, vol 11, pp1–55

Klusman, R. W., Leopold, M. E. and LeRoy, M. P. (2000) 'Seasonal variation in methane fluxes from sedimentary basins to the atmosphere: Results from chamber measurements and modeling of transport from deep sources', *J. Geophys. Res.*, vol 105D, pp24661–24670

Kopf, A. J. (2002) 'Significance of mud volcanism', *Rev. Geophysics*, vol 40, no 1005, pp1–52, doi:10.1029/2000RG000093

Kvenvolden, K. A. (1988) 'Methane hydrate and global climate', *Global Biog. Cycles*, vol 2, pp221–229

Kvenvolden, K. A. and Rogers, B. W. (2005) 'Gaia's breath – global methane exhalations', *Mar. Petrol. Geol.*, vol 22, pp579–590

Kvenvolden, K. A., Lorenson, T. D. and Reeburgh, W. (2001) 'Attention turns to naturally occurring methane seepage', *EOS*, vol 82, p457

Lacroix, A. V. (1993) 'Unaccounted-for sources of fossil and isotopically enriched methane and their contribution to the emissions inventory: A review and synthesis', *Chemosphere*, vol 26, pp507–557

Lassey, K. R., Lowe, D. C. and Smith, A. M. (2007) 'The atmospheric cycling of radiomethane and the "fossil fraction" of the methane source', *Atmos. Chem., Phys.*, vol 7, pp2141–2149

Leifer, I., and Patro, R. K. (2002) 'The bubble mechanism for methane transport from the shallow sea bed to the surface: A review and sensitivity study', *Cont. Shelf Res.*, vol 22, pp2409–2428

Leifer, I., Roberts, D., Margolis, J. and Kinnaman, F. (2006) 'In situ sensing of methane emissions from natural marine hydrocarbon seeps: A potential remote sensing technology', *Earth and Planetary Science Letters*, vol 245, pp509–522

Lelieveld, J., Crutzen, P. J. and Dentener, F. J. (1998) 'Changing concentration, lifetime and climate forcing of atmospheric methane', *Tellus*, vol 50B, pp128–150

Magoon, L. B. and Schmoker, J. W. (2000) 'The Total Petroleum System – the natural fluid network that constraints the assessment units', in *US Geological Survey World Petroleum Assessment 2000. Description and results*, USGS Digital Data Series 60, World Energy Assessment Team, USGS Denver Federal Center, Denver, CO, p31

Mellors, R., Kilb, D. Aliyev, A., Gasanov, A. and Yetirmishli, G. (2007) 'Correlations between earthquakes and large mud volcano eruptions', *J. Geophys. Res.*, vol 112, B04304, doi:10.1029/2006JB004489

Milkov, A. V. (2000) 'Worldwide distribution of submarine mud volcanoes and associated gas hydrates', *Mar. Geology*, vol 167, no 1–2, pp29–42

Milkov, A. V., Sassen, R., Apanasovich, T. V. and Dadashev, F. G. (2003) 'Global gas flux from mud volcanoes: A significant source of fossil methane in the atmosphere and the ocean', *Geoph. Res. Lett.*, vol 30, no 2, 1037, doi:10.1029/2002GL016358

Morner, N. A. and Etiope, G. (2002) 'Carbon degassing from the lithosphere', *Global and Planet. Change*, vol 33, no 1–2, pp185–203

Prather, M., Ehhalt, D., Dentener, F., Derwent, R., Dlugokencky, E. J., Holland, E., Isaksen, I., Katima, J., Kirchhoff, V., Matson, P., Midgley, P. and Wang, M. (2001) 'Atmospheric chemistry and greenhouse gases', in J. T. Houghton, T. Y. Ding, D. J. Griggs, M. Nogeur, P. J. van der Linden, X. Dai, K. Maskell and C. A. Johnson (eds) *Climate Change 2001: The Scientific Basis. Contribution of Working Group I to the Third Assessment Report of the Intergovernmental Panel on Climate Change*, Cambridge University Press, Cambridge, pp239–287

Revil, A. (2002) 'Genesis of mud volcanoes in sedimentary basins: A solitary wave-based mechanism', *Geophys. Res. Lett.*, vol 29, no 12, doi:10.1029/2001GL014465

Ryan, S., Dlugokencky, E. J., Tans, P. P. and Trudeau, M. E. (2006) 'Mauna Loa volcano is not a methane source: Implications for Mars', *Geophys. Res. Lett.*, vol 33, L12301, doi:10.1029/2006GL026223

Sano, Y., Urabe, A., Wakita, H. and Wushiki, H. (1993) 'Origin of hydrogen-nitrogen gas seeps, Oman', *Applied Geochem.*, vol 8, pp1–8

Saunders, D. F., Burson, K. R. and Thompson, C. K. (1999) 'Model for hydrocarbon microseepage and related nearsurface alterations', *AAPG Bulletin*, vol 83, pp170–185

Schimel, D., Alves, D., Enting, I., Heimann, M., Joos, F., Raynaud, D., Wigley, T., Prather, M., Derwent, R., Ehhalt, D., Fraser, P., Sanhueza, E., Zhou, X., Jonas, P., Charlson, R., Rodhe, H., Sadasivan, S., Shine, K. P., Fouquart, Y., Ramaswamy, V., Solomon, S., Srinivasan, J., Albritton, D., Derwent, R., Isaksen, I., Lal, M. and Wuebbles, D. (1996) 'Radiative forcing of climate change', in J. T. Houghton, L. G. M. Filho, B. A. Callander, N. Harris, A. Kattenberg, and K. Maskell (eds) *Climate Change 1995: The Science of Climate Change*, Intergovernmental Panel on Climate Change, Cambridge University Press, Cambridge, pp65–131

Schmale, O., Greinert, J. and Rehder, G. (2005) 'Methane emission from high-intensity marine gas seeps in the Black Sea into the atmosphere', *Geophys. Res. Lett.*, vol 32, L07609, doi:10.1029/2004GL021138

Schumacher, D. and Abrams, M. A. (1996) 'Hydrocarbon migration and its near surface expression', *Amer. Assoc. Petrol. Geol., Memoir*, vol 66, p446

Taran, Y. A. and Giggenbach, W. F. (2003) 'Geochemistry of light hydrocarbons in subduction-related volcanic and hydrothermal fluids', in S. F. Simmons and I. Graham (eds) *Volcanic, Geothermal, and Ore-forming Fluids: Rulers and Witnesses of Processes Within the Earth*, Society of Economic Geologists, Special Publication, no 10, pp61–74

Thielemann, T., Lucke, A., Schleser, G. H. and Littke, R. (2000) 'Methane exchange between coal-bearing basins and the atmosphere: The Ruhr Basin and the Lower Rhine Embayment, Germany', *Org. Geochem.*, vol 31, pp1387–1408

US EPA (United States Environmental Protection Agency) (2010) *Methane and Nitrous Oxide Emissions from Natural Sources*, Report no EPA-430-R-09-025, US Environmental Protection Agency, Office of Atmospheric Programs, Climate Change Division, Washington, DC

Wagner, M., Wagner, M., Piske, J. and Smit, R. (2002) 'Case histories of microbial prospection for oil and gas, onshore and offshore northwest Europe', in D. Schumacher and L. A. LeSchack (eds) *Surface Exploration Case Histories:*

Applications of Geochemistry, Magnetics and Remote Sensing, AAPG Studies in Geology no 48 and SEG Geophys. Ref. Series no 11, pp453–479

Yang, T. F., Yeh, G. H., Fu, C. C., Wang, C. C., Lan, T. F., Lee, H. F., Chen, C.-H., Walia, V. and Sung, Q. C. (2004) 'Composition and exhalation flux of gases from mud volcanoes in Taiwan', *Environ. Geol.*, vol 46, pp1003–1011

5
Termites

David E. Bignell

Introduction

The last 20 years have seen growing interest in the terrestrial carbon cycle in the context of global change. The consequent attempts to fully quantify it, over the whole land surface of the earth, have seen special attention given to greenhouse gas emissions. Social insects have high visibility in many terrestrial biomes, reflecting their large abundance and biomass, with important consequent impacts on ecosystem processes (Hölldobler and Wilson, 2009). Of the social insects, ants and termites clearly outnumber not only bees and wasps, but quite commonly the remaining macroarthropods as well (for example Billen, 1992; Stork and Brendell, 1993). In their classic, but still highly relevant study of central Amazonian rainforest, Fittkau and Klinge (1973) showed that ants and termites together made up more than 15 per cent of all animal biomass. Subsequent work suggests that this dominance is general in the humid tropics and improvements in sampling methods (especially for the more cryptic termites, see Moreira et al, 2008) indicate that the greater contribution is in fact made by termites, which can in exceptional cases reach biomass densities of more than 100g m^{-2}, an impact which could only be matched or exceeded by earthworms (for reviews see Wood and Sands, 1978; Lavelle et al, 1997; Watt et al, 1997; Bignell and Eggleton, 2000; Bignell, 2006). Termite colonies are known to produce a range of trace gases that potentially affect atmospheric chemistry and/or the earth's radiative balance. The gases are CO_2, CH_4, N_2O, H_2, CO and chloroform ($CHCl_3$), but only CO_2 and CH_4 are generated in sufficient quantity to make termites of interest in global budgets (Khalil et al, 1990). Of these, CH_4 emissions attract most attention, as this gas has 25 times the radiative forcing potential of CO_2 over a 100-year time horizon. As a consequence, there have been several quantitative studies of termite assemblages, often accompanied by attempts to estimate their contribution of CH_4 fluxes to the atmosphere.

Biochemistry and microbiology of methane production

A detailed discussion of the process of CH_4 production by microorganisms is beyond the scope of this chapter (see Chapter 2). Excellent recent reviews of the relevant microbiology and microbial physiology can be found in Breznak (2006), Brune (2006), Breznak and Leadbetter (2006) and Purdy (2007). The more general phenomenon of CH_4 production by terrestrial arthropods is reviewed by Hackstein et al (2006). Although the broad basis of lignocellulose digestion by termites has been known for more than 60 years (Hungate, 1946), numerous refinements continue to be added to the picture.

Net methane efflux from termites

Termites (worker caste) isolated in the laboratory have been reported to produce CH_4 at rates ranging from undetectable levels to 1.6µmol g^{-1} h^{-1} (Brauman et al, 1992; Bignell et al, 1997; Nunes et al, 1997; Sugimoto et al, 1998a, 1998b). Zimmerman et al (1982) first drew attention to the potentially large quantities of CH_4 emitted by termites. Their estimates were 75–310Tg yr^{-1}, equivalent to 13–56 per cent of global sources. However, subsequent estimates of annual CH_4 emissions from termites have been rather lower, cf. 10–90Tg (Rasmussen and Khalil, 1983), 10–30Tg (Collins and Wood, 1984), 2–5Tg (Seiler et al, 1984) and 6–42Tg (Fraser et al, 1986). More controversially, while acknowledging the uncertainties Zimmerman et al (1982) predicted that under certain scenarios of land use change, the quantities released could equal those of all other natural sources. Since then the subject has been revisited by numerous authors and almost all of the assumptions made by Zimmerman et al (1982) have been overturned, but their contribution remains pivotal and was the first to give attention to the importance of feeding (= functional) group balances in trace gas emissions. Reviews by Sanderson (1996), Bignell et al (1997) and Sugimoto et al (1998b) summarize the available data on gross CH_4 production, emphasizing the differences in functional group balances across different biogeographical regions and the importance of obtaining accurate local population data before scaling-up. All three reviews give provisional estimates of the gross CH_4 production by termites below, or well below 20 per cent of all global sources, but draw attention to the importance of understanding the interaction between sources and sinks (particularly the local oxidation of CH_4 by soil microorganisms) before a final assessment of net termite contributions can be made.

Clearly, the total produced in a colony depends on the emission rate from individual termites and the colony size. The first can be quite easily determined, while the second is generally estimated by a mixture of guesswork and extrapolations (Eggleton et al, 1996; Bignell et al, 1997; MacDonald et al, 1999). Across the entire spectrum of termite species the available data suggest that emission rates within the same caste are relatively consistent, with higher production in workers (the most abundant caste) than in soldiers, reproductives and nymphs (Sugimoto et al, 1998a). As a consequence, most estimates of colony production are based on measurements of workers alone. Any variation of CH_4 emission by workers from colony to colony is very critical for extrapolations at all spatial scales. In lower termites, extremely large inter-colonial variation has been reported by Wheeler et al (1996), but higher termites appear to be more consistent across a species and even within genera (Sugimoto et al, 1998a).

Early contributions by Brauman et al (1992) and Rouland et al (1993) concentrated on the differences in CH_4 fluxes between species and feeding groups in a sample of 24 types of termite from Africa and North America. On a biomass-specific basis, soil-feeding forms were found to produce more CH_4 than wood-feeding species, with rates exceeding $1.0\mu mol\ g^{-1}\ h^{-1}$ in some cases. This difference has been confirmed in the African assemblage (Nunes et al, 1997) and in Southeast Asia and Australia (Jeeva et al, 1998; Sugimoto et al, 1998a), by studies in which worker caste termites are incubated under standard conditions. Typical values for wood-feeders are in the range negligible to $0.2\mu mol\ g^{-1}\ h^{-1}$, whereas soil-feeders generally cluster between 0.1 and $0.4\mu mol\ g^{-1}\ h^{-1}$, although some higher rates are reported. Values for species of the fungus-growing Macrotermitinae are variable and range from 0.1 to $0.3\mu mol\ g^{-1}\ h^{-1}$; however, some colonies are very large and even at these relatively low biomass-specific rates, can be significant point sources (Tyler et al, 1988; Darlington et al, 1997). But the outstanding CH_4 producers in the Cameroon (Nunes et al, 1997) and Borneo (Jeeva et al, 1998) assemblages are species of the genus *Termes*, with production consistent between 0.5 and $1.0\mu mol\ g^{-1}\ h^{-1}$. Production by other members of the *Termes/Capritermes* clade (sensu Miller, 1991) in Thailand and Northern Australia is also high (Sugimoto et al, 1998a). However, some termite species (especially those feeding on sound wood or relatively undecayed material) produce little or no CH_4. A very high value of $1.6\mu mol\ g^{-1}\ h^{-1}$ is reported for *Incisitermes minor*, but in a single colony only; ten others of the same species produced no CH_4 at all (Wheeler et al, 1996). Sugimoto et al (1998a) also suggested that there could be large variations between colonies in the same species, an observation which is at odds with what is known of the microbiology of the methanogenic process (there would be debilitating redox imbalances in the intestine) and disconcerting for scaling-up exercises. Although not all literature reports can be compared, there appear to be taxonomy-based differences among the subfamilies of Termitidae (Table 5.1), possibly associated with the preponderance of soil-feeders in the Apicotermitinae and Termitinae. In most cases the proportion of total carbon (C) mineralized emitted as CH_4 is small (Bignell et al, 1997). The highest figure in the literature, which shows just over 8 per cent of the total C efflux as CH_4, is for the Australian soil-feeding species *Lophotermes septentrionalis* (Sugimoto et al, 1998a). Four other soil-feeders are reported to produce between 5 per cent and 6 per cent of C efflux as CH_4: *Megagnathotermes sunteri*, *Cubitermes heghi*, *Procubitermes arboricola* (all Termitinae) and *Jugositermes tuberculatus* (Apicotermitinae) (Bignell et al, 1997; Sugimoto et al, 1998a).

Table 5.1 Range of reported values for methane emissions by worker caste termites, grouped by subfamily and feeding type (from Sugimoto et al, 1998b)

Subfamily	Diet	CH_4 emission range, $\mu mol\ g^{-1}\ h^{-1}$	CH_4 emission, average $\mu mol\ g^{-1}\ h^{-1}$	CH_4/CO_2, range
Macrotermitinae	wood, leaf litter	0.02–0.36	0.15 (23)	0.001–0.025
Termitinae (wood-feeders)	wood, grass	0.04–0.17	0.11 (13)	0.0003–0.030
Termitinae (soil-feeders)	soil, humus	0.11–1.09	0.41 (16)	0.026–0.090
Nasutitermitinae	wood, grass	0.11–0.18	0.16 (8)	0.012–0.016
Apicotermitinae	Soil, humus	0.05–0.70	0.28 (7)	0.003–0.054

Note: Arithmetical means of the reported values (number of observations in brackets) and the range of ratios CH_4/CO_2 are also given.
Source: Table from Sugimoto et al (2000), by kind permission of Springer Sciences and Business Media

Most laboratory measurements of CH_4 efflux are made on termites collected from mounds, but an important observation from the Cameroon study is that only about 10 per cent of all termite individuals sampled at any given moment are present in mounds (Eggleton et al, 1996). Many of the termites not in mounds will belong to species that do not build mounds at all, and others will represent foragers temporarily absent from their nests, but in both cases the physiological circumstances may differ from those of termites within mounds, and there may be a different interaction between the gas produced and the immediate surroundings of the termite. This underlines the importance of having a detailed knowledge of assemblage structure and dynamics, as well as a comprehensive set of in vitro measurements of CH_4 effluxes, before scaling-up is attempted.

It has become clear that oxidation of CH_4 in soils by microorganisms is a significant component of the global budget for this gas (IPCC, 1995), accounting for about one quarter of production from all natural sources, and probably the only terrestrial sink of any consequence (Reeburgh et al, 1993). Oxidation is observed across a range of soil environments from tropical rainforest to tundra, but shows sensitivity to disturbance (for example tree-felling or ploughing), an observation that makes the conservation of natural or semi-natural forest soils of particular importance (references in MacDonald et al, 1998). Primary and old growth secondary tropical forests, which represent (in theory) about 8 per cent of the earth's land surface, but also might contain 66–75 per cent of all termite biomass (Bignell et al, 1997) are therefore of greatest interest in the context of CH_4 fluxes. Here there is a potential for oxidation to take place in the soil generally, as well as within nest materials and mound walls.

MacDonald et al (1998, 1999) examined the point-scale relationship between CH_4 production and oxidation in tropical forest soils, concentrating on the role of soil-feeding termites, which in the absence of waterlogging are likely to be the largest source of the gas. Production was patchy, reflecting large variations in the abundance of termites in the soil profile from one sampling point to another. Sink capacity also varied but was most commonly positive, such that undisturbed locations could always in theory oxidize all the CH_4 produced by the associated termites, even where such termites have very high abundance and biomass. Other studies have focused on the role of mound soil as a containment structure and have also shown that oxidation capacity can match the efflux of termite-produced CH_4 (Sugimoto et al, 1998a, 1998b). The studies are important because they evaluate the need to factor local oxidation effects into regional and global scaling-up calculations.

Rates of CH_4 oxidation in soil appear to be primarily controlled by soil physical properties that regulate gas diffusion, such as texture (Dorr et al, 1993), moisture content (Castro et al, 1995) and temperature (Crill, 1991). Inputs of nitrogen (N) (Steudler et al, 1989), pH (Hutsch et al, 1994) and soil organic matter (Czeipel et al, 1995) have also been shown to affect oxidation rates, and soil disturbances such as deforestation and conversion to agriculture consistently decrease the sink capacity (Keller et al, 1990; Lessard et al, 1994). This confirms that the net emission of CH_4 from termites foraging in the litter and surface soil is determined by local oxidation capacity, and that both production and oxidation will be highly sensitive to land use changes. Thus, it is important to know whether termite abundance and soil sink strength have different coefficients with respect to disturbance gradients, as this will directly link land use changes to global trace gas budgets. A pictorial summary of the fluxes in primary forest is given in Figure 5.1.

Scaling-up calculations and global CH_4 budgets

Scaling-up calculations by Martius et al (1996), Sanderson (1996), Bignell et al (1997) and Sugimoto et al (2000) used differing assumptions and principles, but all four agree that CO_2 fluxes by termites are in the range 2–5 per cent of the total from all terrestrial sources. 2 per cent is a large figure for a single insect order with about 0.01 per cent of the global terrestrial species richness, but is nevertheless only a minor source in the context of the whole C budget. With CH_4 emission, it now appears that the net contribution by termites is very small, despite the fact that some trophic groups produce this potent greenhouse gas in large amounts at the point scale. Independent studies using, respectively, static chambers and natural stable isotope ratios showed that mound walls and undisturbed soil between mounds have a sink capacity exceeding production at the landscape level, presumably due to the presence of methylothrophic bacteria (Sugimoto et al, 1998b; MacDonald et al, 1999). Sanderson (1996) estimates the global emission of CH_4 by termites as 19.7 ± 1.5Tg yr^{-1} (approximately 4 per cent of global totals), while Bignell et al (1997)

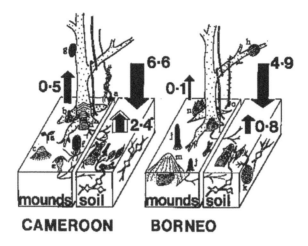

Figure 5.1 Locality CH₄ budgets for near primary and primary forest in Cameroon and Borneo (kg ha⁻¹ yr⁻¹)

Note: For each forest system the left-hand block illustrates the most commonly encountered epigeal termite mounds and arboreal nests. The right-hand block illustrates termite galleries in the soil (in some cases associated with mound-building species). Upward arrows indicate measured CH₄ effluxes (from mounds) and estimated gross CH₄ effluxes (from soil termite populations). Downward arrows indicate the net CH₄ fluxes measured in soils (both sites are net sinks). The illustrated mounds/nests are a) *Astalotermes quietus*, b) *Procubitermes arboricola*, c) *Cephalotermes rectangularis*, d) *Cubitermes* spp, e) *Cubitermes fungifaber*, f) *Thoracotermes macrothorax*, g) *Nasutitermes* spp, h) *Bulbitermes* spp, j) *Dicuspiditermes* spp, k) *Prohamitermes mirabilis*, l) *Procapritermes* spp, m) *Macrotermes gilvus* (strictly not a forest species, but associated with tracks and paths), n) *Hospitalitermes hospitalis*, and o) *Lacessititermes* sp. Effluxes from mounds g, m and n have not been determined and are not in the budgets. Also note, some mounds may be partly or wholly occupied by other species as secondary colonizers.
*The gross CH₄ production by soil termites in Cameroon is the lower estimate of Bignell et al (1997). Higher fluxes are possible, reflecting large inter-annual variations in population size. Not to scale.
Source: Data from Bignell et al (1997), Macdonald et al (1999) and Jeeva (1998). Figure from Sugimoto et al (2000), by kind permission of Springer Sciences and Business Media

give estimates in the range 17–96Tg yr⁻¹. Neither Sanderson (1996) nor Bignell et al (1997) take account of local CH_4 oxidation in their calculations. The arguments given above concerning the sink capacity of tropical forest soils clearly suggest that actual termite-mediated fluxes to the atmosphere in these ecosystems may be small or non-existent. Sugimoto et al (1998b) have made the only calculation of global production of CH_4 that uses the field observations of actual (rather than potential) oxidation rates, which suggested that only 17–47 per cent (30 per cent on average) of the CH_4 produced in epigeal termite mounds is released to the atmosphere. Although mounds can be large point sources, they may nevertheless be small contributors to the gross efflux from all termite sources at landscape level (Sugimoto et al, 1998b).

Comparing the four most comprehensive scaling-up exercises (Martius et al, 1996; Sanderson, 1996; Bignell et al, 1997; Sugimoto et al, 1998b), it is apparent that the main differences in protocols are the factoring-in of CH_4 oxidation and the areas assigned to regional vegetation/biomes types (which then feed into estimations of termite biomass). This is important in the case African rainforests, which have the highest recorded abundance and biomass of termites on earth and are in general (perhaps one should say in principle) dominated by soil-feeding forms with higher rates of CH_4 production. However, estimates of net global production that do not consider oxidation are unrealistic, whatever the accuracy of other assumptions they contain, so the range of 1.47–7.41Tg yr^{-1} given by Sugimoto et al (1998b; see Table 5.2) should be considered the best available. This is much less than the approximately 20Tg yr^{-1} still quoted in some chapters of the most recent report of the IPCC (IPCC, 2007).

Table 5.2 Global estimation of CH_4 emissions by termites

Region and land use	Area (10^6 km^2)	Biomass (g m^{-2}) Min	Max	Emissions (Tg yr^{-1}) Min	Max
Africa					
rainforest	2.219	24	130	0.38	2.20
savanna	6.836	5	20	0.23	0.93
woodland	7.401	5	20	0.08	0.30
cultivated land	3.247	0	10	0	0.07
Indian subcontinent + SE Asia					
rainforest	1.773	5	20	0.06	0.22
savanna	0.476	5	20	0.01	0.03
Woodland	1.277	5	20	0.02	0.09
cultivated land	4.330	0	10	0	0.16
South America					
rainforest	5.318	10	60	0.27	1.59
savanna	2.259	5	20	0.05	0.22
woodland	6.271	5	20	0.15	0.60
cultivated land	3.471	0	10	0	0.17
Australasia + Oceania					
rainforest	0.051	5	20	<0.01	0.01
savanna	0.881	5	20	0.03	0.11
woodland	3.884	5	20	0.12	0.48
cultivated land	1.398	0	10	0	0.09
Middle East					
savanna	0.125	5	20	<0.01	0.01
woodland	2.879	5	20	0.06	0.22
cultivated land	0.707	0	10	0	0.03
Temperate forest	3.863	1	3	0.04	0.07
Total	58.7			1.47	7.41

Source: Table from Sugimoto et al (2000) by kind permission of Springer Sciences and Business Media

Conclusions

Methane is produced as a by-product of fermentation processes in termite intestines, but net effluxes from the insects are often less than predicted because gut structure and physiology favours an alternative process, homeoacetogenesis, for maintaining redox balance (Brune, 2006). Net efflux of CH_4 from soils and termite mounds containing living termites is low, as these materials have a strong sink capacity, at least in undisturbed ecosystems. In general, soil-feeding termites emit more CH_4 than wood-feeders, but soil-feeders are more vulnerable to land-use changes, especially deforestations (Bignell and Eggleton, 2000).

Quantitative assessments of the impact of termites in the biosphere, whether direct or indirect, are hampered by incomplete knowledge of their abundance, biomass and consumption rates, and also by differences in gas emissions between species (Bignell et al, 1997). Assemblage composition (i.e. the number of species, their trophic functional roles and the balance between these) is a crucial factor affecting the contribution of termites to biogeochemical processes and to predictions of the effects of global change on them, but is incompletely documented in many parts of the tropics. There is also incomplete knowledge of the way termite biology and point-scale niche selection influences gas fluxes; for example global budgets for greenhouse gas emissions (IPCC, 1994) require data on net fluxes of CO_2 and CH_4 from termites in their natural settings (in nests and mounds, in galleries, within the soil, within their substrates etc.), as well as laboratory determinations on isolated insects, but such data are few and current funding of projects in this area is low. Further uncertainty arises at the ecosystem level and above, as the exchanges of gases between ground level and the higher atmosphere, and between different parts of mosaic landscapes are scarcely understood, especially in forest or forest-derived systems. However the many uncertainties do not undermine the present view that most CH_4 produced by termites is likely to be oxidized at source.

Early concerns about termite CH_4 emissions in the context of global change (IPCC, 1992, 1994, 1995) stimulated more than a decade of intensive research, from which we have gained much knowledge of termite biology and microbiology, but in the latest report (IPCC, 2007) the issue is downgraded. This reflects the current consensus that contributions to atmospheric CH_4 fluxes are quite small, with a maximum net global emission well under $10Tg\ yr^{-1}$.

References

Bignell, D. E. (2006) 'Termites as soil engineers and soil processors', in H. König and A. Varma (eds) *Intestinal Microorganisms of Termites and other Invertebrates*, Springer-Verlag, Berlin, pp183–220

Bignell, D. E. and Eggleton, P. (2000) 'Termites in ecosystems', in T. Abe, D. E. Bignell and M. Higashi (eds) *Termites: Evolution, Sociality, Symbioses, Ecology*, Kluwer Academic Publishers, Dordrecht, pp363–387

Bignell, D. E., Eggleton, P., Nunes, L. and Thomas, K. L. (1997) 'Termites as mediators of carbon fluxes in tropical forest: Budgets for carbon dioxide and methane emissions', in A. D. Watt, N. E. Stork and M. D. Hunter (eds) *Forests and Insects*, Chapman and Hall, London, pp109–134

Billen, J. (ed) (1992) *Biology and Evolution of Social Insects*, Leuven University Press, Leuven, Belgium

Brauman, A., Kane, M. D., Labat, M. and Breznak, J. A. (1992) 'Genesis of acetate and methane by gut bacteria of nutritionally diverse termites', *Science*, vol 257, pp1384–1387

Breznak, J. A. (2006) 'Termite gut spirochetes' in J. D. Radolph and S. A. Lukehart (eds) *Pathogenic* Treponema: *Molecular and Cellular Biology*, Horizon Scientific Press, Norwich, UK, pp421–443

Breznak, J. A. and Leadbetter, J. (2006) 'Termite gut spirochetes', in M. Dworkin, S. Falkow, E. Rosenberg, K.-H. Schleifer and E. Stackebrandt (eds) *The Prokaryotes*, vol 7, Springer, New York, pp318–329

Brune, A. (2006) 'Symbiotic associations between termites and prokaryotes', in M. Dworkin, S. Falkow, E. Rosenberg, K.-H. Schleifer and E. Stackebrandt (eds) *The Prokaryotes*, vol 7, Springer, New York, pp439–474

Castro, M. S., Steudley, P. A., Melillo, J. M., Aber, J. D. and Bowden, R. D. (1995) 'Factors controlling atmospheric methane consumption by temperate forest soil', *Global Biogeochemical Cycles*, vol 9, pp1–10

Collins, N. M. and Wood, T. G. (1984) 'Termites and atmospheric gas production', *Science*, vol 224, pp84–86

Crill, P. M. (1991) 'Seasonal patterns of methane uptake and carbon dioxide release by a temperate woodland soil', *Global Biogeochemcal Cycles*, vol 5, pp319–334

Czeipel, P. M., Crill, P. M., and Harriss, R. C. (1995) 'Environmental factors influencing the variability of CH_4 oxidation in temperate zone soils', *Journal of Geophysical Research*, vol 100, pp9359–9364

Darlington, J. P. E. C., Zimmerman, P. R., Greenberg, J., Westberg, C. and Bakwin, P. (1997) 'Production of metabolic gases by nests of the termites *Macrotermes jeanneli* in Kenya', *Journal of Tropical Ecology*, vol 13, pp491–510

Dorr, H., Katruff, L. and Levin, I. (1993) 'Soil texture parameterization of the methane uptake in aerated soils', *Chemosphere*, vol 26, pp697–713

Eggleton, P., Bignell, D. E., Sands, W. A., Maudsley, N. A., Lawton, J. H., Wood, T. G. and Bignell, N. C. (1996) 'The diversity, abundance and biomass of termites under differing levels of disturbance in the Mbalmayo Forest Reserve, southern Cameroon', *Philosophical Transactions of The Royal Society of London, Series B*, vol 351, pp51–68

Fittkau, E. J. and Klinge, H. (1973) 'On biomass and trophic structure of the central Amazonian rain forest ecosystem', *Biotropica*, vol 5, pp2–14

Fraser, P. J., Rasmussen, R. A., Creffield, J. W., French, J. R. and Khalil, M. A. K. (1986) 'Termites and global methane – another assessment', *Journal of Atmospheric Chemistry*, vol 4, pp295–310

Hackstein, J. H., van Alen, T. A. and Rosenberg, J. (2006) 'Methane production by terrestrial arthropods', in H. König and A. Varma (eds) *Intestinal Microorganisms of Termites and other Invertebrates*, Springer-Verlag, Berlin, pp155–180

Hölldobler, B. and Wilson, E. O. (2009) *The Superorganism*, W. W. Norton and Company, New York

Hungate, R. E. (1946) 'The symbiotic utilization of cellulose', *Journal of the Elisha Mitchell Scientific Society*, vol 62, pp9–24

Hutsch, B. W., Webster, C. P. and Powlson, D. S. (1994) 'Methane oxidation in soil as affected by landuse, soil pH and nitrogen fertilization', *Soil Biology and Biochemistry*, vol 26, pp1613–1622

IPCC (Intergovernmental Panel on Climate Change) (1992) *Climate Change 1992: The Supplementary Report to the IPCC Scientific Assessment*, J. T. Houghton, B. A. Callandar and S. K. Varney (eds), Cambridge University Press, Cambridge

IPCC (1994) *Radiative Forcing of Climate Change and Evaluation of the IPCC 1992 Emission Scenarios*, J. T. Houghton, L. G. Meira Filho, J. Bruce, L. Hoesung, B. A. Callander, E. Haites, N. Harris, A. Kattenburg and K. Maskell (eds), Cambridge University Press, Cambridge

IPCC (1995) *The Science of Climate Change*, J. T. Houghton, L. G. Meira Filho, B. A. Callander, N. Harris, A. Kettenburg and K. Maskell (eds), Cambridge University Press, Cambridge

IPCC (2007) *Climate Change 2007: The Physical Science Basis. Contribution of Working Group I to the Fourth Assessment Report of the Intergovernmental Panel on Climate Change*, S. Solomon, D. Qin, M. Manning, Z. Chen, M. Marquis, K. B. Averyt, M. Tignor and H. L. Miller (eds), Cambridge University Press, Cambridge and New York

Jeeva, D. (1998) 'Greenhouse gas emission by termites in tropical rain forest of Danum Valley, Sabah, Malaysia', MSc thesis, Universiti Malaysia Sabah, Malaysia

Jeeva, D., Bignell, D. E. Eggleton, P. and Maryati, M. (1998) 'Respiratory gas exchanges of termites from the Sabah (Borneo) assemblage', *Physiological Entomology*, vol 24, pp11–17

Keller, M., Mitre, M. E. and Stallard, R. F. (1990) 'Consumption of atmospheric methane in soils of central Panama: Effects of agricultural development', *Global Biogeochemical Cycles*, vol 4, pp21–27

Khalil, M. A. K., Rasmussen, R. A., French, J. R. J. and Holt, J. A. (1990) 'The influence of termites on atmospheric trace gases: CH_4, CO_2, $CHCl_3$, N_2O, CO, H_2 and light hydrocarbons', *Journal of Geophysical Research*, vol 95, pp3619–3634

Lavelle, P., Bignell, D. E., Lepage, M. Volters, V., Roger, P., Ineson, P., Heal, W. and Dillion, S. (1997) 'Soil function in a changing world: The role of invertebrate ecosystem engineers', *European Journal of Soil Biology*, vol 33, pp159–193

Lessard, R., Rochette, P., Topp, E., Pattey, E., Desjardins, R. L. and Beaumont, G. (1994) 'Methane and carbon dioxide fluxes from poorly drained adjacent cultivated and forest sites', *Canadian Journal of Soil Science*, vol 74, pp139–146

MacDonald, J. A., Eggleton, P., Bignell, D. E. and Forzi, F. (1998) 'Methane emission by termites and oxidation by soils, across a forest disturbance gradient in the Mbalmayo Forest Reserve, Cameroon', *Global Change Biology*, vol 4, pp409–418

MacDonald, J. A., Jeeva, D. Eggleton, P. Davies, R., Bignell, D. E., Fowler, D., Lawton, J. and Maryati, M. (1999) 'The effect of termite biomass and anthropogenic disturbance on the methane budgets of tropical forests in Cameroon and Malaysia', *Global Change Biology*, vol 5, pp869–879

Martius, C., Fearnside, P., Bandeira, A. and Wassmann, R. (1996) 'Deforestation and methane release from termites in Amazonia', *Chemosphere*, vol 33, 517–536

Miller, L. R. (1991) 'A revision of the *Termes/Capritermes* branch of the Termitinae in Australia (Isoptera: Termitidae)', *Invertebrate Taxonomy*, vol 4, pp1147–1282

Moreira, F. M., Huising, J. and Bignell, D. E. (eds) (2008) *A Handbook of Tropical Soil Biology: Sampling and Characterization of Below-ground Biodiversity*, Earthscan, London

Nunes, L., Bignell, D. E., Lo, N. and Eggleton, P. (1997) 'On the respiratory quotient (RQ) of termite', *Journal of Insect Physiology*, vol 43, pp749–758

Purdy, K. J. (2007) 'The distribution and diversity of *Euryarchaeota* in termite guts', *Advances in Applied Microbiology*, vol 62, pp63–80

Rasmussen, R. A. and Khalil, M. A. K. (1983) 'Global production of methane by termites', *Nature*, vol 301, pp700–702

Reeburgh, W. S., Whalen, S. C. and Alperin, M. J. (1993) 'The role of methylotrophy in the global CH_4 budget', in J. C. Murrell and D. Kelley (eds) *Microbial Growth on C-1 Compounds*, Intercept, Andover, UK, pp1–14

Rouland, C., Brauman, A., Labat, M. and Lepage, M. (1993) 'Nutritional factors affecting methane emissions from termites', *Chemosphere*, vol 26, pp617–622

Sanderson, M. G. (1996) 'Biomass of termites and their emissions of methane and carbon dioxide', *Global Biogeochemical Cycles*, vol 10, pp543–557

Seiler, W., Conrad, R. and Scharffe, D. (1984) 'Field studies of methane emissions from termite nests into the atmosphere and measurement of methane uptake by tropical soils', *Journal of Atmospheric Chemistry*, vol 1, pp171–186

Steudler, P. A., Bowden, R. D., Melillo, J. M. and Aber, J. D. (1989) 'Influence of nitrogen fertilization on methane uptake in temperate forest soils', *Nature*, vol 341, pp314–315

Stork, N. E. and Brendell, M. D. J. (1993) 'Arthropod abundance in lowland rainforest of Seram' in J. Proctor and I. Edwards (eds) *Natural History of Seram*, Intercept, Andover, UK, pp115–130

Sugimoto, A., Inoue, T., Tayasu, I, Miller, L., Takeichi, S. and Abe, T. (1998a) 'Methane and hydrogen production in a termite-symbiont system', *Ecological Research*, vol 13, pp241–257

Sugimoto, A., Inoue, T., Kirtibur, N. and Abe, T. (1998b) 'Methane oxidation by termite mounds estimated by the carbon isotopic composition of methane', *Global Biogeochemical Cycles*, vol 12, pp595–605

Sugimoto, A., Bignell, D. E. and MacDonald, J. A. (2000) 'Global impact of termites on the carbon cycle and atmospheric trace gases' in T. Abe, D. E. Bignell and M. Higashi (eds) *Termites: Evolution, Sociality, Symbioses, Ecology*, Kluwer Academic Publishers, Dordrecht, pp409–435

Tyler, S. C., Zimmerman, P., Cumberbatch, C., Greenberg, J. P., Westberg, C. and Darlington, J. P. E. C. (1988) 'Measurements and interpretation of $\delta^{13}C$ of methane from termites, rice paddies, and wetlands in Kenya', *Global Biogeochemical Cycles*, vol 2, pp341–355

Watt, A. D., Stork, N. E., Eggleton, P., Srivastava, D., Bolton. B., Larsen, T. B., Brendell, M. J. D. and Bignell, D. E. (1997) 'Impact of forest loss and regeneration on insect abundance and diversity', in A. D. Watt, N. E. Stork and M. D. Hunter (eds) *Forests and Insects*, Chapman and Hall, London, pp271–284

Wheeler, G. S., Tokoro, M., Scheffrahn, R. H. and Su, N. Y. (1996) 'Comparative respiration and methane production rates in Nearctic termites', *Journal of Insect Physiology*, vol 42, pp799–806

Wood, T. G. and Sands, W. A. (1978) 'The role of termites in ecosystems', in M. V. Brian (ed) *Production Ecology of Ants and Termites*, Cambridge University Press, Cambridge, pp245–292

Zimmerman, P. R., Greenberg, J. P., Wandiga, S. O. and Crutzen, P. J. (1982) 'Termites, a potentially large source of atmospheric methane', *Science*, vol 218, pp563–565

6
Vegetation

Andy McLeod and Frank Keppler

Introduction

Sources of atmospheric CH_4 in the biosphere have until recently been attributed to originate from strictly anaerobic microbial processes in wetland soils and rice paddies (Chapters 3 and 8), the guts of termites and ruminants (Chapters 5 and 9), human and agricultural waste (Chapter 10), and from biomass burning (Chapter 7), fossil fuel mining (Chapter 12) and geologic sources including mud volcanoes and seeps (Chapter 4). However, in early 2006, Keppler et al published a surprising report of direct CH_4 emission from vegetation foliage under aerobic conditions. Their study enclosed samples of detached leaves, air-dried leaves, intact plants and the plant structural component pectin in CH_4-free air inside closed vials or chambers and measured the build-up of CH_4 in the enclosure. This revealed rates of CH_4 emission from a range of air-dried tree and grass leaves from C3 and C4 plants in the range 0.2–3 nanograms (ng) per gram dry weight per hour (g^{-1} d.wt. h^{-1}) at 30°C but increasing to much higher rates of 12–370ng g^{-1} d.wt. h^{-1} for intact plants. Methane emission rates increased by a factor of three to five when chambers were exposed to natural sunlight. The emissions of air-dried leaves appeared to be non-enzymatic as they increased over the range 30–70°C (Figure 6.1) and above the threshold of 50–60°C (Berry and Raison, 1982) at which plant enzymes are denatured. Furthermore, air-dried leaves that were sterilized by prior exposure to gamma radiation emitted the same amount of CH_4 as the untreated leaves suggesting that microbial production was not involved.

Although these rates of emission were small, Keppler et al (2006) also completed a rough extrapolation to estimate a total annual global emission of CH_4 from this living vegetation source by using mean sunlit and dark emission rates for leaf biomass scaled by day length, duration of growing season and total net primary productivity in each biome. Their estimate of 62–236Tg (1Tg = 10^{12}g) CH_4 yr^{-1}, with the largest contribution of 46–169Tg CH_4 yr^{-1} from tropical forests and grassland, was observed to equate to 10-40 per cent of the known annual CH_4 source strength (IPCC, 2007). Plant litter was estimated to contribute 0.5–6.6Tg CH_4 yr^{-1}. Consequently, these first observations of

Keppler et al (2006) caused intense interest, considerable debate and some scepticism among the scientific community and the media (Schiermeier, 2006a, 2006b) leading to further experimental studies and a wider consideration of their implications for the global CH_4 budget and greenhouse gas mitigation options (Lowe, 2006; NIEPS, 2006).

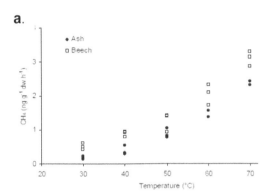

Figure 6.1 (a) Release of CH_4 from air-dried leaves of ash (*Fraxinus excelsior*) and beech (*Fagus sylvatica*) in the temperature range 30–70°C under aerobic conditions; (b) Closed leaf chambers used for whole plant measurement of CH_4 emissions

Source: (a) adapted from Keppler et al (2006)

Experimental laboratory studies

The earliest laboratory study reporting an emission of CH_4 from leaves was conducted in the late 1950s at the Academy of Sciences of Georgia (Tbilisi) on emissions of volatile organic compounds (VOCs) from leaves of willow and poplar tress (Sanadze and Dolidze, 1960). In that study, mass spectrometric analysis was used to scan the volatile emissions of leaves incubated in 1.5 litre glass containers. Based on their mass spectra they concluded that plants released CH_4 as well as ethane, propane, isoprene and several other VOCs, but their subsequent work focused on isoprene and provided no explanation for their detection of CH_4 or any follow-up studies on CH_4 release.

It was not until the study of Keppler et al (2006), described above, that the scientific community began to examine the question of vegetation as a direct source of CH_4 in some detail. The first subsequent experimental study by Dueck et al (2007) grew six plant species hydroponically from seed for nine weeks inside a hermetically sealed plant growth chamber provided with ^{13}C-labelled CO_2 in order to create isotopically labelled plant material. Shoots of four species were then sealed into a continuous-flow gas exchange cuvette with a visible light (300 or 600µmol m^{-2} s^{-1}) and a corresponding air temperature of

25 or 35°C, respectively. They observed CH_4 emissions $-10ng\ g^{-1}\ h^{-1}$ to $42ng\ g^{-1}\ h^{-1}$, some 18 times lower than the average rates reported by Keppler et al (2006), and which were not statistically different from zero. In a further experiment, they found no release of ^{13}C-labelled CH_4 into the sealed growth chamber from the labelled plant material over six days. Both experiments led to their conclusion that there was no evidence for substantial aerobic CH_4 emission by terrestrial plants.

Beerling et al (2008) measured CH_4 and CO_2 exchange rates of corn (*Zea mays*) and tobacco (*Nicotiana tabacum*) using a high-accuracy gas analysis system inside a controlled environment room. They found no evidence for aerobic CH_4 emissions during alternating three-hour periods of $700\mu mol\ m^{-2}\ s^{-1}$ visible light and darkness over a 12-hour experiment at 25°C. They concluded that there was no evidence for a link between aerobic CH_4 emissions from leaves and photosynthetic or respiratory metabolism but noted that an emission might be linked to a non-enzymatic process driven by non-photosynthetic radiation such as ultraviolet (UV) wavelengths which would not have been detected by their experimental system. It was not until researchers used irradiation containing a UV wavelength component, that several observations of CH_4 release by vegetation under aerobic conditions were demonstrated (McLeod et al, 2008; Vigano et al, 2008).

Vigano et al (2008) measured the effect of UV radiation and elevated temperature on the emission of CH_4 from dry and detached fresh leaves from over 20 species and plant structural components including pectin, lignin and cellulose. They used a range of lamps to provide exposures to variable amounts of UV-A (320–400 nanometres (nm)), UV-B (290–320nm) and UV-C (<290nm) radiation, mostly at least five times higher than ambient spectrally unweighted UV irradiances. They demonstrated that CH_4 emissions were linearly related to UV irradiance and were almost instantaneous after irradiation, indicating a direct photochemical process. By conducting one experiment with dry grass for 35 days and other experiments using both ambient and CH_4-free air, they were able to provide some evidence that the CH_4 originated within the plant material and emissions were not explained by desorption from plant surfaces. They also used a dry leaf of ^{13}C-labelled wheat (*Triticum aestivum*) from the study of Dueck et al (2007) and were able to demonstrate that ^{13}C-labelled CH_4 was emitted at $32ng\ g^{-1}$ leaf dry weight h^{-1} under an unweighted UV irradiance of three times the typical tropical conditions. Their CH_4 emissions rates from UV-A and UV-B lamp irradiation of plant material ranged from zero to $393ng\ g^{-1}$ dry weight h^{-1} (with all but one value for cotton flowers being below $200ng\ g^{-1}$ dry weight h^{-1}). The highest emissions were detected at UV levels that exceeded typical ambient levels but the study clearly indicated a mechanism for aerobic CH_4 release from plant material under some experimental conditions.

McLeod et al (2008) also performed experiments to expose detached fresh leaves to UV radiation from filtered lamps and pectin-impregnated glass fibre sheets to both lamps and natural sunlight. They found a linear response of CH_4

emission from pectin, up to 750ng g^{-1} h^{-1}, with each lamp type and tested a range of common and idealized spectral weighting functions. They identified a function that decayed one decade in 80nm wavelength, which gave a significant linear regression between weighted UV and CH_4 for all lamp sources and sunlight. These irradiances, which included sunlight in Edinburgh during September (Figure 6.2), were within the range of ambient UV

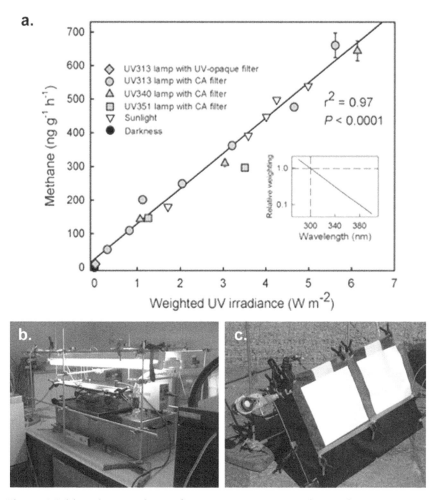

Figure 6.2 (a) Methane production from citrus pectin at 30°C showing linear regression with ultraviolet irradiance weighted with the inset spectral weighting function, using lamps filtered with cellulose diacetate (CA), a UV-opaque filter or sunlight in Edinburgh (55°55′N, 3°10′W) between 6 and 21 September 2006; (b) Equipment used for the temperature-controlled laboratory irradiation of pectin-impregnated glass fibre sheets; (c) Equipment used for the temperature-controlled exposure of pectin-impregnated sheets to natural sunlight

Source: (a) adapted from McLeod et al (2008)

exposures, especially in tropical regions. Methane emissions from pectin were virtually eliminated by chemical removal of methyl ester groups from the pectin molecule or by adding a scavenger of singlet oxygen to the pectin. The latter observation was the first indication that reactive oxygen species (ROS) might be involved in a mechanism for aerobic CH_4 production in vegetation. Their experiments also demonstrated that UV irradiation of dry pectin not only produced CH_4 but also ethene, ethane and CO_2 and the study included UV irradiation of freshly detached leaves of tobacco (*Nicotiana tabacum*) that released CH_4 (and also ethene and ethane) at a rate of 12ng g^{-1} leaf dry weight per hour over 45 hours. Although these studies were performed using pure pectin or detached fresh leaves they clearly demonstrated a mechanism for aerobic CH_4 production driven by levels of UV radiation, including sunlight, within the ambient range.

Methane emissions under aerobic conditions were investigated in detached leaves and stems of 44 indigenous species of the Inner Mongolian Steppe using gas-tight serum bottles by Wang et al (2008). The study included ten herbaceous hydrophytes (wetland-adapted plants) from low-lying areas and 34 xerophytes (arid-adapted plants) including shrubs and herbs from upland areas. Plants were sampled in the early morning, separated into stems and leaves and samples sealed in triplicate serum bottles for incubation in CH_4-free air and in darkness for 10–20 hours at 20–22°C. Gas samples were withdrawn (and replaced with CH_4-free air) by syringe and the CH_4 flux calculated from the initial rate of change of CH_4 concentration. The study found that nine species emitted CH_4 in the range 0.5–13.5ng g^{-1} d.wt. h^{-1} but 80 per cent of the species tested emitted no detectable CH_4. The 20 per cent proportion of species emitting CH_4 were about the same for hydrophytes, xerophytes, C3 plants and C4 plants but whereas 78 per cent of shrubs emitted CH_4, only 6 per cent of herbaceous plants did so. They observed the highest emission rates of 6.8–13.5ng CH_4 g^{-1} d.wt. h^{-1} in two hydrophytes – *Glyceria spiculosa* and *Scirpus yagara* – and these emitted from stems and not from detached leaves. The xerophytes examined included seven out of nine shrub species that emitted CH_4 from detached leaves but not stems, while none of the herbaceous xerophytes emitted CH_4. The authors also examined the carbon isotope ratio ($\delta^{13}C$ value) of the emitted CH_4 in order to distinguish between a plant and a soil-derived source.

Kirschbaum and Walcroft (2008) performed laboratory studies to investigate whether reported CH_4 emissions might have been caused by CH_4 desorption from sample surfaces after prior exposure to higher concentrations and to examine CH_4 emissions from living detached leaves and intact plant materials. They tested detached leaves of wormwood (*Artemesia absinthum*), yarrow (*Achillea millefolium*), dandelion flowers (*Taraxacum officinale*), broadleaf (*Griselina littoralis*), five finger (*Pseudopanax arboreus*) and a mixture of local grasses for six days sealed inside 5.7 litre Plexiglas chambers exposed indoors to low visible (and no UV) light (5µmol quanta m^{-2} s^{-1} from fluorescent lamps) and flushed at the start with CH_4-free air. They also used intact seedlings of corn (*Zea mays*) grown in pots containing vermiculite as an inert rooting medium. Measured rates of CH_4 emission were zero or very small ranging from -0.25 ± 1.1ng CH_4 kg^{-1} d.wt. s^{-1} for *Z. mays* to 0.1 ± 0.08ng CH_4 kg^{-1} d.wt. s^{-1} for mixed detached grasses. These values are much lower than those reported by Keppler et al (2006). As organic materials have a high adsorption capacity for CH_4 and its desorption was a possible experimental artefact that might explain observed vegetation emissions. Kirschbaum and Walcroft (2008) also conducted experiments with stacks of cellulose filter papers (as organic adsorbers). These were exposed first to ambient CH_4 concentration and then sealed inside chambers for six days and the time-course of CH_4 release determined. They found no significant desorption of CH_4 and concluded that desorption is not a quantitatively important artefact contributing to observed aerobic CH_4 fluxes.

Experiments on CH_4 emissions from plants were also conducted by Nisbet et al (2009) who examined genome sequences of plant material and found no evidence that enzymes necessary for classical methanogenesis pathways found in microorganisms were present in plants. They conducted experiments that demonstrated the influence of CH_4 dissolved in soil water (described below) and also measured the release of CH_4 from plant material enclosed in flasks. They grew thale cress (*Arabidopsis thaliana*) and the green alga (*Chlamydomonas reinhardtii*) on an agar medium under aseptic conditions in 2 litre flasks and found no increase in CH_4 concentration above ambient with a throughflow of 5mL min^{-1}. They also examined detached leaves of *Z. mays* and whole plants of rice (*Oryza sativa*) grown on vermiculite in closed 15 litre glass flasks under low levels of visible light (cool-white fluorescent lamps, 180µmol m^{-2} s^{-1}) and found no detectable CH_4 emissions. Their conclusion from these experiments was that plants grown under controlled conditions in the laboratory under artificial light are not capable of producing CH_4.

The effects of temperature and UV radiation on non-plant material and pectin were also demonstrated by Bruhn et al (2009). They exposed detached leaves of six species, fruit tissues and purified pectin inside glass vials to a range of temperatures in darkness, to visible light (400–700nm photosynthetically active radiation (PAR) at 400µmol photons m^{-2} s^{-1} and to UV-A and UV-B radiation from two types of lamp. They found that detached green leaves of *Betula populifolia* released CH_4 under PAR irradiation at 31.7ng g^{-1} d.wt. h^{-1} at 4°C over 24 hours and 4.5ng g^{-1} d.wt. h^{-1} at 30°C throughout one week. Dry pectin also released CH_4 at 80°C in darkness at rates comparable to those observed by Keppler et al (2006) but also at similar rates at a much lower temperature of 37°C when dissolved in water. At a high temperature of 80°C, all tested plant material released CH_4 but with two orders of magnitude variation between species.

Bruhn et al (2009) also found that CH_4 emission from twigs of *Picea abies* and from pectin was stimulated by UV radiation and the effect of UV-B wavelengths was greater that UV-A (Figure 6.2). The UV-B irradiance had a linear relationship with CH_4 emission, corresponding to the observations of

McLeod et al (2008). By digesting pectin with the enzyme pectin methyl esterase prior to temperature or irradiation treatments, they demonstrated a reduction in CH_4 emissions thus supporting the observations of Keppler et al (2006) and McLeod et al (2008) that methyl groups of pectin are a source of CH_4 from vegetation under aerobic conditions.

Qaderi and Reid (2009) recently reported effects of temperature, UV-B radiation and water stress on CH_4 emissions (and plant growth and CO_2 exchange) from the leaves of six crops: faba bean (*Vicia faba*), sunflower (*Helianthus annuus*), pea (*Pisum sativum*), canola (*Brassica napus*), barley (*Hordeum vulgare*) and wheat (*T. aestivum*). They grew one-week-old seedlings for one week in controlled environment growth chambers under two temperature regimes (24°C day/20°C night and 30°C day/26°C night), three levels of UV-B radiation using filtered lamps (zero, ambient, enhanced) and watered either to excess (well watered) or to wilting point (water stressed). Emissions from freshly detached leaves into CH_4-free air inside plastic syringes were measured over the range 57–210ng CH_4 g^{-1} d.wt. h^{-1}. Significantly higher CH_4 emissions were observed from plants subject to higher temperature (1.14 times higher) or water stress (1.21 times higher) compared to plants grown under lower temperature or watered to excess. The zero and enhanced levels of UV-B increased CH_4 emissions compared to the ambient UV-B treatment. However, unlike previous studies (McLeod et al, 2008; Vigano et al, 2008; Bruhn et al, 2009) in which emissions were concurrent with UV exposure, the measurements of Qaderi and Reid (2009) were after cessation of the stress treatments, including UV, and may therefore represent different aspects of the plant response. They also found that CH_4 emissions from attached leaves of *P. sativum* were 1.89 times higher than for detached leaves and that emissions increased with incubation time over four hours, suggesting that CH_4 was being produced by the plants and not simply diffusing from CH_4 stored in the leaves.

Although some of the studies outlined above clearly demonstrated that UV irradiation and elevated temperature can lead to CH_4 formation in plant foliage and from pectin, further understanding of the process was only revealed by studies using stable isotopes and biochemical analyses described later in this chapter.

Global vegetation emissions and related uncertainties

The first extrapolations from laboratory measurements to the global scale (Keppler et al, 2006) suggested that vegetation could constitute a substantial fraction (62–242Tg yr^{-1}) of the total global emissions of CH_4 (see Introduction). This large figure was derived by using mean sunlit and dark emission rates for leaf biomass scaled by day length, duration of growing season and total net primary productivity (NPP) in each biome. Alternative extrapolations of the same data were subsequently published that accounted for differences in foliage turnover rates between biomes and significantly lowered estimates of the global CH_4 emission from vegetation (Houweling et al, 2006; Kirschbaum et al, 2006; NIEPS, 2006; Parsons et al, 2006; Butenhoff and Khalil, 2007; Ferretti et al, 2007; Megonigal and Güenther, 2008). A comprehensive overview of the ranges of estimated global emissions of CH_4 from vegetation was given by Megonigal and Güenther (2008) and some of the approaches of the revised estimates (Table 6.1) are discussed in more detail below.

Kirschbaum et al (2006) first indicated that the upscaling approach applied by Keppler et al (2006) contained some methodological inconsistencies. In their alternative approach, they used two different methods for scaling CH_4 emissions from plants to the global rate. In their first estimate, Kirschbaum et al (2006) used an approach similar to that of Keppler et al (2006), but based their estimate on leaf mass in different biomes instead of NPP. This approach provided a global CH_4 emission of 36 (range 15–60) Tg yr^{-1}, with about half of the emissions attributed to tropical forests. The second approach related CH_4 emissions to photosynthesis and used independent estimates of photosynthesis as the basis for upscaling, which resulted in a global emission rate of 9.6Tg yr^{-1} which is lower than the leaf-mass-based estimate. Thus both approaches of Kirschbaum et al (2006) suggested substantially lower global emissions from vegetation than Keppler et al (2006).

Biome leaf mass was also used as a scaling factor by Parsons et al (2006). Their calculation of global CH_4 emissions from plants (up to 52Tg yr^{-1}) reduced the original estimate of Keppler et al (2006) by 72 per cent. They also showed that the reduction varied across all biomes, because of the variable relationship between standing biomass and NPP. In some cases, the value of NPP can be similar to the standing biomass, however, in biomes with low turnover rates, such as tropical forests, NPP is substantially smaller than standing biomass. In this context they raised the question of whether non-leafy tissues release CH_4, and if so, then the potential emission from systems such as tropical forests would increase enormously.

A bottom-up estimate to model monthly plant emissions on a 0.5 × 0.5° grid was conducted by Butenhoff and Khalil (2007), using the temperature- and sunlight-dependent emission rates determined by Keppler et al (2006). In contrast to the latter study, sunlight exposure was determined not only by day length, but also by cloud cover and canopy shading. Instead of NPP they used foliage biomass. They concluded that CH_4 emissions from the terrestrial plant community were likely to be in the range 20–69Tg yr^{-1} and varied largely due to a sensitivity to the prescribed temperature dependence of the plant emission rate, which is still unknown.

In order to constrain the potential magnitude of global CH_4 emissions from upland plants, Megonigal and Güenther (2008) used a foliar VOC emissions model called MEGAN (Model of Emissions of Gases and Aerosols from Nature) to incorporate certain canopy and physical processes that were not considered by the estimates of Kirschbaum et al (2006) and Parsons et al (2006). In particular, they used the temperature responses reported by Keppler et al (2006) and accounted for the effects of self-shading within the plant

Table 6.1 Estimates of global aerobic methane emissions by vegetation

Scaling method	Global methane production (Tg yr^{-1})	Reference
Sunlit and dark leaf emission rate scaled by day length, season length and biome net primary production. Mean rate of 149Tg yr^{-1} based on emission rates 374ng g^{-1} d.wt. h^{-1} (sunlight) and 119ng g^{-1} d.wt. h^{-1} (no sun). Range of 62–236Tg yr^{-1} was based on emission rates 198–598ng g^{-1} d.wt. h^{-1} (sunlight) and 30–207ng g^{-1} d.wt. h^{-1} (no sun)	62–236	Keppler et al (2006)
Leaf emission rates of Keppler et al (2006) scaled by biome leaf biomass – mean 36Tg yr^{-1} (range 15–60Tg yr^{-1}); or by leaf photosynthesis, 10Tg yr^{-1}	10–60	Kirschbaum et al (2006)
Leaf emission rates of Keppler et al (2006) scaled by biome leaf biomass. Leafy biomass, 42Tg yr^{-1}; non-leafy biomass, 11Tg yr^{-1}	53	Parsons et al (2006)
Atmospheric transport model, isotope ratios, mass balance. Pre-industrial plausible value 85Tg yr^{-1} to maximum present day upper limit 125 Tg yr^{-1}	85–125	Houweling et al (2006)
Leaf emission rates of Keppler et al (2006) scaled using model of cloud cover and canopy shading. Scaled using Leaf Area Index (LAI), 36Tg yr^{-1}; scaled using foliage biomass, 20Tg yr^{-1}, maximum expected 69Tg yr^{-1}	20–69	Butenhoff and Khalil (2007)
Mass balance, ice core isotope ratios using: Pre-industrial, 'best estimate' 0–46Tg yr^{-1}, 'maximum estimate' 9–103Tg yr^{-1}; Modern source, 'best estimate' 0–176Tg yr^{-1}, 'maximum estimate' 0–213Tg yr^{-1}	0–213	Ferretti et al (2007)
Global VOC emissions model assuming VOCs and CH$_4$ have similar biochemical origin. Range dependent on land cover and weather data	34–56	Megonigal and Güenther (2008)

Source: Adapted from Megonigal and Güenther (2008) and Keppler et al (2009)

canopy. Furthermore they applied MEGAN with the assumption that the mechanism of CH$_4$ production shared similar features with the biochemical pathways for VOC production, such as methanol. The parameterization of light and temperature in the MEGAN model was similar to the global model of aerobic CH$_4$ emissions developed by Butenhoff and Khalil (2007). The global distribution of CH$_4$ emissions from foliage simulated with MEGAN agreed well with the predictions of Keppler et al (2006) and the observations of Frankenberg et al (2005). However, the annual global CH$_4$ emission from living vegetation estimated with MEGAN was significantly lower, ranging from 34 to 56Tg yr^{-1}, depending on the land cover and weather data used to drive the model. This figure is close to the alternative extrapolations provided by Parsons et al (2006) and Kirschbaum et al (2006), and the global model developed by Butenhoff and Khalil (2007).

Houweling et al (2006) used atmospheric transport model simulations with and without vegetation emissions and compared them with background CH_4 concentrations, values of stable carbon isotope ratios of CH_4 and satellite measurements. For the present-day atmospheric CH_4 concentration, they suggested an upper limit of 125Tg yr^{-1} for vegetation emissions, while analysis of pre-industrial CH_4, pointed to 85Tg yr^{-1} as a more plausible maximum for plant emissions. These estimates are 36–90 per cent of the maximum estimate reported by Keppler et al (2006).

Another stable isotope approach for revised global limits of aerobic CH_4 emissions was provided by Ferretti et al (2007) who examined ice core records of atmospheric CH_4 concentration and stable carbon isotope ratios over the last 2000 years. Their 'best estimate' of a top-down approach suggested that global plant emissions must be much lower than proposed by Keppler et al (2006) during the last 2000 years and are likely to lie in the range 0–46Tg yr^{-1} and 0–176Tg yr^{-1} during the pre-industrial and modern eras, respectively. They further suggested that CH_4 emissions from plants are not essential to reconcile either the pre-industrial or the modern CH_4 budgets. However, the high ^{13}C content of CH_4 before AD1500, as recovered from ice cores (Ferretti et al, 2005), is hard to reconcile with the knowledge that pre-industrial emissions were dominated by wetland emissions that are isotopically depleted in ^{13}C. The initial hypothesis that pre-industrial anthropogenic biomass burning caused the high ^{13}C levels (Ferretti et al, 2005) has been further questioned by recent data revealing an even higher ^{13}C content in the early Holocene (Schaefer et al, 2006).

The revised estimates of the proposed aerobic plant source of CH_4 discussed above were in the range of 0–176Tg yr^{-1} and lower than the first global estimates provided by Keppler et al (2006). Most of these revised estimates were low enough (0–69Tg yr^{-1}) to be accommodated within the uncertainty in the global CH_4 budget. However, all estimates were based on the initial experimental laboratory data presented by Keppler et al (2006). More reliable global estimates of vegetation CH_4 emissions should be possible when more experimental data, including measurements from the field and the laboratory, have been collected, as well as an understanding of the mechanisms for CH_4 formation from vegetation.

Earth observation and canopy flux measurement

Atmospheric CH_4 concentrations have been studied from space using satellite-mounted instruments that allow a global spatial and temporal analysis and this has raised a number of uncertainties about the possible magnitude of vegetation sources. Frankenberg et al (2005) used data from the SCIAMACHY (scanning imaging absorption spectrometer for atmospheric chartography) instrument on board the European Space Agency satellite Envisat to examine trace gas concentrations including CH_4. They found that CH_4 concentrations were greater than expected over tropical rainforests when compared with simulated concentration distributions derived from a global chemistry-transport model using ground-based inventories. They suggested potential sources for this discrepancy including wetlands, biomass burning, termites and cattle but included a hitherto unknown CH_4 source that might be directly related to tropical forests. Keppler et al (2006) subsequently suggested that their experimental observations of aerobic CH_4 emissions from vegetation were a potential explanation for the discrepancy. Significantly greater tropical emissions than expected were also reported from these satellite observations in an extended analysis by Bergamaschi et al (2007) and Schneising et al (2009). More recent studies have shown that the CH_4 data retrieval was positively biased in tropical regions as a result of spectroscopic interference by water vapour, but source inversions based upon an updated data retrieval method still point to substantial tropical CH_4 emissions (Frankenberg et al, 2008).

In response to the experimental observations of Keppler et al (2006), Crutzen et al (2006) re-analysed data reported by Scharffe et al (1990) from a 1988 study of CH_4 concentrations in the atmospheric boundary layer above the savanna climate region of Venezuela. Scharffe et al (1990) used chamber methods to demonstrate that their forest soils were a substantial sink for CH_4, while savanna soils were a source, and then compared these values to the accumulation of CH_4 in the nocturnal boundary layer. Nocturnal accumulation was more than ten times larger than the soil emissions and they attributed this to dispersed sources of flooded soil and termites. However, Crutzen et al (2006) suggested that this additional CH_4 source may have largely come from savanna and forest vegetation and could be responsible for ~3–60Tg yr^{-1} CH_4 emissions. Although their estimate was criticized for assumptions and uncertainties about the height of the nocturnal boundary layer and uniform mixing, their observations provided some support for the observations of Keppler et al (2006) on the global role of vegetation emissions.

Do Carmo et al (2006) measured concentration profiles of CH_4 and CO_2 in the canopy layer of undisturbed upland forests at three sites in the Brazilian Amazon region. They detected night-time increases in both CH_4 and CO_2 concentration at all sites and during both wet and dry seasons. They concluded that their data strongly suggest a widespread source of CH_4 in the canopy layer of upland Amazon forest sites that would contribute an annual flux from the Amazon region of 4–38Tg yr^{-1}. They avoided measuring at sites subject to inundation and chose locations on topographically high ground, which gave no support for a wetland origin of their nocturnal CH_4 emissions. Consequently, their observations provided a possible explanation for the anomalously high CH_4 levels detected over tropical regions by satellite observations described above (Frankenberg et al, 2005; Bergamaschi et al, 2007) and with a possible aerobic CH_4 emission from tropical forest vegetation.

Another field study (Sinha et al, 2007) presented the results of high-frequency CH_4 measurements in the boreal forests of Finland and the tropical forests of Suriname, in April–May 2005 and October 2005, respectively. Two weeks of continuous CH_4 concentrations from a boreal site were used with a simple boundary layer model to estimate CH_4 emissions from the surrounding area. The average median concentrations during a typical diurnal cycle were 1.83µmol mol^{-1} and 1.74µmol mol^{-1} for the boreal and tropical forest respectively, with remarkable similarity in the time series of both diurnal profiles. Night-time CH_4 emissions from the boreal forest, calculated from the increase of CH_4 during the night and measured nocturnal boundary layer heights, yielded a global flux of 45.5±11Tg CH_4 yr^{-1} for the boreal forests, which would represent ~8 per cent of the global CH_4 budget. Sinha et al (2007) highlighted the importance of the boreal forest for the global CH_4 budget. However, the upscaling approach as well as the accuracy of the applied analytical system was criticized by several researchers.

Miller et al (2007) collected flask samples using an aircraft and constructed a four-year record of the vertical profile of CH_4 concentrations up to 3600m above sea level at two sites in eastern and central Amazonia. As these were direct measurements, they were not influenced by any spectroscopic interference that impacted the satellite data described above. They observed large enhancements in CH_4 concentration that were not seen at the background monitoring sites of the tropical Atlantic Ocean and calculated an upwind CH_4 emission from tropical forests averaging 27mg m^{-2} day^{-1}. This was not explained by any known individual CH_4 source (wetlands, plants, fire, termites) and atmospheric consumption by the OH radical, or by the sum of all sources – including an estimate of 4mg m^{-2} day^{-1} for plant emissions estimated from the laboratory measurements of Keppler et al (2006) and 5mg m^{-2} day^{-1} unknown night-time emissions, perhaps from plants, observed by Do Carmo et al (2006). Consequently, their airborne measurements indicated a large CH_4 source from the tropical Amazon region and that the individual sources of this CH_4 have yet to be fully explained.

Bowling et al (2009) conducted a study of CH_4 flux from a 110-year-old high-elevation conifer forest in the Rocky Mountains of the US where the dominant species were lodgepole pine (*Pinus contorta*), Engelmann spruce (*Picea engelmannii*) and subalpine fir (*Abies lasiocarpa*). The site elevation of 3050m meant that trees were naturally exposed to high UV irradiances (>2W m^{-2} unweighted UV). The study used micrometeorological techniques over a six-week summer period to measure vertical profiles of CH_4 and CO_2 above and within the canopy and in the subcanopy air space near the ground and it confirmed that the forest soils were a persistent CH_4 sink throughout the study. Although they experienced variability in their CH_4 concentration measurements due to topographic wind systems and urban emissions, they found no evidence for substantial emissions of CH_4 from canopy foliage but could not eliminate the possibility of a weak canopy source in these conifer species.

Recent developments of fast response CH_4 analysers used with eddy covariance techniques (for example Smeets et al, 2009) are now considered capable of detecting even the lower range of CH_4 fluxes from tropical forests observed by Do Carmo et al (2006), Miller et al (2007) and Sinha et al (2007) described above but studies have yet to be completed. However, a full understanding of the processes in a range of ecosystems may still require the use of gas exchange chambers in the field as described below.

Field studies: Gas exchange chambers

One of the first field studies to evaluate CH_4 emissions from vegetation in the field was conducted in 1990 by Sanhueza and Donoso (2006) on a soil–grass system dominated by *Trachypogon* species in the tropical savanna of Venezuela. They used a stainless steel and glass chamber placed onto a soil frame in four unperturbed plots of vegetation and three plots where standing vegetation was clipped and plant litter removed. Methane was measured intermittently during one hour by a small recirculation of chamber air through a gas chromatograph during two periods of the wet season (rainy and less rainy) in October and November respectively. Although the vegetation was exposed to sunlight before the chamber was put in place, it was covered by opaque aluminium foil during measurements. They found that these wet season soils consumed CH_4 with a flux of 4.7ng m^{-2} s^{-1}, but in comparison the soil–vegetation–litter flux of 6ng m^{-2} s^{-1} suggested a net emission from the vegetation and leaf litter of 10.7ng m^{-2} s^{-1}. They concluded by calculating that savanna grasses could make a modest contribution (~1 per cent) to global CH_4 emissions but this was based on measurements in the absence of sunlight.

Two field studies in China also examined CH_4 emissions from plant communities inside closed gas exchange chambers. Cao et al (2008) conducted a three-year study on three plant communities in alpine ecosystems of the Qinghai-Tibet Plateau. A 'grass community', comprising a herbaceous layer only, was examined in a *Kobresia humilis* meadow and a 'shrub' and 'grass' community studied in a *Potentilla* meadow, containing both shrub and herbaceous vegetation, in areas both with and without the shrub *Potentilla fruticosa*. Methane emissions were monitored inside Plexiglas chambers located over 50 × 50cm plots and covered with foam and white cloth to reduce temperature rise in sunlight. The experiment compared emissions from replicate chambers over plots of intact vegetation with those over plots from which above-ground plant parts and live roots had been removed. Methane concentrations were determined between 9am and 10am at a range of daily intervals throughout the year and emission rates calculated from the difference between chambers with intact vegetation and those containing only bare soil. The study observed that bare soils were a net sink for atmospheric CH_4 with a seasonal variation but that the 'grass' community vegetation was emitting CH_4 at the rate of 2.6 and 7.8µg m^{-2} h^{-1} in the *Kobresia* meadow and *Potentilla* meadow respectively. On a biomass basis, this corresponded to 68.3 and 22.6ng CH_4 g^{-1} d.wt. h^{-1} respectively. However, the 'shrub' community of the

Potentilla meadow consumed atmospheric CH_4 at a rate of 5.8μg m^{-2} h^{-1}. These observations led the authors to suggest that because *Kobresia* meadows occupy >60 per cent of the alpine meadows of the Tibetan plateau, they would make a significant regional contribution of 0.13Tg yr^{-1} to the global CH_4 budget.

Notably, this study was conducted inside white opaque chambers (Cao et al, 2008) and, consequently, the vegetation experienced no UV radiation and a reduced level of diffuse visible radiation. The CH_4 source was assumed from the difference in CH_4 concentrations between vegetated and bare soil chambers and there was no direct and conclusive evidence that the apparent CH_4 emissions originated from the vegetation itself. The observed CH_4 could have had a soil origin either via the transpiration stream or other aspects of the applied treatments influencing the net soil flux. Indeed, a third study of alpine meadows and oat pasture in the same area has subsequently suggested that the vegetation was not necessarily the true source of the reported CH_4 emission. Wang et al (2009a) enclosed four vegetation types in the same *Kobresia* meadow as Cao et al (2008): natural alpine meadow; restored alpine meadow; bare soil with roots; and annual oats. They used closed, opaque, stainless steel chambers over two years and conducted parallel studies of water-filled pore space and temperature in the soil. The study established a significant relationship between CH_4 consumption rate with both soil moisture and soil temperature. This led the authors to suggest that diffusive transport of CH_4 through the soil gas phase would be greater when soil moisture was low allowing CH_4 to reach methanotrophic microorganisms that might also respond positively to temperature. Consequently, they suggested that apparent CH_4 production by vegetation, when calculated by comparison with bare soil, might represent differences in soil temperature and water-filled pore space arising from the removal of plants and not represent a true vegetation source. The question of true aerobic production of CH_4 by the vegetation in these studies therefore remains open and further studies covering soil and radiation conditions are necessary to evaluate these observations (Wang et al, 2009a).

Plant- 'mediated' methane emissions

The possibility that experimental observations of aerobic CH_4 emissions from plants could have resulted from soil-derived CH_4 either via internal air spaces or dissolved in the transpiration stream was considered by several researchers after the initial observations of Keppler et al (2006). Such vegetation-mediated sources of soil CH_4 are distinct from the generation of CH_4 inside plant tissues discussed above.

The transport of CH_4 from anaerobic processes in soil to the atmosphere via internal air spaces, such as aerenchyma, is well known in vascular plants (especially grasses and sedges) of wetlands and rice paddies (Chapters 3 and 8). However, the extent of this process in other vegetation types has received less attention. Fluxes of CH_4 from the stems of two-year-old seedlings of black alder (*Alnus glutinosa*), a wetland tree species, were examined by Rusch and

Rennenberg (1998). In controlled experiments they found that CH_4 (and N_2O) was emitted from the bark of stems when the CH_4 concentration in the root zone was above ambient concentrations. Methane efflux decreased with stem height and was consistent with diffusion through air spaces in the tissue as the mechanism of transport. Similar conclusions were reached following field observations of CH_4 emissions from the stem surfaces of mature *Fraxinus mandshurica* var. japonica trees in a Japanese floodplain forest by Terazawa et al (2007). They used steel chambers attached to stems at two heights to quantify CH_4 efflux and measured substantial CH_4 emissions that, when multiplied by tree density, were almost equivalent to the average CH_4 flux rate from the atmosphere to the soil at the site. As the CH_4 concentration in upper soil layers was below ambient concentration, they concluded that much higher CH_4 concentrations in groundwater were driving a concentration gradient resulting in CH_4 transport from submerged soil layers to the atmosphere via air-permeable tissues.

The transport of microbially generated CH_4 to the atmosphere from deeper soil layers via the transpiration stream has also been considered as a possible explanation of apparent plant CH_4 emissions in experimental studies and field observations. Kirschbaum et al (2007) considered a number of potential artefacts in experimental studies including the adsorption of CH_4 onto cell surfaces and the release of solubilized CH_4 in plant lipid and liquid phases including the transpiration stream.

Nisbet et al (2009) conducted two experiments with basil (*Occimum basilicum*) and celery (*Apium graveolens*) that demonstrated that plants equilibrated with CH_4-saturated soil water and subsequently sealed inside closed chambers for ~16 hours generate an elevated CH_4 concentration in the enclosure. They therefore asserted that Keppler et al (2006) primarily measured transpired CH_4 and that when this CH_4 is removed (as in Dueck et al, 2007) no CH_4 emission is observed. However, this overlooks the pre-exposure of leaf samples by Keppler et al (2006) for one hour in CH_4-free air, when considerable dissolved CH_4 would have diffused out from the leaf samples. Kirschbaum et al (2007) calculated the quantity of CH_4 physically held in living leaves, taking into account the water/air and lipid/air equilibrium partition coefficients of CH_4. They concluded that release of CH_4 physically held in leaves is likely to have made only a small contribution to the observed fluxes into CH_4-free air reported by Keppler et al (2006), equivalent to only 50–150 seconds at the rates of emission that were observed over several hours. Consequently, CH_4 dissolved in the transpiration water does not explain many of the observations of foliar CH_4 release. However, it remains a possibility that intact plants in the field with a continuous transpiration stream could transport microbially produced CH_4 from deep sources to the atmosphere.

Vegetation-mediated fluxes from soil should be evaluated carefully in experimental studies and modelling of terrestrial CH_4 sources, and warrant further investigations in order to improve estimates of global emissions, particularly in ecosystems with high or variable water tables.

Verification of plant-derived methane using stable isotope techniques

In recent years stable isotope analysis has become a powerful tool particularly for environmental scientists to track the complimentary processes of methanogenesis and methanotrophy and to study the global CH_4 cycle. Most chemical and biochemical reactions show a slight preference for one isotope of an element over another, usually because of their different masses. The partitioning of the light and heavy isotopes of hydrogen and carbon and the resultant isotope signatures ($\delta^{13}C$ and δ^2H values) can be diagnostic for the identification of CH_4 origin and pathway. Furthermore, isotope-labelling experiments of potential organic precursor molecules are often applied to finally verify hypothetic reaction pathways.

Based on their first results, Keppler et al (2006) suggested the possibility of the involvement of the methyl moiety of the esterified carboxyl group (methoxyl group) of pectin – an important plant structural cell wall component – in CH_4 formation. However, other researchers suggested that the observed release of CH_4 might be caused by adsorption and desorption processes of ambient air (for example Kirschbaum et al, 2006). To exclude this possibility, Keppler et al (2008) employed stable isotope analysis together with pectin and methyl polygalacturonate, both containing trideuterium-labelled methyl groups, to demonstrate that methoxyl groups in plant pectin are indeed precursors of CH_4. A strong deuterium signal in the emitted CH_4 was observed from these labelled polysaccharides under UV irradiation and heating. These results provided unambiguous evidence that methoxyl groups of pectin can act as a source of atmospheric CH_4 under aerobic conditions. Although the authors did not suggest a reaction pathway, their study was an important first step in gaining more information about potential plant precursors involved in CH_4 formation. The same conclusion was reached by McLeod et al (2008) and Bruhn et al (2009), who found that CH_4 production stopped when methoxyl groups of pectin were removed prior to their experiments.

Following the observation that dry and fresh leaves as well as several structural plant components such as pectin and lignin emit CH_4 upon irradiation with UV light (Keppler et al, 2008; McLeod et al, 2008; Vigano et al, 2008; Messenger et al, 2009b). Vigano et al (2009) investigated the source isotope signatures of the CH_4 emitted from a range of dry natural plant leaves and structural compounds. Their data showed that UV-induced CH_4 from organic matter is strongly depleted in both ^{13}C and ^2H (deuterium) compared to the bulk biomass. It was also verified that the ^{13}C and ^2H values of CH_4 does not depend on UV intensity, i.e. the same isotope signature was derived using a total unweighted UV-B radiation of 4W m^{-2} and 20W m^{-2}. This is a surprising result because both ^{13}C- and ^2H-depleted CH_4 is generally viewed as a clear isotope signature for microbial production occurring, for example in wetlands and in the rumen of cows, i.e. that microorganisms preferentially use the lighter isotopes. However, Vigano et al (2009) clearly demonstrated that isotopically light CH_4 is not an unambiguous fingerprint for microbial sources, but can also be generated photochemically from plant matter.

An elegant isotope method to confirm CH_4 formation from living plants was provided by Brüggemann et al (2009). They showed that plants of grey poplar (*Populus* × *canescens*, syn. *Populus tremula* × *Populus alba*) derived from cell cultures under sterile conditions released ^{13}C-labelled CH_4 after feeding the plants with labelled $^{13}CO_2$. The rapid transfer of ^{13}C label from assimilated $^{13}CO_2$ to $^{13}CH_4$ suggested that freshly synthesized, non-structural products of photosynthesis also contribute to aerobic plant CH_4 emissions. They detected CH_4 emissions under low levels of visible light (100µmol m^{-2} s^{-1} photosynthetic flux density) and 27°C:24°C (day:night) temperatures in the range 0.16–0.7ng CH_4 g^{-1} d.wt h^{-1}, which were small compared to those found by Keppler et al (2006) but notably in the absence of UV irradiation. Molecular biological analysis proved the absence of microbial contamination with known methanogenic microorganisms and ruled out the possibility that CH_4 emission from the poplar shoot cultures could be of microbial origin.

Environmental stress factors and methane formation in foliage

As part of a study on the role of UV radiation in foliar emissions of CH_4, Messenger et al (2009a, 2009b) investigated the role of ROS. By using ROS scavengers and ROS generators they demonstrated that the hydroxyl radical and singlet oxygen (but not hydrogen peroxide or the superoxide radical) were involved in CH_4 release from plant pectin. This supports the original suggestion of Sharpatyi (2007) that CH_4 formation in plants may be due to a free radical mechanism. ROS are formed in plants in response to a range of biotic and abiotic stress factors and as part of cellular signalling processes during growth, hormone action or programmed cell death (Apel and Hirt, 2004) and these might also lead to some CH_4 formation. In line with this hypothesis, McLeod et al (2008) demonstrated some CH_4 emission (and also ethane and ethene) from UV-irradiated pectin and from tobacco leaves using a bacterial pathogen and a chemical generator of ROS. Most recently, Wang et al (2009b) have shown that physical injury (by cutting) causes CH_4 emission from vegetation and Qaderi and Reid (2009) have reported that leaf CH_4 emissions may increase under temperature, water and UV radiation stress conditions. The possibility of widespread CH_4 formation in foliage is also indicated by the occurrence and distribution of methanotrophic bateria. In general, methanotrophs and methylotrophs are capable of oxidizing C1 compounds such as CH_4 and methanol. Methanotrophic bacteria that oxidize CH_4 as a sole source of carbon and energy have been identified in soil, freshwater and marine environments (Hanson and Hanson, 1996) and from the aerenchyma of wetland plants (King, 1996; Gilbert et al, 1998; Raghoebarsing et al, 2005). The last might be expected given the role of air spaces in transport of soil-derived CH_4 to the atmosphere in these species (Chapter 3). However, methanotrophs have also been identified in bud and leaf tissue of linden and spruce trees (Doronina et al, 2004), knotgrass (Pirttilä et al, 2008) and stems of maize (*Z. mays*) (Seghers et al, 2004). This raises the question of the source of CH_4 in these plant tissues and the wider occurrence of methanotrophs in other plant species and tissues that may be an indicator of co-evolution and provide a useful indicator of CH_4 availability in plants. The range of diverse processes and environmental factors that may influence vegetation CH_4 emissions requires further investigation to evaluate the mechanisms involved and their significance for global emissions in a range of species and vegetation types.

Overview and global significance

Until recently, only incomplete combustion processes were considered to form CH_4 in the presence of oxygen. However, a large number of studies have now reported aerobic CH_4 emissions from vegetation. The preliminary estimates of Keppler et al (2006) that plant emissions of CH_4 might contribute 10–40 per cent of total global emissions led to considerable debate about current uncertainties in the global CH_4 budget. The authors did not suggest that there were major policy implications from their results but the press and some review articles (Lowe, 2006; Schiermeier, 2006b) suggested that the benefits of forests for CO_2 sequestration and carbon storage might be substantially reduced. Subsequent analyses, described above, led to refinements of the global upscaling calculation and a reduction in the estimated range of global CH_4 emissions from vegetation. Even at an early stage, two reports by expert groups (Bergamaschi et al, 2006; NIEPS, 2006) concluded that the CH_4 produced by plants would be only a small fraction of the global CH_4 budget and likely to be insignificant in the accounting of benefits from carbon sequestration. No change was recommended in current or proposed policies for land use projects designed for sequestration of carbon but a need for flexible carbon offset policies that allow the incorporation of updates in scientific understanding was noted (NIEPS, 2006). The original findings by Keppler et al (2006) were set in context by the original authors in 2007 when Keppler and Röckmann (2007) stated that their calculations showed that the climatic benefits gained by establishing new forests to absorb CO_2 would far exceed the relatively small negative effect of adding more CH_4 to the atmosphere.

However, the collective evidence from independent observations from several research fields now shows that plants do produce CH_4 under aerobic conditions and that this may not be an exotic process, but widespread in nature. The contribution of vegetation emissions to the global total is currently unknown but varies between plant species and between different studies so that research is needed to assess the nature and causes of this variability and to determine the significance of these emissions at the cellular, organism, ecosystem and global scale. An urgent challenge is to test the new hypotheses about aerobic CH_4 formation from different biomolecules and cellular/tissue structures and to draw a comprehensive picture of possible molecular sources and their specific importance in plants. The results published to date imply that aerobic CH_4 formation may be an integral part of cellular responses towards changes in oxidative status present in all eukaryotes and that it is highly variable in time and source strength. Some CH_4 emitted by plants may originate from soil processes via the transpiration stream or physical diffusion though plant tissues. However, CH_4 may also be formed from biomolecules induced by ambient UV light, increased temperatures, or hypoxia, or from more general physiological processes. It appears that plant CH_4 generation may also be linked to environmental stress, and research is needed to estimate the contribution of stress-induced emissions to the global CH_4 budget, not forgetting the consumption of CH_4 by methanotrophs. Atmospheric scientists should revisit the biogenic sources of CH_4 (including wetlands and plants) in view of possible global change feedbacks. These may include links with stratospheric ozone depletion, changing humidity and temperature regimes, rising CO_2 concentrations, land use change and ecosystem responses, which are relevant for both modern and palaeo-climatic changes. A broad range of field measurements at different scales and satellite measurements at higher resolution are needed in order to feed numerical modelling studies.

The observation that CH_4 formation under aerobic conditions occurs in plants is robust but the magnitude and significance for the global CH_4 budget requires further evaluation. The topic presents new, interesting and very challenging avenues for future research in plant and environmental sciences. It will require a considerable effort by researchers from different disciplines to quantify these emissions and complete the detail of the vegetation contribution to biogeochemical cycling of CH_4 and its importance for our atmosphere and climate.

References

Apel, K. and Hirt, H. (2004) 'Reactive oxygen species: Metabolism, oxidative stress, and signal transduction', *Annual Review of Plant Biology*, vol 55, pp373–399

Beerling, D. J., Gardiner, T., Leggett, G., Mcleod, A. and Quick, W. P. (2008) 'Missing methane emissions from leaves of terrestrial plants', *Global Change Biology*, vol 14, no 8, pp1821–1826

Bergamaschi, P., Dentener, F., Grassi, G., Leip, A., Somogyi, Z., Federici, S., Seufert, G. and Raes, F. (2006) 'Methane emissions from terrestrial plants', European Commission, DG Joint Research Centre, Institute for Environment and Sustainability, Ispra, Italy

Bergamaschi, P., Frankenberg, C., Meirink, J. F., Krol, M., Dentener, F., Wagner, T., Platt, U., Kaplan, J. O., Korner, S., Heimann, M., Dlugokencky, E. J. and Goede, A. (2007) 'Satellite chartography of atmospheric methane from SCIAMACHY on board ENVISAT: 2. Evaluation based on inverse model simulations', *Journal of Geophysical Research – Atmospheres*, vol 112, no D2, D02304

Berry, J. A. and Raison, J. K. (1982) 'Responses of macrophytes to temperature', in O. L. Lange, P. S. Nobel, C. B. Osmond and H. Ziegler (eds) *Physiological Plant*

Ecology: I. Responses to the Physical Environment, Encyclopedia of Plant Physiology, New Series Vol 12A, Springer-Verlag, Berlin, Heidelberg, New York

Bowling, D. R., Miller, J. B., Rhodes, M. E., Burns, S. P., Monson, R. K. and Baer, D. (2009) 'Soil and plant contributions to the methane flux balance of a subalpine forest under high ultraviolet irradiance', *Biogeosciences Discussions*, vol 6, pp4765–4801

Brüggemann, N., Meier, R., Steigner, D., Zimmer, I., Louis, S. and Schnitzler, J. P. (2009) 'Nonmicrobial aerobic methane emission from poplar shoot cultures under low-light conditions', *New Phytologist*, vol 182, no 4, pp912–918

Bruhn, D., Mikkelsen, T. N., Øbro, J., Willats, W. G. T. and Ambus, P. (2009) 'Effects of temperature, ultraviolet radiation and pectin methyl esterase on aerobic methane release from plant material', *Plant Biology*, vol 11, pp43–48

Butenhoff, C. L. and Khalil, M. A. K. (2007) 'Global methane emissions from terrestrial plants', *Environmental Science and Technology*, vol 41, no 11, pp4032–4037

Cao, G., Xu, X., Long, R., Wang, Q., Wang, C., Du, Y. and Zhao, X. (2008) 'Methane emissions by alpine plant communities in the Qinghai–Tibet Plateau', *Biology Letters*, vol 4, no 6, pp681–684

Crutzen, P. J., Sanhueza, E. and Brenninkmeijer, C. A. M. (2006) 'Methane production from mixed tropical savanna and forest vegetation in Venezuela', *Atmospheric Chemistry and Physics Discussions*, vol 6, pp3093–3097

Do Carmo, J. B., Keller, M., Dias, J. D., De Camargo, P. B. and Crill, P. (2006) 'A source of methane from upland forests in the Brazilian Amazon', *Geophysical Research Letters*, vol 33, L04809

Doronina, N. V., Ivanova, E. G., Suzina, N. E. and Trotsenko, Y. A. (2004) 'Methanotrophs and methylobacteria are found in woody plant tissues within the winter period', *Microbiology*, vol 73, no 6, pp702–709

Dueck, T. A., De Visser, R., Poorter, H., Persijn, S., Gorissen, A., De Visser, W., Schapendonk, A., Verhagen, J., Snel, J., Harren, F. J. M., Ngai, A. K. Y., Verstappen, F., Bouwmeester, H., Voesenek, L. and Van Der Werf, A. (2007) 'No evidence for substantial aerobic methane emission by terrestrial plants: A ^{13}C-labelling approach', *New Phytologist*, vol 175, no 1, pp29–35

Ferretti, D. F., Miller, J. B., White, J. W. C., Etheridge, D. M., Lassey, K. R., Lowe, D. C., Meure, C. M. M., Dreier, M. F., Trudinger, C. M., Van Ommen, T. D. and Langenfelds, R. L. (2005) 'Unexpected changes to the global methane budget over the past 2000 years', *Science*, vol 309, no 5741, pp1714–1717

Ferretti, D. F., Miller, J. B., White, J. W. C., Lassey, K. R., Lowe, D. C. and Etheridge, D. M. (2007) 'Stable isotopes provide revised global limits of aerobic methane emissions from plants', *Atmospheric Chemistry and Physics*, vol 7, pp237–241

Frankenberg, C., Meirink, J. F., Van Weele, M., Platt, U. and Wagner, T. (2005) 'Assessing methane emissions from global space-borne observations', *Science*, vol 308, no 5724, pp1010–1014

Frankenberg, C., Bergamaschi, P., Butz, A., Houweling, S., Meirink, J. F., Notholt, J., Petersen, A. K., Schrijver, H., Warneke, T. and Aben, I. (2008) 'Tropical methane emissions: A revised view from SCIAMACHY onboard ENVISAT', *Geophysical Research Letters*, vol 35, no 15, L15811

Gilbert, B., Assmus, B., Hartmann, A. and Frenzel, P. (1998) 'In situ localization of two methanotrophic strains in the rhizosphere of rice plants', *FEMS Microbiology Ecology*, vol 25, no 2, pp117–128

Hanson, R. S. and Hanson, T. E. (1996) 'Methanotrophic bacteria', *Microbiological Reviews*, vol 60, no 2, pp439–471

Houweling, S., Rockmann, T., Aben, I., Keppler, F., Krol, M., Meirink, J. F., Dlugokencky, E. J. and Frankenberg, C. (2006) 'Atmospheric constraints on global emissions of methane from plants', *Geophysical Research Letters*, vol 33, no 15, L15821

IPCC (Intergovernmental Panel on Climate Change) (2007) *Climate Change 2007: The Physical Science Basis Contribution of Working Group I to the Fourth Assessment Report of the Intergovernmental Panel on Climate Change*, S. Solomon, D. Qin, M. Manning, Z. Chen, M. Marquis, K. B. Averyt, M. Tignor and H. L. Miller (eds), Cambridge University Press, Cambridge and New York

Keppler, F. and Röckmann, T. (2007) 'Methane, plants and climate change', *Scientific American*, vol 296, no 2, pp52–57

Keppler, F., Hamilton, J. T. G., Brass, M. and Röckmann, T. (2006) 'Methane emissions from terrestrial plants under aerobic conditions', *Nature*, vol 439, no 7073, pp187–191

Keppler, F., Hamilton, J. T. G., McRoberts, W. C., Vigano, I., Brass, M. and Röckmann, T. (2008) 'Methoxyl groups of plant pectin as a precursor of atmospheric methane: Evidence from deuterium labelling studies', *New Phytologist*, vol 178, pp808–814

Keppler, F., Boros, M., Frankenberg, C., Lelieveld, J., McLeod, A., Pirttilä, A. M., Röckmann, T. and Schnitzler, J. P. (2009) 'Methane formation in aerobic environments', *Environmental Chemistry*, vol 6, pp459–465

King, G. M. (1996) 'In situ analyses of methane oxidation associated with the roots and rhizomes of a bur reed, *Sparganium eurycarpum*, in a Maine wetland', *Applied and Environmental Microbiology*, vol 62, no 12, pp4548–4555

Kirschbaum, M. U. F. and Walcroft, A. (2008) 'No detectable aerobic methane efflux from plant material, nor from adsorption/desorption processes', *Biogeosciences*, vol 5, no 6, pp1551–1558

Kirschbaum, M. U. F., Bruhn, D., Etheridge, D. M., Evans, J. R., Farquhar, G. D., Gifford, R. M., Paul, K. I. and Winters, A. J. (2006) 'A comment on the quantitative significance of aerobic methane release by plants', *Functional Plant Biology*, vol 33, no 6, pp521–530

Kirschbaum, M. U. F., Niinemets, U., Bruhn, D. and Winters, A. J. (2007) 'How important is aerobic methane release by plants?', *Functional Plant Science and Technology*, vol 1, pp138–145

Lowe, D. C. (2006) 'Global change: A green source of surprise', *Nature*, vol 439, no 7073, pp148–149

McLeod, A. R., Fry, S. C., Loake, G. J., Messenger, D. J., Reay, D. S., Smith, K. A. and Yun, B. W. (2008) 'Ultraviolet radiation drives methane emissions from terrestrial plant pectins', *New Phytologist*, vol 180, no 1, pp124–132

Megonigal, J. P. and Güenther, A. B. (2008) 'Methane emissions from upland forest soils and vegetation', *Tree Physiology*, vol 28, pp491–498

Messenger, D. J., McLeod, A. R. and Fry, S. C. (2009a) 'Reactive oxygen species in aerobic methane formation from vegetation', *Plant Signaling and Behavior*, vol 4, no 7, pp1–2

Messenger, D. J., McLeod, A. R. and Fry, S. C. (2009b) 'The role of ultraviolet radiation, photosensitizers, reactive oxygen species and ester groups in mechanisms of methane formation from pectin', *Plant Cell and Environment*, vol 32, no 1, pp1–9

Miller, J. B., Gatti, L. V., D'amelio, M. T. S., Crotwell, A. M., Dlugokencky, E. J., Bakwin, P., Artaxo, P. and Tans, P. P. (2007) 'Airborne measurements indicate large methane emissions from the eastern Amazon basin', *Geophysical Research Letters*, vol 34, no 10, L10809

NIEPS (Nicholas Institute for Environmental Policy Solutions) (2006) 'Do recent scientific findings undermine the climate benefits of carbon sequestration in Forests? An expert review of recent studies on methane emissions and water tradeoffs', Duke University, Nicholas Institute for Environmental Policy Solutions, Durham, NC

Nisbet, R. E. R., Fisher, R., Nimmo, R. H., Bendall, D. S., Crill, P. M., Gallego-Sala, A. V., Hornibrook, E. R. C., Lopez-Juez, E., Lowry, D., Nisbet, P. B. R., Shuckburgh, E. F., Sriskantharajah, S., Howe, C. J. and Nisbet, E. G. (2009) 'Emission of methane from plants', *Proceedings of the Royal Society B – Biological Sciences*, vol 276, no 1660, pp1347–1354

Parsons, A. J., Newton, P. C. D., Clark, H. and Kelliher, F. M. (2006) 'Scaling methane emissions from vegetation', *Trends in Ecology and Evolution*, vol 21, no 8, pp423–424

Pirttilä, A. M., Hohtola, A., Ivanova, E. G., Fedorov, D. N. F., Doronina, N. V. and Trotsenko, Y. A. (2008) 'Identification and localization of methylotrophic plant-associated bacteria', in S. Sorvari and A. M. Pirttilä (eds) *Prospects and Applications for Plant Associated Microbes, A Laboratory Manual: Part A, Bacteria*, Biobien Innovations, Turku, Finland, pp218–224

Qaderi, M. M. and Reid, D. M. (2009) 'Methane emissions from six crop species exposed to three components of global change: Temperature, ultraviolet-B radiation and water stress', *Physiologia Plantarum*, vol 137, no 2, pp139–147

Raghoebarsing, A. A., Smolders, A. J. P., Schmid, M. C., Rijpstra, W. I. C., Wolters-Arts, M., Derksen, J., Jetten, M. S. M., Schouten, S., Damste, J. S. S., Lamers, L. P. M., Roelofs, J. G. M., Den Camp, H. and Strous, M. (2005) 'Methanotrophic symbionts provide carbon for photosynthesis in peat bogs', *Nature*, vol 436, no 7054, pp1153–1156

Rusch, H. and Rennenberg, H. (1998) 'Black alder (*Alnus glutinosa* (L.) Gaertn.) trees mediate methane and nitrous oxide emission from the soil to the atmosphere', *Plant and Soil*, vol 201, no 1, pp1–7

Sanadze, G. A. and Dolidze, G. M. (1960) 'About chemical nature of volatile emissions released by leaves of some plants', *Reports of Academy of Sciences of USSR*, vol 134, no 1, pp214–216

Sanhueza, E. and Donoso, L. (2006) 'Methane emission from tropical savanna *Trachypogon* sp. Grasses', *Atmospheric Chemistry and Physics*, vol 6, pp5315–5319

Schaefer, H., Whiticar, M. J., Brook, E. J., Petrenko, V. V., Ferretti, D. F. and Severinghaus, J. P. (2006) 'Ice record of delta C-13 for atmospheric CH_4 across the Younger Dryas-Preboreal transition', *Science*, vol 313, no 5790, pp1109–1112

Scharffe, D., Hao, W. M., Donoso, L., Crutzen, P. J. and Sanhueza, E. (1990) 'Soil fluxes and atmospheric concentration of CO and CH_4 in the northern part of the

Guayana shield, Venezuela', *Journal of Geophysical Research – Atmospheres*, vol 95, no D13, pp22475–22480

Schiermeier, Q. (2006a) 'Methane finding baffles scientists', *Nature*, vol 439, p128

Schiermeier, Q. (2006b) 'The methane mystery', *Nature*, vol 442, pp730–731

Schneising, O., Buchwitz, M., Burrows, J. P., Bovensmann, H., Bergamaschi, P. and Peters, W. (2009) 'Three years of greenhouse gas column-averaged dry air mole fractions retrieved from satellite – Part 2: Methane', *Atmospheric Chemistry and Physics*, vol 9, pp443–465

Seghers, D., Wittebolle, L., Top, E. M., Verstraete, W. and Siciliano, S. D. (2004) 'Impact of agricultural practices on the *Zea mays* L. endophytic community', *Applied and Environmental Microbiology*, vol 70, no 3, pp1475–1482

Sharpatyi, V. A. (2007) 'On the mechanism of methane emission by terrestrial plants', *Oxidation Communications*, vol 30, no 1, pp48–50

Sinha, V., Williams, J., Crutzen, P. J. and Lelieveld, J. (2007) 'Methane emissions from boreal and tropical forest ecosystems derived from in-situ measurements', *Atmospheric Chemistry and Physics Discussions*, vol 7, pp14011–14039

Smeets, C. J. P. P., Holzinger, R., Vigano, I., Goldstein, A. H. and Röckmann, T. (2009) 'Eddy covariance methane measurements at a Ponderosa pine plantation in California', *Atmospheric Chemistry and Physics Discussions*, vol 9, pp5201–5229

Terazawa, K., Ishizuka, S., Sakatac, T., Yamada, K. and Takahashi, M. (2007) 'Methane emissions from stems of *Fraxinus mandshurica* var. japonica trees in a floodplain forest', *Soil Biology and Biochemistry*, vol 39, no 10, pp2689–2692

Vigano, I., Holzinger, R., Van Weelden, H., Keppler, F., McLeod, A. and Röckmann, T. (2008) 'Effect of UV radiation and temperature on the emission of methane from plant biomass and structural components', *Biogeosciences*, vol 5, pp937–947

Vigano, I., Röckmann, T., Holzinger, R., Van Dijk, A., Keppler, F., Greule, M., Brand, W. A., Geilmann, H. and Van Weelden, H. (2009) 'The stable isotope signature of methane emitted from plant material under UV irradiation', *Atmospheric Environment*, doi:10.1016/j.atmosenv.2009.07.046

Wang, Z. P., Han, X. G., Wang, G. G., Song, Y. and Gulledge, J. (2008) 'Aerobic methane emission from plants in the Inner Mongolia Steppe', *Environmental Science and Technology*, vol 42, no 1, pp62–68

Wang, S., Yang, X., Lin, X., Hu, Y., Luo, C., Xu, G., Zhang, Z., Su, A., Chang, X., Chao, Z. and Duan, J. (2009a) 'Methane emission by plant communities in an alpine meadow on the Qinghai-Tibetan Plateau: A new experimental study of alpine meadows and oat pasture', *Biology Letters*, vol 5, no 4, pp535–538

Wang, Z. P., Gulledge, J., Zheng, J. Q., Liu, W., Li, L. H. and Han, X. G. (2009b) 'Physical injury stimulates aerobic methane emissions from terrestrial plants', *Biogeosciences*, vol 6, no 4, pp615–621

7
Biomass Burning

Joel S. Levine

Introduction

Biomass burning or the burning of vegetation is the burning of living and dead vegetation and includes human-initiated burning and natural lightning-induced burning. The majority of the biomass burning, primarily in the tropics (perhaps as much as 90 per cent), is believed to be human-initiated for land clearing and land use change. Natural fires triggered by atmospheric lightning only account for in the order of about 10 per cent of all fires (Andreae, 1991). As will be discussed, a significant amount of biomass burning occurs in the boreal forests of Russia, Canada and Alaska.

Biomass burning is a significant source of gases and particulates to the regional and global atmosphere (Crutzen et al, 1979; Seiler and Crutzen, 1980; Crutzen and Andreae, 1990; Levine et al, 1995). The bulk of the world's biomass burning occurs in the tropics – in the tropical forests of South America and Southeast Asia and in the savannas of Africa and South America. Biomass burning is truly a multidisciplinary subject, encompassing the following areas: fire ecology, fire measurements, fire modelling, fire combustion, remote sensing, fire combustion gaseous and particulate emissions, the atmospheric transport of these emissions and the chemical and climatic impacts of these emissions. Over the last few years, a series of dedicated books have documented much of our current understanding of biomass burning in different ecosystems. These volumes include: Goldammer (1990), Levine (1991), Crutzen and Goldammer (1993), Goldammer and Furyaev (1996), Levine (1996a, 1996b), van Wilgen et al (1997), Kasischke and Stocks (2000), Innes et al (2000) and Eaton and Radojevic (2001).

Simpson et al (2006) found that biomass burning had a significant influence on the large global CH_4 pulses observed in 1998 and 2002–2003. Simpson et al (2006) attributed the large global increase of methane in 1997 to extensive and widespread surface vegetation and below-surface peat burning in Indonesia in 1997 (Levine, 1999) and in the boreal forests of Russia in 1998 and the global CH_4 increase from 2000 to 2003 (with a peak in 2003) is attributed to enhanced boreal fire emissions that occurred during this time.

The important role of the world's boreal forest fires in global climate change and in the emissions of CH_4 and other combustion products from above-ground vegetation and below-ground peat fires in Indonesia will be discussed in more detail. As we will see later in this chapter, the emission ratio of CH_4 from the burning of underground peat is three times larger than the emission ratio of CH_4 from burning of above-ground vegetation.

Global impacts of biomass burning

On an annual global basis, biomass burning is a significant source of gases and particulates to the atmosphere. The gaseous and particulate emissions produced during biomass burning are dependent on the nature of the biomass matter, which is a function of the ecosystem and the temperature of the fire, which is also ecosystem dependent. In general, biomass is composed mostly of carbon (about 45 per cent by weight) and hydrogen and oxygen (about 55 per cent by weight), with trace amounts of nitrogen (0.3 to 3.8 per cent by mass), sulphur (0.1 to 0.9 per cent), phosphorus (0.01 to 0.3 per cent), potassium (0.5 to 3.4 per cent) and still smaller amounts of chlorine, and bromine (Andreae, 1991).

During complete combustion, the burning of biomass matter produces CO_2 and water vapour as the primary products, according to the reaction:

$$CH_2O + O_2 \rightarrow CO_2 + H_2O$$

where CH_2O represents the approximate average chemical composition of biomass matter. In the more realistic case of incomplete combustion in cooler and/or oxygen-deficient fires, i.e. the smoldering phase of burning, carbon is released in the forms of CO, CH_4, non-methane hydrocarbons (NMHCs), and various partially oxidized organic compounds, including aldehydes, alcohols, ketones and organic acids, and particulate black (soot) carbon. Nitrogen is present in biomass mostly as amino groups ($R-NH_2$) in the amino acids of proteins. During combustion the nitrogen is released by pyrolytic decomposition of the organic matter and partially or completely oxidized to various volatile nitrogen compounds, including molecular nitrogen (N_2), nitric oxide (NO), nitrous oxide (N_2O), ammonia (NH_3), hydrogen cyanide (HCN), cyanogen (NCCN), organic nitriles (acetonitrile, CH_3CN; acrylonitrile, CH_2CHCN; and propionitrile CH_3CH_2CN) and nitrates. The sulphur in biomass is organically bound in the form of sulphur-containing amino acids in proteins. During burning the sulphur is released mostly in the form of sulphur dioxide (SO_2) and smaller amounts of carbonyl sulphide (COS) and non-volatile sulphate (SO_4^-). About one half of the sulphur in the biomass matter is left in the burn ash, whereas, very little of the fuel nitrogen is left in the ash.

Laboratory biomass burning experiments conducted by Lobert et al (1991) have identified the carbon (Table 7.1) and nitrogen compounds released to the atmosphere by burning. The major gases produced during the biomass burning

Table 7.1 Carbon and gases produced during biomass burning

Compound	Mean emission factor relative to the fuel C (per cent)
Carbon dioxide (CO_2)	82.58
Carbon monoxide (CO)	5.73
Methane (CH_4)	0.424
Ethane (CH_3CH_3)	0.061
Ethene ($CH_2=CH_2$)	0.123
Ethine ($CH\equiv CH$)	0.056
Propane (C_3H_8)	0.019
Propene (C_3H_6)	0.066
n-butane (C_4H_{10})	0.005
2-butene (cis) (C_4H_8)	0.004
2-butene (trans) (C_4H_8)	0.005
i-butene, i-butene ($C_4H_8 + C_4H_8$)	0.033
1,3-butadiene(C_4H_6)	0.021
n-pentane (C_5H_{12})	0.007
Isoprene (C_5H_8)	0.008
Benzene (C_6H_6)	0.064
Toluene (C_7H_8)	0.037
m-, p-xylene (C_8H_{10})	0.011
o-xylene (C_8H_{10})	0.006
Methyl chloride (CH_3Cl)	0.010
NMHC (As C) (C_2 to C_8)	1.18
Ash (As C)	5.00
Total Sum C	94.92 (including ash)

Source: Lobert et al (1991)

process include CO_2, CO, CH_4, oxides of nitrogen (NO_x = NO + NO_2), and ammonia (NH_3). As well as CO_2 and CH_4 acting as direct greenhouse gases, CO, CH_4 and the oxides of nitrogen also lead to the photochemical production of ozone (O_3) in the troposphere. In the troposphere, ozone is harmful to both vegetation and humans at concentrations not far above the global background levels. Nitric oxide leads to the chemical production of nitric acid (HNO_3) in the troposphere. Nitric acid is the fastest-growing component of acidic precipitation. Ammonia is the only basic gaseous species that neutralizes the acidic nature of the troposphere.

Particulates – small (usually about 10μm or smaller) solid particles, such as smoke or soot particles – are also produced during the burning process and released into the atmosphere. These solid particulates absorb and scatter incoming sunlight and hence impact the local, regional and global climate. In addition, these particulates (specifically particulates 2.5μm or smaller) can lead to various human respiratory and general health problems when inhaled. The gases and particulates produced during biomass burning lead to the formation of 'smog.' The word 'smog' was coined as a combination of smoke and fog and is now used to describe any smoky or hazy pollution in the atmosphere.

Gaseous and particulate emissions produced during biomass burning and released into the atmosphere impact the local, regional and global atmosphere and climate in several different ways, including:

1 Biomass burning is a significant global source of CO_2 and CH_4. Both gases are greenhouse gases that lead to global warming.
2 Biomass burning is a significant global source of carbon monoxide, CH_4, non-CH_4 hydrocarbons and oxides of nitrogen. These gases lead to the photochemical production of ozone in the troposphere. Tropospheric ozone is a pollutant and irritant and has a negative impact on plant, animal and human life.
3 Methyl chloride and methyl bromide, while only released in trace amounts during biomass burning, have a negative impact on stratospheric ozone. Methyl chloride and methyl bromide produced during biomass burning will become even more important in the future, as human-produced sources of chlorine and bromine are phased out as a result of the Montreal Protocol banning stratospheric ozone-destroying chlorine- and bromine-containing gases.
4 Particulates produced during biomass burning absorb and scatter incoming solar radiation, which impact climate. In addition, biomass burning-produced particulates lead to reduced atmospheric visibility. Recent research suggests that atmospheric particulates produced during biomass burning may directly enter the stratosphere via strong vertical thermal convective currents produced during the fire.
5 Particulates produced during biomass burning become cloud condensation nuclei (CCN) and impact the formation and distribution of clouds.
6 Particulates produced during biomass burning, particularly particulates of 10μm or less in diameter, lead to severe respiratory problems when inhaled.

The geographical distribution of biomass burning

The locations of biomass burning are varied and include tropical savannas, tropical, temperate and boreal forests, and agricultural lands after the harvest. The burning of fuelwood for domestic use is another source of biomass burning. Global estimates of the annual amounts of biomass burning from these sources are estimated in Table 7.2 (Andreae, 1991).

Table 7.2 Global estimates of annual amounts of biomass burning and of the resulting release of carbon to the atmosphere

Source	Biomass burned (Tg dm yr^{-1})	Carbon released (Tg C yr^{-1})
Savanna	3690	1660
Agricultural waste	2020	910
Fuelwood	1430	640
Tropical forests	1260	570
Temperate/boreal forests	280	130
World totals	8680	3910

Note: dm = dry matter (biomass matter).
Source: Andreae (1991)

As already noted, biomass matter is composed of about 45 per cent carbon by weight. Table 7.2 gives estimates of the carbon released (Tg C yr^{-1}) by the burning of this biomass (the total biomass burned is multiplied by 45 per cent to determine the amount of carbon released into the atmosphere during burning). Combining estimates of the total amount of biomass matter burned per year (Table 7.2) with measurements of the gaseous and particulate emissions from biomass burning (Table 7.1) permits estimates of the global production and release into the atmosphere of gases and particulates from biomass burning, which will be discussed in more detail.

Biomass burning in the boreal forests

In the past, it was generally assumed that biomass burning was primarily a tropical phenomenon. This is because most of the information that we have on the geographical and temporal distribution of biomass burning is largely based on tropical burning. Very little information was available on the geographical and temporal distribution of biomass burning in the boreal forests, which cover about 25 per cent of the world's forests. To illustrate how our knowledge of the geographical extent of burning in the world's boreal forests has increased in recent years, consider the following.

Early estimates based on surface fire records and statistics suggested that 1.5 million hectares (1 hectare = 2.47 acres) of boreal forests burn annually (Seiler and Crutzen, 1980). Later studies, based on more comprehensive surface fire records and statistics, indicated that earlier values underestimated burning in the world's boreal forests and that an average of 8 million hectares burned annually during the 1980s, with great year-to-year fluctuations (Stocks et al, 1993). One of the largest fires ever measured occurred in the boreal forests of the Heilongjiang Province of Northeastern China in May 1987. In less than four weeks, more than 1.3 million hectares of boreal forest were burned (Levine et al, 1991; Cahoon et al, 1994). At the same time, extensive fire activity occurred across the Chinese border in Russia, particularly in the area east of Lake Baikal between the Amur and Lena rivers. Estimates based on National Oceanic and Atmospholic Administration (NOAA) Advanced Very High Resolution Radiometer (AVHRR) imagery indicate that 14.4 million hectares (35.7 million acres) in China and Siberia were burned in 1987 (Cahoon et al, 1994), dwarfing earlier estimates of boreal forest fire burned area.

While 1987 was an extreme fire year in eastern Asia, the sparse database may suggest a fire trend. Is burning in the boreal forests increasing with time, or are satellite measurements providing more accurate data? Satellite measurements are certainly providing a more accurate assessment of the extent and frequency of burning in the world's boreal forests. As global warming continues, predicted warmer and dryer conditions in the world's boreal forests will result in more frequent and larger fires and greater production of CO_2 and CH_4 by these fires, such increased burning could therefore represent a significant positive climate change feedback.

Calculations using the satellite-derived burn area and measured emission ratios of gases for boreal forest fires indicate that the Chinese and Siberian fires of the 1987 contributed about 20 per cent of the total CO_2 produced by savanna burning, 36 per cent of the total CO produced by savanna burning, and 69 per cent of the total CH_4 produced by savanna burning (Cahoon et al, 1994). Since savanna burning represents the largest component of tropical burning in terms of the vegetation consumed by fire (Table 7.2), it is apparent that the atmospheric emissions from boreal forest burning must be included in global emissions budgets.

There are several reasons that burning in the world's boreal forests is very important:

1. The boreal forests are very susceptible to global warming. Small changes in the surface temperature can significantly influence the ice and snow albedo feedback (a change in how reflective the land surface is). Thus, infrared absorption processes by fire-produced greenhouse gases, as well as fire-induced changes in surface albedo and infrared emissivity in the boreal forest regions, are more environmentally significant than in the tropics.
2. In the world's boreal forests, global warming will result in warmer and drier conditions. This in turn may result in enhanced frequency of fire and an accompanying enhancement in the emissions of CO_2 and CH_4 that will then amplify the greenhouse effect.
3. Fires in the boreal forests are the most energetic in nature. The average fuel consumption per unit area in the boreal forest is in the order of 25,000 kg ha^{-1}, which is about an order of magnitude greater than in the tropics. Large boreal forest fires typically spread very quickly, most often as 'crown fires', causing the burning of the entire tree up to and including the crown. Large boreal forest fires release enough energy to generate convective smoke columns that routinely reach well into the upper troposphere, and on occasion, may directly penetrate across the tropopause into the

stratosphere. The tropopause is at a minimum height over the world's boreal forests. As an example, a 1986 forest fire in northwestern Ontario (Red Lake) generated a convective smoke column 12 to 13km in height, penetrating across the tropopause into the stratosphere (Stocks and Flannigan, 1987). There is a strong link between boreal forest fires in Canada and eastern Russia in 1998 and increased stratospheric aerosols during the same period (Fromm et al, 2000).

4   The cold temperature of the troposphere over the world's boreal forests results in low levels of tropospheric water vapour. The deficiency of tropospheric water vapour and the scarcity of incoming solar radiation over most of the year result in very low photochemical production of the hydroxyl radical over the boreal forests. The OH radical is the overwhelming chemical scavenger in the troposphere and controls the atmospheric lifetime of many tropospheric gases, including CH_4. The very low concentrations of the OH radical over the boreal forests result in enhanced atmospheric lifetimes for most tropospheric gases, including CH_4 produced by biomass burning. Hence, gases produced by burning, such as CO, CH_4 and the oxides of nitrogen, will have enhanced atmospheric lifetimes over the boreal forest.

New information about burning in the world's boreal forests, based on satellite measurements, was reported by Kasischke et al (1999). Some of the findings reported in this study are summarized here:

1   Fires in the boreal forest covering at least 100,000 hectares are not uncommon.
2   In the boreal forests of North America, most fires (>90 per cent) are crown fires. The remainder are surface fires. Crown fires consume much more fuel (30 to 40 tonnes of biomass material per hectare burned) than surface fires (8 to 12 tonnes of biomass material per hectare burned).
3   The fire record for North America over the past three decades clearly shows the episodic nature of fire in the boreal forests. Large fire years occur during extended periods of drought, which allow naturally ignited fires (i.e. lightning-ignited fires) to burn large areas. Since 1970, the area burned during six episodic fire years in the North American boreal forest was 6.2 million hectares per year, while 1.5 million hectares burned per year in the remaining years. There is evidence that a similar episodic pattern of fire may also exist in the Russia boreal forest.
4   The fire data in the North American boreal forest show a significant increase in the annual area burned over the past three decades, with an average 1.5 million hectares per year burning during the 1970s and 3.2 million hectares per year burning during the 1990s. This increase in burning corresponds to rises of 1.0 to 1.6°C over the same period (Hansen et al, 1996). The projected 2 to 4°C increase in temperature due to projected increases in greenhouse gases during the 21st century may result in high levels of fire activity throughout the world's boreal forests in the future (Stocks et al, 1993).

5 During typical years in the boreal forests, the amounts of biomass consumed during fire ranges between 10 and 20 tonnes per hectare. During the drought years with episodic fires, the amounts of biomass consumed during biomass burning may be as high as 50 to 60 tonnes per hectare. Assuming that biomass is about 50 per cent carbon by mass, such amounts would release 450 to 600Tg C globally. These amounts are considerably higher than the often-quoted value for total carbon released by biomass burning in the world's boreal and temperate forests of 130Tg C globally (Andreae, 1991; see Table 7.2).

Calculation of gaseous and particulate emissions from burning

To assess both the environmental and health impacts of biomass burning, information is needed on the gaseous and particulate emissions produced during the fire and released into the atmosphere. The calculation of gaseous emissions from vegetation and peat fires can be calculated using a form of an expression from Seiler and Crutzen (1980) for each burning ecosystem/terrain:

$$M = A \times B \times E \tag{1}$$

where M = total mass of vegetation or peat consumed by burning (tonnes), A = area burned (km^2), B = biomass loading (tonnes km^{-2}), and E = burning efficiency (dimensionless). The total mass of carbon (M(C)) released to the atmosphere during burning is related to M by the following expression:

$$M(C) = C \times M \text{ (tonnes of carbon)} \tag{2}$$

where C is the mass percentage of carbon in the biomass. For tropical vegetation, C = 0.45 (Andreae, 1991); for peat, C = 0.50 (Yokelson et al, 1996). The mass of CO_2 (M(CO_2)) released during the fire is related to M(C) by the following expression:

$$M(CO_2) = CE \times M(C) \tag{3}$$

The combustion efficiency (CE) is the fraction of carbon emitted as CO_2 relative to the total carbon compounds released during the fire. For tropical vegetation fires, CE = 0.90 (Andreae, 1991); for peat fires, CE = 0.77 (Yokelson et al, 1997). The biomass load range and the burning efficiency for tropical ecosystems are summarized in Table 7.3.

Once the mass of CO_2 produced by burning is known, the mass of any other species, X_i (M(X_i)), produced by burning and released to the atmosphere can be calculated with knowledge of the CO_2-normalized species emission ratio (ER(X_i)). The emission ratio is the ratio of the production of species X_i to the production of CO_2 in the fire. The mass of species, X_i, is related to the mass of CO_2 by the following expression:

Table 7.3 Biomass load range and burning efficiency in tropical ecosystems

Vegetation type	Biomass load range (tonnes km^{-2})	Burning efficiency
Peat[1]	97,500	0.50
Tropical rainforests[2]	5000–55,000	0.20
Evergreen forests	5000–10,000	0.30
Plantations	500–10,000	0.40
Dry forests	3000–7000	0.40
Fynbos	2000–4500	0.50
Wetlands	340–1000	0.70
Fertile grasslands	150–550	0.96
Forest/savanna mosaic	150–500	0.45
Infertile savannas	150–500	0.95
Fertile savannas	150–500	0.95
Infertile grasslands	150–350	0.96
Shrublands	50–200	0.95

Source: From Scholes et al (1996) except [1] Brunig (1977) and Supardi et al (1993); [2] Brown and Gaston (1996)

$$M(X_i) = ER(X_i) \times M(CO_2) \text{ (tonnes of element } X_i) \qquad (4)$$

where X_i = CO, CH$_4$, NO$_x$, NH$_3$, and O$_3$. It is important to re-emphasize that O$_3$ is not a direct product of biomass burning. However, O$_3$ is produced via photochemical reactions of CO, CH$_4$ and NO$_x$, all of which are produced directly by biomass burning. Hence, the mass of ozone resulting from biomass burning may be calculated by considering the ozone precursor gases produced by biomass burning. Values for emission ratios for tropical forest fires and peat fires are summarized in Table 7.4.

To calculate the total particulate matter (TPM) released from tropical forest fires and peat fires, the following expression is used (Ward, 1990):

$$\text{TPM} = M \times P \text{ (tonnes of carbon)} \qquad (5)$$

where P is the conversion of biomass matter or peat matter to particulate matter during burning. For the burning of tropical vegetation, P = 20 tonnes of TPM per kilotonne of biomass consumed by fire; for peat burning, we assume P = 35 tonnes of TPM per kilotonne of organic soil or peat consumed by fire (Ward, 1990).

Arguably, the major uncertainties in the calculation of gaseous and particulate emissions resulting from fires involve poor or incomplete

Table 7.4 Emission ratios for tropical forest fires and peat fires

Species	Tropical forest fires (%)	Reference	Peat fires (%)	Reference
CO_2	90.00	Andreae (1991)	77.05	Yokelson et al (1997)
CO	8.5	Andreae et al (1988)	18.15	Yokelson et al (1997)
CH_4	0.32	Blake et al (1996)	1.04	Yokelson et al (1997)
NO_x	0.21	Andreae et al (1988)	0.46	Derived from Yokelson et al (1997)
NH_3	0.09	Andreae et al (1988)	1.28	Yokelson et al (1997)
O_3	0.48	Andreae et al (1988)	1.04	Derived from Yokelson et al (1997)
TPM[1]	20 t kt^{-1}	Ward (1990)	35t kt^{-1}	Ward (1990)

Note: [1] Total particulate matter emission ratios are in units of t kt^{-1} (tonnes of total particulate matter per kilotonne of biomass or peat material consumed by fire).

information about four fire and ecosystem parameters: (1) the area burned (A); (2) the ecosystem or terrain that burned, i.e. forests, grasslands, agricultural lands, peatlands, etc.; (3) the biomass loading (B), i.e. the amount of biomass per unit area of the ecosystem prior to burning; and (4) the fire efficiency (C), i.e. the amount of biomass in the burned ecosystem that was actually consumed by burning.

A case study of biomass burning: The 1997 wildfires in Southeast Asia

Extensive and widespread tropical forest and peat fires swept throughout Kalimantan and Sumatra, Indonesia, between August and December 1997 (Brauer and Hisham-Hishman, 1988; Hamilton et al, 2000). The fires resulted from burning for land clearing and land use change. However, the severe drought conditions resulting from El Niño caused small land-clearing fires to become large uncontrolled wildfires. Based on satellite imagery, it has been estimated that a total of 45,600km² burned on Kalimantan and Sumatra between August and December 1997 (Liew et al, 1998). The gaseous and particulate emissions produced in these fires and released into the atmosphere reduced atmospheric visibility, impacted the composition and chemistry of the atmosphere, and affected human health. Some of the consequences of the fires in Southeast Asia were: (1) more than 200 million people were exposed to high levels of air pollution and particulates produced during the fires; (2) more than 20 million smoke-related health problems were recorded; (3) fire-related damage cost in excess of US$4 billion; (4) on 26 September 1997, a commercial airliner (Garuda Airlines Airbus 300-B4) crashed in Sumatra due to very poor visibility due to smoke from the fires on landing with 234 passengers killed; (5) on 27 September 1997, two ships collided at sea due to poor visibility in the Strait of Malacca, off the coast of Malaysia, with 29 crew members killed.

International concern about the environmental and health impacts of these fires was great. Three different agencies of the United Nations organized workshops and reports on the environmental and health impacts of these fires: the World Meteorological Organization (WMO) 'Workshop on Regional Transboundary Smoke and Haze in Southeast Asia', Singapore, 2–5 June 1998, the World Health Organization (WHO) 'Health Guidelines for Forest Fires Episodic Events', Lima, Peru, 6–9 October 1998, and the United Nations Environmental Programme (UNEP) Report on 'Wildland Fires and the Environment: A Global Synthesis', published in February 1999 (Levine et al, 1999). The Indonesian fires also formed the basis of an article in *National Geographic* magazine, entitled 'Indonesia's plague of fire' (Simons, 1998).

Indonesia ranks third, after Brazil and the Democratic Republic of the Congo (formerly Zaire), in its area of tropical forest. Of Indonesia's total land area of 1.9 million km^2, current forest cover estimates range from 0.9 to 1.2 million km^2, or 48 to 69 per cent of the total. Forests dominate the landscape of Indonesia (Makarim et al, 1998). Large areas of Indonesian forests burned in 1982 and 1983. In Kalimantan alone, the fires burned from 2.4 to 3.6 million ha of forests (Makarim et al, 1998). It is interesting to note that there is an uncertainty of 1.2 million ha.

Liew et al (1998) analysed 766 Satellite Pour l'Observation de la Terre (SPOT) 'quicklook' images with almost complete coverage of Kalimantan and Sumatra from August to December 1997. They estimate the burned area in Kalimantan to be 30,600km^2 and the burned area in Sumatra to be 15,000km^2, for a total burned area of 45,600km^2 (this is equivalent to the combined areas of the states of Rhode Island, Delaware, Connecticut and New Jersey, in the US). The estimate of Liew et al (1998) represents only a lower limit estimate of the area burned in Southeast Asia in 1997, since the SPOT data only covered Kalimantan and Sumatra and did not include fires on the other Indonesian islands of Irian Jaya, Sulawesi, Java, Sumbawa, Komodo, Flores, Sumba, Timor and Wetar, or the fires in the neighbouring countries of Malaysia and Brunei.

What is the nature of the ecosystem/terrain that burned in Kalimantan and Sumatra?
In October 1997, NOAA satellite monitoring produced the following distribution of fire hot spots in Indonesia (UNDAC, 1998): agricultural and plantation areas: 45.95 per cent; bush and peat soil areas: 24.27 per cent; productive forests: 15.49 per cent; timber estate areas: 8.51 per cent; protected areas: 4.58 per cent; and transmigration sites: 1.20 per cent (the three forest/timber areas add up to a total of 28.58 per cent of the area burned). While the distribution of fire hot spots is not an actual index for area burned, the NOAA satellite-derived hot spot distribution is quite similar to the ecosystem/terrain distribution of burned area deduced by Liew et al (1998) based on SPOT images of the actual burned areas: agricultural and plantation

areas: 50 per cent; forests and bushes: 30 per cent; and peat swamp forests: 20 per cent. Since the estimates of burned ecosystem/terrain of Liew et al (1998) are based on actual SPOT images of the burned area, their estimates were adopted in our calculations.

What is the biomass loading for the three terrain classifications identified by Liew et al (1998)?
Values for biomass loading or fuel load for various tropical ecosystems are summarized in Table 7.3. The biomass loading for tropical forests in Southeast Asia ranges from 5000 to 55,000 tonnes km^{-2}, with a mean value of 23,000 tonnes km^{-2} (Brown and Gaston, 1996). However, in our calculations we have used a value of 10,000 tonnes km^{-2} to be conservative. The biomass loading for agricultural and plantation areas (mainly rubber trees and oil palms) of 5000 tonnes km^{-2} is also a conservative value (Liew et al, 1998). Nichol (1997) has investigated the peat deposits of Kalimantan and Sumatra and used a biomass loading value of 97,500 tonnes km^{-2} (Supardi et al, 1993) for the dry peat deposits 1.5m thick as representative of the Indonesian peat in her study. Brunig (1997) gives a similar value for peat biomass loading.

The combustion efficiency for forests is estimated at 0.20 and for peat is estimated at 0.50 (Levine and Cofer, 2000). Values for emission ratios for tropical forest fires and peat fires are summarized in Table 7.4. Inspection of Table 7.4 indicates that the emission ratio of CH_4 from the burning of underground peat is three times larger than the emission ratio of CH_4 from burning of above-ground vegetation. Based on the discussions presented in this section, the values for burned area, biomass loading and combustion efficiency used in the calculations are summarized in Table 7.5.

Table 7.5 Parameters used in calculations

1	Total area burned in Kalimantan and Sumatra, Indonesia in 1997: 45,600km^2				
2	Distribution of burned areas, biomass loading and combustion efficiency				
	A	Agricultural and plantation areas	50%	5000 tonnes km^{-2}	0.20
	B	Forests and bushes	30%	10,000 tonnes km^{-2}	0.20
	C	Peat swamp forests	20%	97,500 tonnes km^{-2}	0.50

Results of calculations: Gaseous and particulate emissions from the fires in Kalimantan and Sumatra, Indonesia, August to December 1997
The calculated gaseous and particulate emissions from the fires in Kalimantan and Sumatra, from August to December 1997, are summarized in Table 7.6 (Levine, 1999) (it is important to keep in mind that wildfires continued throughout Southeast Asia from January through to April 1998 and that the fires covered much more of the region than Kalimantan and Sumatra).

Table 7.6 Gaseous and particulate emissions from the fires in Kalimantan and Sumatra in 1997

	Agricultural/ plantation fire emissions	Forest fire emissions	Peat fire emissions	Total fire emissions
CO_2	9.234 (4.617–13.851)	11.080 (5.54–16.62)	171.170 (85.585–256.755)	191.485 (95.742–287.226)
CO	0.785 (0.392–1.177)	0.942 (0.471–1.413)	31.067 (15.533–46.600)	32.794 (16.397–49.191)
CH_4	0.030 (0.015–0.045)	0.035 (0.017–0.052)	1.780 (0.89–2.67)	1.845 (0.922–2.767)
NO_x	0.023 (0.011–0.034)	0.027 (0.013–0.040)	0.921 (0.460–1.381)	0.971 (0.485–1.456)
NH_3	0.010 (0.005–0.015)	0.012 (0.006–0.018)	2.563 (1.281–3.844)	2.585 (1.292–3.877)
O_3	0.177 (0.088–0.265)	0.213 (0.106–0.319)	6.710 (3.35–10.06)	7.100 (3.55–10.65)
TPM	0.460 (0.23–0.69)	0.547 (0.273–0.820)	15.561 (7.780–23.341)	16.568 (8.284–24.852)

Note: For total burned area = 45,600km². For each species, the best estimate emission value is on first line and the range of emission values in parenthesis under best guess (see text for discussion of emission estimate range and uncertainty calculations). Units of emissions: million metric tonnes (Mt) of C for CO_2, CO, and CH_4; Mt of N for NO_x and NH_3; Mt of O_3 for O_3; MtC of particulates; 1Mt = 10^{12} grams = 1Tg.
Source: Levine (1999)

For each of the seven species listed, the emissions due to agricultural/plantation burning (A), forest burning (F), and peat burning (P) are given. The total (T) of all three components (A+F+P) is also given. The 'best estimate' total emissions are: CO_2: 191.485 million Mt of C (Tg C); CO: 32.794Tg C; CH_4: 1.845Tg C; NO_x: 5.898Tg N; NH_3: 2.585Tg N; O_3: 7.100Tg O_3; and total particulate matter: 16.154Tg C.

The CO_2 emissions from these fires are about 2.2 per cent of the global annual net emission of CO_2 from all sources (see Table 7.2 for global annual production of CO_2, which is ~8700Tg C). The percentage for other gases produced by these fires compared to the global annual production from all sources is: CO: 2.98 per cent; CH_4: 0.48 per cent; oxides of nitrogen: 2.43 per cent; ammonia: 5.87 per cent; and TPM: 1.08 per cent.

However, it is important to re-emphasize that these emission calculations represent lower limit values since the calculations are only based on burning in Kalimantan and Sumatra in 1997. The calculations do not include burning in Java, Sulawesi, Irian Jaya, Sumbawa, Komodo, Flores, Sumba, Timor and Wetar in Indonesia or in neighbouring Malaysia and Brunei.

It is interesting to compare the gaseous and particulate emissions from the 1997 Kalimantan and Sumatra fires with those from the Kuwait oil fires of 1991, described as a 'major environmental catastrophe'. Laursen et al (1992) have calculated the emissions of CO_2, CO, CH_4, NO_x and particulates from the Kuwait oil fires in units of Mt per day. The Laursen et al (1992) calculations are summarized in Table 7.7. To compare these calculations with the calculations for Kalimantan and Sumatra (Table 7.7), we have normalized our calculations by the total number of days of burning. The SPOT images (Liew et al, 1998) covered a period of five months (August–December 1997) or about 150 days. For comparison with the Kuwait fire emissions, we divided our calculated emissions by 150 days. The gaseous and particulate emissions from the fires in Kalimantan and Sumatra significantly exceeded the emissions from the Kuwait oil fires. The 1997 fires in Kalimantan and Sumatra were evidently a significant source of gaseous and particulate emissions to the local, regional and global atmospheres.

Table 7.7 Comparison of gaseous and particulate emissions: The Indonesian fires and the Kuwait oil fires

Species	Indonesian fires	Kuwait oil fires
CO_2	1.28×10^6	5.0×10^5
CO	2.19×10^5	4.4×10^3
CH_4	1.23×10^4	1.5×10^3
NO_x	6.19×10^3	2.0×10^2
Particulates	1.08×10^5	1.2×10^4

Note: Units of emissions: Mt per day of C for CO_2, CO and CH_4; Mt per day of N for NO_x; Mt per day for particulates
Source: Data on Kuwait oil fires from Laursen et al (1992)

Global estimates and conclusions

Based on Equations 1–4 (pages 104–105), a number of investigators have calculated the global production of CH_4 due to biomass burning. Several of these studies are summarized in Table 7.8. Inspection of Table 7.8 indicates that estimates for the global annual production of CH_4 from biomass burning ranges from about 11 to 80Tg CH_4 yr^{-1}, accounting for between 2 per cent and 14 per cent of the global CH_4 emissions each year (assuming a total global annual production of 582Tg CH_4 yr^{-1} (IPCC, 2007). As stated earlier, the warmer and drier conditions projected for many areas as a result of climate change in the 21st century may result in enhanced fire frequency, intensity and area of burning. All else being equal, this may therefore result in a significant increase in global emissions of both CH_4 and CO_2 from biomass burning in coming decades.

Table 7.8 Production of methane by biomass burning: Global estimates

Methane emissions in Tg CH_4 yr^{-1} (range of estimate)	Reference
(11–53)	Crutzen and Andreae (1990)
38	Andreae (1991)
51.9 (27–80)	Levine et al (2000)
39	Andreae and Merlet (2001)
50 (27–80)	Wuebbles and Hayhoe (2002)
12.32	Hoelzemann et al (2004)
32.2	Ito and Penner (2004)

References

Andreae, M. O. (1991) 'Biomass burning: Its history, use, and distribution and its impact on environmental quality and global climate', in J. S. Levine (ed) *Global Biomass Burning: Atmospheric, Climatic, and Biospheric Implications*, MIT Press, Cambridge, MA, pp3–21

Andreae, M. O. and Merlet, P. (2001) 'Emission of trace gases and aerosols from biomass burning', *Global Biogeochemical Cycles*, vol 15, pp955–966

Andreae, M. O., Browell, E. V., Garstang, M., Gregory, G. L., Harriss, R. C., Hill, G. F., Jacob, D. J., Pereira, M. C., Sachse, G. W., Setzer, A. W., Silva Dias, P. L., Talbot, R. W., Torres, A. L., and Wofsy, S. C. (1988) 'Biomass burning emission and associated haze layers over Amazonia', *Journal of Geophysical Research*, vol 93, pp1509–1527

Blake, N. J., Blake, D. R., Sive, B. C., Chen, T.-Y., Rowland, F. S., Collins, J. E., Sachse, G. W. and Anderson, B. E. (1996) 'Biomass burning emissions and vertical distribution of atmospheric methyl halides and other reduced carbon gases in the South Atlantic Region', *Journal of Geophysical Research*, vol 101, pp24, 151–24, 164

Brauer, M. and Hisham-Hishman, J. (1988) 'Fires in Indonesia: Crisis and reaction', *Environmental Science and Technology*, vol 32, pp404A–407A

Brown, S. and Gaston, G. (1996) 'Estimates of biomass density for tropical forests', in J. S. Levine (ed), *Biomass Burning and Global Change, Volume 1*, MIT Press, Cambridge, MA, pp133–139

Brunig, E. F. (1997) 'The tropical rainforest – A wasted asset or an essential biospheric resource?', *Ambio*, vol 6, pp187–191

Cahoon, D. R., Stocks, B. J., Levine, J. S., Cofer, W. R. and J. M. Pierson (1994) 'Satellite analysis of the severe 1987 forest fires in northern China and southeastern Siberia', *Journal of Geophysical Research*, vol 99, pp18627–18638

Crutzen, P. J. and Andreae, M. O. (1990) 'Biomass burning in the tropics: Impact on atmospheric chemistry and biogeochemical cycles', *Science*, vol 250, pp1669–1678

Crutzen, P. J. and Goldammer, J. G. (1993) *Fire in the Environment: The Ecological,*

Atmospheric, and Climatic Importance of Vegetation Fires, John Wiley and Sons, Chichester

Crutzen, P. J., Heidt, L. E., Krasnec, J. P., Pollock, W. H. and Seiler, W. (1979) 'Biomass burning as a source of atmospheric gases CO, H_2, N_2O, NO, CH_3Cl, COS', *Nature*, vol 282, pp253–256

Eaton, P. and Radojevic, M. (2001) *Forest Fires and Regional Haze in Southeast Asia*, Nova Science Publishers, Huntington, NY

Fromm, M., Alfred, J., Hoppel, K., Hornstein, J., Bevilacqua, R., Shettle, E., Servranckx, R., Li, Z. and Stocks, B. (2000) 'Observations of boreal forest fire smoke in the stratosphere by POAM III, SAGE II, and lidar in 1998', *Geophysical Research Letters*, vol 27, pp1407–1410

Goldammer, J. G. (1990). *Fire in the Tropical Biota: Ecosystem Processes and Global Challenges*, Springer-Verlag, Berlin

Goldammer, J. G. and Furyaev, V. V. (1996) *Fire in Ecosystems of Boreal Eurasia*, Kluwer Academic Publishers, Dordrecht, The Netherlands

Hamilton, M. S., Miller, R. O. and Whitehouse, A. (2000) 'Continuing fire threat in Southeast Asia', *Environmental Science and Technology*, vol 34, pp82A–85A

Hansen, J., Ruedy, R., Sato, M. and Reynolds, R. (1996) 'Global surface air temperature in 1995: Return to pre-Pinatubo level', *Geophysical Research Letters*, vol 23, pp1665–1668

Hoelzemann, J. J., Schultz, M. G., Brasseur, G. P. and Granier, C. (2004) 'Global wildland emission model (GWEM): Evaluating the use of global area burnt satellite data', *Journal of Geophysical Research*, vol 109, D14S04, doi:10.1029/2003JD0036666

Innes, J. L., Beniston, M. and Verstraet, M. M. (2000) *Biomass Burning and its Interrelationships with the Climate System*, Kluwer Academic Publishers, Dordrecht

IPCC (2007) *Climate Change 2007: The Physical Science Basis. Contribution of Working Group I to the Fourth Assessment Report of the Intergovernmental Panel on Climate Change*, S. Solomon, D. Qin, M. Manning, Z. Chen, M. Marquis, K. B. Averyt, M. Tignor and H. L. Miller (eds), Cambridge University Press, Cambridge, UK and New York, NY

Ito, A. and Penner, J. E. (2004) 'Global estimates of biomass burning emissions based on satellite imagery for the year 2000', *Journal of Geophysical Research*, vol 109, D14S05, doi:10.1029/2003JD004423

Kasischke, E. S. and Stocks, B. J. (2000) *Fire, Climate Change, and Carbon Cycling in the Boreal Forest*, Ecological Studies Volume 138, Springer-Verlag, New York

Kasischke, E. S. Bergen, K., Fennimore, R., Sotelo, F., Stephens, G., Janetos, A. and Shugart, H. H. (1999) 'Satellite imagery gives clear picture of Russia's boreal forest fires', *EOS, Transactions of the American Geophysical Union*, vol 80, pp141–147

Laursen, K. K., Ferek, R. J. and Hobbs P. V. (1992) 'Emission factors for particulates, elemental carbon, and trace gases from the Kuwait oil fires', *Journal of Geophysical Research*, vol 97, pp14491–14497

Levine, J. S. (1991) *Global Biomass Burning: Atmospheric, Climatic, and Biospheric Implications*, MIT Press, Cambridge, MA

Levine, J. S. (1996a) *Biomass Burning and Global Change: Remote Sensing, Modeling and Inventory Development, and Biomass Burning in Africa*, MIT Press, Cambridge, MA

Levine, J. S. (1996b) *Biomass Burning and Global Change: Biomass Burning in South America, Southeast Asia, and Temperate and Boreal Ecosystems, and the Oil Fires of Kuwait*, MIT Press, Cambridge, MA

Levine, J. S. (1999) 'The 1997 fires in Kalimantan and Sumatra Indonesia: Gaseous and particulate emissions', *Geophysical Research Letters*, vol 26, pp815–818

Levine, J. S. and Cofer, W. R. (2000) 'Boreal forest fire emissions and the chemistry of the atmosphere', in E. S. Kasischke and B. J. Stocks (eds) *Fire, Climate Change and Carbon Cycling in the North American Boreal Forests*, Ecological Studies Series, Springer-Verlag, New York, pp31–48

Levine, J. S., Cofer, W. R. and Pinto, J. P. (2000) 'Biomass Burning', in M. A. K. Khalil (ed) *Atmospheric Methane: Its Role in the Global Environment*, Springer-Verlag, New York, pp190–201

Levine, J. S., Cofer, W. R., Winstead, E. L., Rhinehart, R. P., Cahoon, D. R., Sebacher, D. I., Sebacher, S. and Stocks, B. J. (1991) 'Biomass burning: Combustion emissions, satellite imagery, and biogenic emissions', in J. S. Levine (ed) *Global Biomass Burning: Atmospheric, Climatic, and Biospheric Implications*, MIT Press, Cambridge, MA, pp264–272

Levine, J. S., Cofer, W. R., Cahoon, D. R. and Winstead, E. L. (1995) 'Biomass burning: A driver for global change', *Environmental Science and Technology*, vol 29, pp120A–125A

Levine, J. S., Bobbe, T., Ray, N., Witt, R. G. and Singh A. (1999) *Wildland Fires and the Environment: A Global Synthesis*, Environment Information and Assessment Technical Report 99-1, The United Nations Environmental Programme, Nairobi, Kenya

Liew, S. C., Lim, O. K., Kwoh, L. K. and Lim, H. (1998) 'A study of the 1997 fires in South East Asia using SPOT quicklook mosaics', paper presented at the 1998 International Geoscience and Remote Sensing Symposium, 6–10 July, Seattle, Washington

Lobert, J. M., Scharffe, D. H., Hao, W.-M., Kuhlbusch, T., Seuwen, A. R., Warneck, P. and Crutzen, P. J. (1991) 'Experimental evaluation of biomass burning emissions: Nitrogen and carbon containing compounds', in J. S. Levine (ed) *Global Biomass Burning: Atmospheric, Climatic, and Biospheric Implications*, MIT Press, Cambridge, MA, pp289–304

Makarim, N., Arbai, Y. A., Deddy, A. and Brady, M. (1998) 'Assessment of the 1997 land and forest fires in Indonesia: National coordination', *International Forest Fire News*, United Nations Economic Commission for Europe and the Food and Agriculture Organization of the United Nations, Geneva, Switzerland, No 18, January, pp4–12

Nichol, J. (1997) 'Bioclimatic impacts of the 1994 smoke haze event in Southeast Asia', *Atmospheric Environment*, vol 44, pp1209–1219

Scholes, R. G., Kendall, J. and Justice, C. O. (1996) 'The quantity of biomass burned in southern Africa', *Journal of Geophysical Research*, vol 101, pp23, 667–23, 676

Seiler, W. and Crutzen, P. J. (1980) 'Estimates of gross and net fluxes of carbon between the biosphere and the atmosphere from biomass burning', *Climatic Change*, vol 2, pp207–247

Simons, L. M. (1998) 'Indonesia's plague of fire', *National Geographic*, vol 194, no 2, pp100–119

Simpson, I. J., Rowland, F. S., Meinardi, S. and Blake, D. R. (2006) 'Influence of biomass burning during recent fluctuations in the slow growth of global tropospheric methane', *Geophysical Research Letters*, vol 33, L22808, doi:10,1029/2006GL027330

Stocks, B. J. and Flannigan, M. D. (1987) 'Analysis of the behavior and associated weather for a 1986 northeastern Ontario wildfire: Red Lake #7.' *Proceedings of the Ninth Conference on Fire and Forest Meteorology*, San Diego, CA, American Meteorological Society, Boston, MA, pp94–100.

Stocks, B. J., Fosberg, M. A., Lyman, T. J., Means, L., Wotton, B. M., Yang, Q., Jin, J.-Z., Lawrence, K., Hartley, G. J., Mason, J. G. and McKenney, D. T. (1993) 'Climate change and forest fire potential in Russian and Canadian boreal forests', *Climatic Change*, vol 38, pp1–13

Supardi, A., Subekty, D. and Neuzil, S. G. (1993) 'General geology and peat resources of the Siak Kanan and Bengkalis Island peat deposits, Sumatra, Indonesia', in J. C. Cobb and C. B. Cecil (eds) *Modern and Ancient Coal Forming Environments*, Society of America Special Paper, Volume 86, pp45–61

UNDAC (United Nations Disaster Assessment and Coordination Team) (1998) 'Mission on Forest Fires, Indonesia, September–November 1997', *International Forest Fire News*, United Nations Economic Commission for Europe and the Food and Agriculture Organization of the United Nations, Geneva, Switzerland, No. 18, pp13–26

van Wilgen, B. W., Andreae, M. O., Goldammer, J. G. and Lindesay, J. A. (1997) *Fire in Southern African Savannas: Ecological and Atmospheric Perspectives*, Witwatersrand University Press, Johannesburg

Ward, D. E. (1990) 'Factors influencing the emissions of gases and particulate matter from biomass burning', in J. G. Goldammer (ed) *Fire in the Tropical Biota: Ecosystem Processes and Global Challenges*, Springer-Verlag, Berlin, Ecological Studies, vol 84, pp418–436

Wuebbles, D. J. and Hayhoe, K. (2002) 'Atmospheric methane and global change', *Earth Science Reviews*, vol 57, pp177–210

Yokelson, R. J., Griffith, D. W. T. and Ward, D. E. (1996) 'Open-path Fourier transform infrared studies of large-scale laboratory biomass fires', *Journal of Geophysical Research*, vol 101, pp21067–21080

Yokelson, R. J., Susott, R., Ward, D. E., Reardon J. and Griffith, D. W. T. (1997) 'Emissions from smouldering combustion from biomass measured by open-path fourier transform infrared spectroscopy', *Journal of Geophysical Research*, vol 102, pp18865–18877

8
Rice Cultivation

Franz Conen, Keith A. Smith and Kazuyuki Yagi

Introduction

More than a third of all CH_4 emissions come from soils, as a result of the microbial breakdown of organic compounds in strictly anaerobic conditions. This process occurs in natural wetlands (Chapter 3), in flooded rice fields and in landfill sites rich in organic matter (Chapter 11), as well as in the gut of some species of soil-dwelling termites (Chapter 5). Rice fields make a significant contribution to anthropogenic CH_4 emissions, which are responsible for the concentration in the atmosphere more than doubling since the pre-industrial era, when, on the basis of evidence from ice core analysis, it was only about 0.7µmol mol^{-1} (Prather et al, 1995). Early estimates suggested that rice production contributes about a quarter of the total anthropogenic CH_4 source and is of similar strength to the ruminant source or the energy sector (Fung et al, 1991; Hein et al, 1997). However, estimates have declined substantially with time, to values mostly between 25 and 50Tg CH_4 year^{-1}. Higher values generally originate from inverse modelling of observed fluctuations in atmospheric CH_4 mixing ratios (top-down method) (for example Hein et al, 1997; Chen and Prinn, 2006), whereas lower values are obtained by scaling up field observations (bottom-up method) (Figure 8.1).

As far as the bottom-up estimates are concerned, the trend towards smaller emission estimates reflects the increasing level of our understanding of the processes governing emissions, aided by a growing number of measurements and modelling exercises. It is not related to any changes that may have occurred over the same period in actual cultivation methods or cultivated area. The experimental and modelling studies show that emissions may be affected by the continuity, or lack of it, of the flooding regime, the extent of incorporation of organic residues into the soil, the general level of productivity of the crop, the cultivar used and other factors. Improved understanding of the fundamental processes controlling CH_4 production, and knowledge about how they are influenced by agricultural management, not only gives greater confidence in the emission estimates, but also indicates possible ways to reduce the emissions.

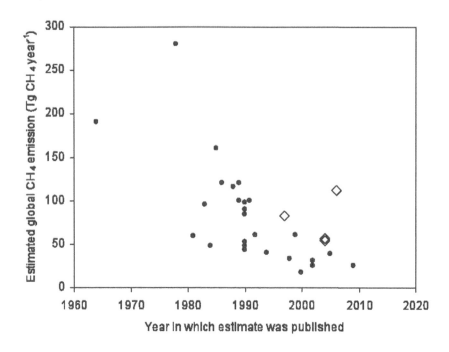

Figure 8.1 Decrease over time in estimates of global CH$_4$ emissions from rice fields.

Note: Large open symbols represent estimates from global inverse modelling (top-down method).
Source: Adapted from Sass (2002), with inclusion of the estimates by Fung et al (1991), Hein et al (1997), Olivier et al (1999, 2005), Scheehle et al (2002), Wang et al (2004), Mikaloff Fletcher et al (2004), Chen and Prinn (2006) and Yan et al (2009)

Rice production

Rice is grown under a variety of climatic, soil and hydrological conditions in the world, from northeastern regions of China (53°N) to southern regions of Australia (35°S) and from sea level to altitudes of more than 2500m. It grows well in flood-prone areas of South and Southeast Asia (in as much as 5m of floodwater) and in drought-prone upland areas of Asia, South America and Africa (Neue and Sass, 1994). Nowadays, more than 90 per cent of the harvest area of rice is located in monsoon Asian countries and rice supports two thirds of the people living there as a staple food. In several Asian languages the words for food and rice, or for rice and agriculture, are the same, indicating its overwhelming importance for human survival over millennia.

In order to meet increasing demand for rice, the harvest area in the world has steadily expanded from 84 to 154 million hectares (Mha) between 1935 and 2005, corresponding to an annual increase of about 1 per cent (Figure 8.2). This rapid increase in the harvest area implies increased emissions of CH$_4$ during the last 70 years. In addition, introducing high-yielding varieties, together with new cultivation technologies, has significantly increased rice

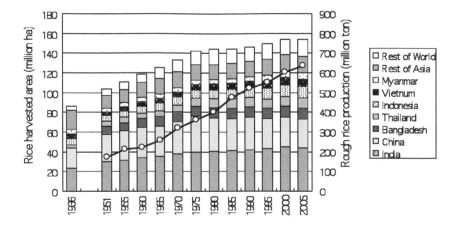

Figure 8.2 Global changes in rice harvest area (bars) and rough rice production (circles)
Source: Data from International Rice Research Institute (IRRI) World Rice Statistics

yields, resulting in more than a threefold increase in world production during the last 50 years. It has been suggested that the increase in the yield has additionally increased CH_4 emissions because of accelerating carbon turnover in the rice–soil system, caused by adding more organic matter to the soil in the form of crop residues (Kimura et al, 2004). The rate of global CH_4 emissions from rice fields will probably increase further in the next decade, as there is an estimated global need for an additional 50 million tonnes of rough rice by 2015 (about 9 per cent of current production) in order to meet expected consumption rates (International Rice Research Institute, 2006).

Rice farming provides a livelihood to hundreds of millions of small farmers, challenged by the possibility of floods, droughts, pests and other threats to their crop. The major goal in their enterprise is to secure the crop yield necessary to sustain their families. Reducing CH_4 emissions is of little concern to them in this situation. Nevertheless, a number of traditional management practices do curb CH_4 emissions (although this is not their primary purpose), providing a 'win-win' outcome rather than a conflict between different economic, environmental and social goals. These issues are discussed in a later section.

Biogeochemistry of methane production

Methane production in natural wetlands, and in rice paddies, is a process occurring in strictly reduced (anoxic) conditions (see also Chapter 2). The creation of these conditions is controlled by both chemical and microbiological soil properties (Conrad, 1989a, 1993; Neue and Roger, 1993). Aerobic, drained soils become completely anoxic after flooding because of the barrier to entry of atmospheric oxygen presented by the water layer: the rate of diffusion of oxygen and other gases through water is about 10^{-4} times that in the gas phase. The only regions where anoxia does not prevail are at the soil–floodwater interface and the zones around plant roots (Conrad, 1993). Respiration by microorganisms and plant roots rapidly depletes the remaining oxygen in the system and then other chemical species – nitrate, manganese(IV), iron(III), sulphate and carbon dioxide – act in turn as alternative electron acceptors, and are consequently reduced by microbial activity (for example Ponnamperuma, 1972, 1981; Peters and Conrad, 1996). Organic compounds such as humic acids may also act as electron acceptors (Lovley et al, 1996).

In most paddy soils the concentration of sulphate in the pore water increases for a few days before sulphate reduction starts (Ponnamperuma, 1981; Yao et al, 1999). According to Yao et al (1999), the increase is due to the release of sulphate adsorbed onto ferric iron minerals such as goethite either by reduction of bicarbonate or Fe(III). In acidic soils, clays and hydrous oxides of aluminium strongly sorb sulphate and release it when the pH increases after flooding.

The sequential reduction of the various electron acceptors takes place in the order indicated by their redox potentials, i.e. as predicted by thermodynamic theory (Ponnamperuma, 1972, 1981; Zehnder and Stumm, 1988). After their reduction, methanogenesis becomes possible. Hydrogen is a key substrate for this process. It is produced from the degradation of organic substances (Conrad, 1996a) and is rapidly consumed as an electron donor in various redox reactions, so that the turnover time of H_2 is very short (Conrad et al, 1989b; Yao et al 1999). At the end of these reduction processes, the H_2 partial pressure increases, allowing hydrogenotrophic methanogenesis to begin. Yao et al (1999) reported that this took place at partial pressures between 1 and 23Pa. A steady-state concentration of H_2 is determined by the relative rates of H_2 production and H_2 consumption. The thermodynamic conditions for hydrogenotrophic methanogenesis are described in detail by Yao and Conrad (1999).

Acetate is the other compound acting as a major driver of methanogenesis. Like H_2, it is produced from organic substances. Yao et al (1999) found that acetate turnover time in paddy soil was much longer than that of H_2, typically in the range of hours or even days, and that CH_4 production was initiated at concentrations of 30–8000µM, and at pH values between 6.0 and 7.5 and a redox potential (Eh value) of between −80 and +250mV. These Eh values at the onset of methanogenesis were higher than the −150mV reported by Wang et al (1993) and Masscheleyn et al (1993), but experiments with cultures of methanogenic bacteria show that O_2 has a much more adverse effect on methanogenic activity than high redox potentials and that methanogens are able to initiate CH_4 production in the absence of O_2 at higher Eh values: up to +420mV (Fetzer and Conrad, 1993) and 0 to +100 mV (Garcia et al, 1974; Peters and Conrad, 1996; Ratering and Conrad, 1998). Yao et al (1999)

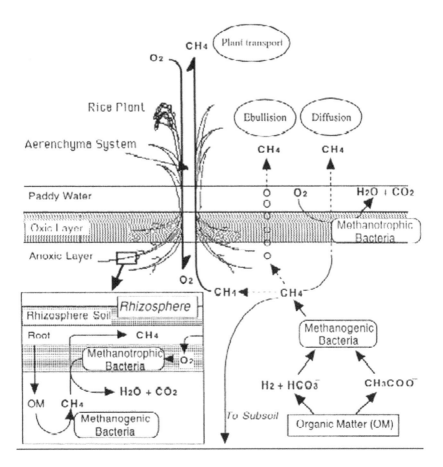

Figure 8.3 Production, oxidation and emission of methane in rice paddy fields
Source: Yagi et al (1997)

concluded that redox potential is not a good indicator for the onset of soil methanogenesis, and should only be used as an indicator when the soil and its CH_4 production behaviour have been carefully characterized (for example Yagi et al, 1996; Sigren et al, 1997).

The emission pathways of CH_4 which are accumulated in flooded paddy soils are: diffusion into the floodwater, loss through ebullition, and diffusive transport through the aerenchyma system of rice plants, as shown in Figure 8.3 (Yagi et al, 1997). In temperate rice fields, more than 90 per cent of CH_4 emissions take place through plants (Schütz et al, 1989). In tropical rice fields, by contrast, significant amounts of CH_4 may evolve by ebullition, in particular during the early part of the season and when organic inputs are high (Denier van der Gon and Neue, 1995).

Methane oxidation

Part of the CH_4 produced by methanogens is consumed by CH_4-oxidizing bacteria, or methanotrophs. It is known that microbially mediated CH_4 oxidation, in particular aerobic CH_4 oxidation, ubiquitously occurs in soils and aquatic environments, where it modulates actual CH_4 emission (for example Conrad, 1996b). In rice paddy fields, it is possible that a part of the CH_4 produced in anaerobic soil layers is oxidized in aerobic layers such as the surface soil–water interface and the rhizosphere of rice plants (Gilbert and Frenzel, 1995), and the net emission will be positive or negative depending on the relative magnitudes of methanogenesis and methanotrophy, respectively.

Early work reported that up to 80 per cent of the CH_4 produced in rice fields was oxidized before it could escape to the atmosphere (for example Sass et al, 1991; Tyler et al, 1997), but the introduction of difluoromethane as a selective inhibitor of CH_4 oxidation has yielded results that suggest that this fraction is 40 per cent or less, depending on the stage of the growing season (Krüger et al, 2001).

In aerobic soil systems, the oxidation of atmospheric CH_4 is inhibited by the addition of nutrients such as nitrogen, but in the rice paddy system, the reverse is true, at least for the first part of the season. Krüger and Frenzel (2003) concluded that any agricultural treatment improving the N supply to rice plants would also be favourable for the CH_4-oxidizing bacteria. However, N fertilization had only a transient influence and was counterbalanced in the field by an elevated CH_4 production.

Isotopic studies

The production pathways to CH_4 – via acetate or via H_2 and CO_2 – have different isotopic signatures, mainly in the $\delta^{13}C$ of the CH_4 produced. Sugimoto and Wada (1993) showed that the fermentation process of acetotrophic bacteria has only a small fractionation effect on the $^{13}C/^{12}C$ ratio, which means that the $\delta^{13}C$ of the CH_4 is only 11 parts per mil lower than that of the methyl group of acetate. The latter is in the range of that of the original organic matter, which is mainly derived from the rice plant, a C3 plant with a typical signature of –28 parts per mil. Methane from CO_2 reduction is much more depleted in $\delta^{13}C$. Reviewing the literature then available, Marik et al (2002) reported that a $\delta^{13}C$ for CH_4 produced via this pathway of as low as –68 parts per mil can be obtained, and Krüger et al (2002) reported values of –69 to –87 parts per mil. The ratio between deuterium (D) and hydrogen (H) (the D/H ratio) also changes during methanogenesis, with CH_4 δD values ranging between –330 and –360 parts per mil (Marik et al, 2002).

Both molecular diffusion and CH_4 oxidation result in an enrichment of the remaining CH_4 in both ^{13}C and D. The fractionation factor for $\delta^{13}C$ is 1.019 for diffusion, and for oxidation it ranges from 1.003 to 1.025 (Coleman et al, 1981; Happell et al, 1994). Deuterium in CH_4 has the same factor as ^{13}C for diffusion, and for oxidation it ranges from 1.044 to 1.325 (Coleman et al,

1981; Bergamaschi and Harris, 1995). The interplay of production, diffusive transport and oxidation means that the isotopic signature of the CH_4 emitted from rice fields changes during the growing season (Tyler et al, 1994; Bergamaschi, 1997; Marik et al, 2002).

Impact of rice cultivation systems

Rice is a unique crop in that it is highly adaptable and can be grown in very diverse environments (Yoshida, 1981). The ecosystems within which rice is grown may be characterized by seasonal change of temperature, rainfall pattern, depth of flooding and drainage, and by the adaptation of rice to these agroecological factors. In addition, the degree of water control available is a useful tool to classify rice ecosystems because it characterizes management design for improving productivity (Huke and Huke, 1997) and also the conditions for CH_4 emissions (Sass et al, 1992; Yagi et al, 1996; Wassmann et al, 2000b). The major rice ecosystems are classified as irrigated, rainfed, deepwater and upland. Irrigated rice can be further subdivided into continuously and intermittently flooded systems, and rainfed rice into regular, drought-prone and deepwater systems, according to the flooding patterns during the cultivation period.

The available database indicates that the CH_4 emission per unit area and season follows the order: continuously flooded irrigated rice ≥intermittently flooded irrigated rice ≥deepwater rice >regular (flood-prone) rainfed rice >drought-prone rainfed rice (Table 8.1). Upland rice is not a source of CH_4, since it is grown in aerated soils that never become flooded for any significant period of time. However, this ranking only provides an initial assessment of the emission potentials that can locally be superseded by crop management favouring or lowering actual emission rates (Wassmann et al, 2000a, 2000b). The flooding pattern before the cultivation period significantly influences the emission rates (Fitzgerald et al, 2000; Cai et al, 2003). Differences in residue recycling, organic amendments, scheduled short aeration periods, soils, fertilization and rice cultivars are major additional causes for variations of CH_4 fluxes in rice fields. Various organic amendments incorporated into rice soils, either of endogenous (straw, green manure etc.) or exogenous origin (compost, farmyard manure etc.), increases CH_4 emissions (Schütz et al, 1989; Yagi and Minami, 1990; Sass et al, 1991). The impact of organic amendments on CH_4 emissions depends on type and amount of the applied material which can be described by a dose response curve (Denier van der Gon and Neue, 1995; Yan et al, 2005). Lowest CH_4 fluxes are recorded in fields with low residue recycling, multiple aeration periods, poor soils and low fertilization with resulting poor rice growth and low yields. The source strength of rainfed rice is most uncertain because of its high variability in all factors controlling CH_4 emissions.

Table 8.1 Ratio of the areas of irrigated rice fields subject to various water regimes

Country	Continuous flooding	Single drainage	Multiple drainage	Source
India	0.30	0.44	0.26	ALGAS report[a]
Indonesia	0.43	0.22	0.35	ALGAS report
Vietnam	1	0	0	ALGAS report
China	0.2	0	0.8	Li et al, 2002
Japan, Korea, Bangladesh	0.2	0	0.8	Assumed to be the same as China
Other monsoon Asian countries	0.43	0.22	0.35	Assumed to be the same as Indonesia
Other countries	0.3	0.44	0.36	Assumed to be the same as India

Note: [a] ALGAS = Asia Least Cost Greenhouse Gas Abatement Strategy. Reports were downloaded from www.adb.org/REACH/algas.asp, the website of the Asian Development Bank (ADB).
Source: Yan et al (2009)

Role of organic matter and nutrients

In rice cultivation, as in any other form of agriculture, it is necessary to sustain soil fertility by returning plant nutrients to the soil. Plant residues, green manure from intercrops or the aquatic plant *Azolla* and its associated N-fixing blue-green alga *Anabaena azollae*, human faeces and animal manure are the most important forms of organic fertilizers used in rice crops. Since the 'green revolution' in the 1960s introduced new varieties with increased yields and nutrient requirements, these organic amendments have been generally complemented in many regions by mineral forms of fertilizer. A major difference between organic and mineral fertilizers, in the context of methanogenesis, is that organic fertilizers contain, in addition to plant nutrients, energy sources (for example carbohydrates) that stimulate soil microbial activity. The energy content in organic amendments declines rapidly with time during aerobic decomposition. Applied freshly, however, organic materials lead to large CH_4 emissions. For example straw, incorporated shortly before flooding, produces three to four times larger emissions compared to when it is incorporated 30 or more days before flooding and is partly decomposed before conditions turn anoxic (Yan et al, 2005). Globally, off-season incorporation of rice straw has the potential to mitigate emissions of about 4Tg CH_4 each year (Yan et al, 2009).

Composting organic fertilizers off-site can be even more effective, however, it involves additional labour to transport material from and to the field. Whether decomposed on-site or composted off-site, plant nutrients, including a large proportion of the nitrogen, are retained while the energy content is reduced. In contrast, burning of straw removes all energy but also almost all the nitrogen, while other nutrients are still retained. Straw burning is already practised on 30 per cent of the rice growing area (Yan et al, 2009). However, in some areas the burning of straw is prohibited. Where few arguments other than legal restrictions are preventing straw from being burned, easing those restrictions may be the easiest mitigation option.

Role of plant physiology – varietal differences

Energy supply to microorganisms and anoxic conditions controlling methanogenesis may also be influenced by the growing rice plant. Some energy is supplied to CH_4-generating organisms in the form of dissolved organic carbon released by the growing plant roots into the surrounding soil. Influenced by plant physiology, release rates seem to be variety-dependent (Sigren et al, 1997; Inubushi et al, 2003; Lou et al, 2008). Varieties may also differ in the degree to which they promote oxygen diffusion into the root environment (Satpathy et al, 1998). Greater rates of oxygen diffusion from roots into the soil could stimulate CH_4 oxidation and inhibit CH_4 production by increasing the oxygen pressure at the soil–root interface. Among ten high-yielding varieties tested in India, fourfold differences in seasonal CH_4 emissions were observed between varieties producing similar grain yields, while no correlation between emissions and yield was observed (Satpathy et al, 1998) (Figure 8.4).

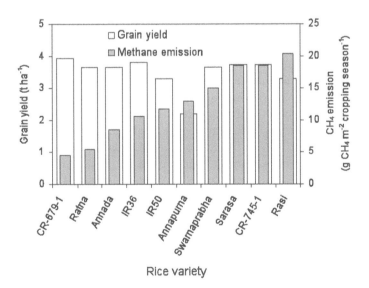

Figure 8.4 Variation in relationship between yield and CH_4 emissions, for ten high-yielding Indian rice varieties

Source: Based on data from Satpathy et al (1998)

In the US, Huang et al (1997) found a positive correlation between grain yield and CH_4 emissions among ten varieties tested under the same conditions. However, a high emission was not always found to accompany high yield. Also in the US, a positive correlation was found between plant height and CH_4 emission (Ding et al, 1999), and semi-dwarf varieties were found to emit 36 per cent less CH_4 than tall ones (Lindau et al, 1995). In the largest study of this kind, involving a total of 29 varieties, results were inconsistent (Wassmann et al, 2002). Within a given season differences were small; compared during nine seasons under varying environmental conditions, most were insignificant. Wassmann et al (2002) concluded that the variety-specific differences were small compared to other factors, varied between seasons and were too elusive for solid classification of varieties with respect to their CH_4 mitigation potential. However, since there are hundreds of high-yielding rice varieties in use and new ones are constantly being bred, this is certainly not the last word on this issue, and efforts to identify mechanisms responsible for variety-dependent CH_4 emissions are continuing.

Impact of water management

As indicated above, CH_4 production is a process occurring under anaerobic conditions in strictly reduced (anoxic) conditions. For such conditions to establish, soils usually have to be flooded or completely waterlogged for at least several days without interruption. During drier periods oxygen enters the soil, redox potentials rapidly increase again and CH_4 production ceases. This is often the case in rainfed rice production, where CH_4 emissions are on average only about one third of those in irrigated systems (Abao et al, 2000; Setyanto et al, 2000; Yan et al, 2005). A traditional management practice in irrigated rice paddies is drainage on one or more occasions during the growing season. Drainage is much applied in Japan and China to enhance yields (Greenland, 1997) and is also popular in northern India (Jain et al, 2000), whereas in Vietnam, for example, continuous flooding is the norm (Table 8.1).

An additional benefit of drainage is that of disturbing the life cycle of water-dependent vectors of human disease (malaria, Japanese encephalitis and others) (Greenland, 1997). To relieve the mosquito-induced stress on local inhabitants, intermittent drainage is prescribed for fields surrounding towns in the rice-growing area of northern Italy (S. Russo, personal communication). Estimated CH_4 reductions compared to continuous flooding are between 7 and 80 per cent (Wassmann et al, 2000b, and references therein) and 26–46 per cent (Zheng et al, 2000). Intermittent irrigation can be as effective (Husin et al, 1995; Yagi et al, 1996), or even more effective than mid-season drainage alone (Lu et al, 2000). Based on an analysis by Yan et al (2005) of more than 1000 seasonal measurements from over 100 sites, the 2006 IPCC *Guidelines for National Greenhouse Gas Inventories* (IPCC, 2007a) adopt values for the reduction in CH_4 emissions from a single mid-season drainage of 40 per cent and for multiple drainages of 48 per cent, compared to continuous flooding.

Some of the gains achieved by drainage in terms of global warming potential can be offset by enhanced N_2O emissions (Cai et al, 1997; Akiyama et al, 2005). The IPCC guidelines estimate that, on average, 0.31 per cent of the nitrogen fertilizer applied to rice paddies is emitted as N_2O (IPCC, 2007a). This emission factor was based on an analysis conducted by Akiyama et al (2005), in which they calculated a N_2O emission factor of 0.22 per cent for continuously flooded rice paddies and an emission factor of 0.37 per cent for intermittently irrigated rice paddies. Yan et al (2009) estimated that 27 million hectares of the global rice area is continuously flooded. Assuming an average fertilizer application rate of 150kg N ha^{-1}, if these continuously flooded rice fields were all drained more than once during the rice-growing season the N_2O emission from rice fields would increase by approximately 9.5Gg. Even though the GWP of 1kg of N_2O is approximately 12 times higher that of 1kg of CH_4 (IPCC, 2007b), they calculate that the increased GWP resulting from this extra N_2O emission would be only approximately 2.7 per cent of the reduction in GWP that would result from the 4.14Tg reduction in CH_4 emissions, and therefore draining the fields is beneficial in terms of net climate forcing. However, it should be emphasized that only where fields are currently under continuous flooding and where flooding can be controlled, is drainage a mitigation option. Yan et al (2009) estimate that the global mitigation potential through water management is the same as that associated with off-season straw incorporation (i.e. 4Tg CH_4 yr^{-1}).

The effects of both applying rice straw to the land outside the growing season, and drainage during the season, are shown in Figures 8.5 and 8.6, and Table 8.2.

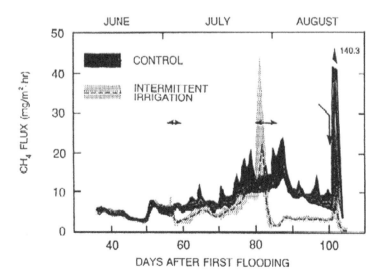

Figure 8.5 Effect of water management on CH_4 emission from a rice paddy field

Note: The arrows indicate the period of mid-season drainage in the intermittent irrigation plot and the timing of final drainage in both of the plots.
Source: Yagi et al (1997)

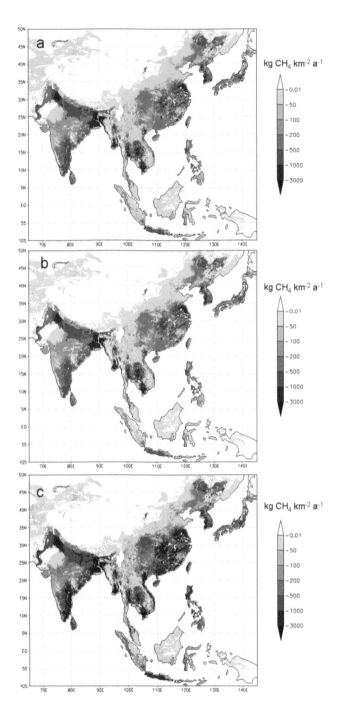

Figure 8.6 Distribution of potential mitigating effects by (a) applying rice straw off-season where possible; (b) draining all continuously flooded rice fields; and (c) adopting both options

Note: Negative values indicate an emission reduction.
Source: Yan et al (2009)

Table 8.2 Mitigation potential (per cent) of methane emission from rice cultivation in major rice-producing countries by applying rice straw off-season where possible, draining all continuously flooded rice fields, and adopting both options simultaneously

Country	Rice straw off-season	Draining rice field	Both options
China	12.8	15.6	26.4
India	16.3	13.6	27.5
Bangladesh	22.4	4.4	25.9
Indonesia	8.4	21.7	28.6
Vietnam	5.7	36.6	40.7
Myanmar	15.9	19.8	33.2
Thailand	20.2	4.7	24.2
The Philippines	9.0	22.7	30.0
Pakistan	25.1	28.7	46.7
Japan	33.6	15.6	43.9
US	35.2	21.8	49.3
Cambodia	27.9	6.6	33.4
South Korea	26.7	12.0	35.5
Nepal	19.0	16.7	32.6
Nigeria	19.6	6.3	24.7
Sri Lanka	18.5	24.5	38.8
Brazil	27.7	17.0	39.9
Madagascar	22.7	2.8	24.8
Malaysia	16.4	23.5	36.6
Laos	21.7	5.2	26.0
Globe	16.1	16.3	30.1

Source: Yan et al (2009)

National and global emissions assessments

The Task Force on National Greenhouse Gas Inventories (TFI) of the IPCC produces from time to time guidelines for the preparation of the inventories of national greenhouse gas emissions that have to be submitted annually by the signatories to the United Nations Framework Convention on Climate Change (UNFCCC). Emissions from agriculture are inherently more difficult to assess than those from other economic sectors, for example energy production or transport, no matter whether they relate to nitrous oxide or CH_4 emissions, because of the variability of environmental conditions in which they occur – many of the factors involved have been discussed above. In many countries there are no – or insufficient – experimental data to determine directly the emissions of CH_4 from rice-growing systems, and so methodologies and a series of 'default' factors have been developed to provide guidance for calculation.

The latest IPCC Guidelines (IPCC, 2007a) provide two basic equations (Equations 6 and 7) for calculating national inventories of CH_4 emissions from rice cultivation. Because the natural conditions and agricultural management of rice production may be highly variable within a country, it is recommended that this variability should be accounted for by disaggregating national total harvested area into sub-units (for example harvested areas under different water regimes or organic amendment: conditions 'i, j and k' in Equation 6 below). Then, total annual emissions are equal to the sum of emissions from each sub-unit of harvested area:

$$\text{Emissions from rice cultivation} = \Sigma_{ijk} (EF_{ijk} \cdot t_{ijk} \cdot A_{ijk}) \tag{6}$$

where EF_{ijk} is a daily emission factor, t_{ijk} is the rice cultivation period in days, and A_{ijk} is the annual harvested area of rice. Emissions for each sub-unit are adjusted by multiplying a baseline default emission factor for continuously flooded fields without organic amendments (EF_c) by various scaling factors as shown in Equation 7:

$$EF_i = EF_c \cdot SF_w \cdot SF_p \cdot SF_o \cdot SF_{s,r} \tag{7}$$

where EF_i is the adjusted daily emission factor for a particular harvested area and SF_w, SF_p, SF_o and $SF_{s,r}$ are scaling factors for the water regime during the cultivation period, that in the pre-season, the organic amendment applied, and others (soil type, rice cultivar, etc.), respectively.

A value for the default baseline emission factor EF_c of 1.30 (with error range of 0.80–2.20) kg CH_4 ha^{-1} day^{-1} is provided by IPCC (2007a); it was estimated via a statistical analysis of available field measurement data (Yan et al, 2005). Table 8.3 shows examples of the default scaling factors to account for the differences in water regime during the cultivation period.

Conclusions and future outlook

Rice is the staple food of the humid monsoon regions of the world, and the need to collectively organize water and crop management to maximize its production has been the foundation of several great cultures. Millennia-old traditions in rice cultivation survive, but necessarily interact with specific crop requirements, a complex biogeochemistry and more recent developments in plant breeding, fertilization and pest management. Increasing populations require matching increases in food supply and security of this supply. All this makes rice cultivation a particularly complex source of CH_4 to understand and to tackle, given that farmers' priorities will inevitably be associated with

Table 8.3 Default CH_4 emission scaling factors for water regimes during the cultivation period relative to a continuously flooded irrigated field

Water regime	Scaling factor (SF_w)	Error range
Upland	0	–
Irrigated		
Continuously flooded	1	0.79–1.26
Intermittently flooded – single aeration	0.60	0.46–0.80
Intermittently flooded – multiple aeration	0.52	0.41–0.66
Rainfed and deepwater		
Regular rainfed	0.28	0.21–0.37
Drought prone	0.25	0.18–0.36
Deepwater	0.31	ND

Note: ND = not determined.
Source: Yan et al (2005); IPCC (2007a)

production, to feed their families, and therefore that environmental considerations must inevitably get less attention.

Understanding the relationships between the variety of conditions under which rice is grown and CH_4 emissions has vastly improved during the past two decades. Water and residue management are now seen as the two main factors controlling the turnover of organic matter under anaerobic conditions and thus the CH_4 emissions that are dependent on that turnover. These factors are also seen as the ones where changes in current practice may result in the most substantial emissions reductions. Globally, emissions could be reduced by about 30 per cent if continuously flooded fields were drained at least once during the season, and if rice straw was applied to land off-season. However, nothing happens without a cause. The variety of current practices is certainly sustained for many practical reasons, but also by conventions for which the initial cause may no longer be known. To convince millions of rice farmers to change crop management is probably one of the more demanding options to reduce global CH_4 emissions.

AR4 of the IPCC suggested that agricultural mitigation options often have co-benefits with improved crop productivity and environmental quality (Smith et al, 2007). Improved water management with mid-season drainage is one promising option in this regard because it can enhance rice yield while reducing CH_4 emissions. In general, any efforts to mitigate CH_4 emissions from rice cultivation through changes in agricultural practices should attempt to achieve synergy between climate change policies and sustainable development.

References

Abao, E. B. Jr., Bronson, K. F., Wassmann, R. and Singh, U. (2000) 'Simultaneous records of methane and nitrous oxide emissions in rice-based cropping systems under rainfed conditions', *Nutrient Cycling in Agroecosystems*, vol 58, pp131–139

Akiyama, H., Yagi, K. and Yan, X. Y. (2005) 'Direct N_2O emissions from rice paddy fields: Summary of available data', *Global Biogeochemical Cycles*, vol 19, article GB1005

Bergamaschi, P. (1997) 'Seasonal variation of stable hydrogen and carbon isotope ratios in methane from a Chinese rice paddy', *Journal of Geophysical Research*, vol 102, pp25383–25393

Bergamaschi, P. and Harris, G. W. (1995) 'Measurements of stable isotope ratios ($^{13}CH_4/^{12}CH_4$; $^{12}CH_3D/^{12}CH_4$) in landfill methane using a tunable diode laser absorption spectrometer', *Global Biogeochemical. Cycles*, vol 9, pp439–447

Cai, Z. C., Xing, G. X., Yan, X. Y., Xu, H., Tsuruta, H., Yagi, K. and Minami, K. (1997) 'Methane and nitrous oxide emissions from rice paddy fields as affected by nitrogen fertilisers and water management', *Plant and Soil*, vol 196, pp7–14

Cai, Z. C., Tsuruta, H., Gao, M., Xu, H. and Wei, C. F. (2003) 'Options for mitigating methane emission from a permanently flooded rice field', *Global Change Biology*, vol 9, pp37–45

Chen, Y.-H. and Prinn, R. G. (2006) 'Estimation of atmospheric methane emission between 1996–2001 using a 3-D global chemical transport model', *Journal of Geophysical Research*, vol 111, article D10307

Coleman, D. D., Risatti, J. B. and Schoell, M. (1981) 'Fractionation of carbon and hydrogen isotopes by methane-oxidizing bacteria', *Geochimica et Cosmochimica Acta*, vol 45, pp1033–1037

Conrad, R. (1989a) 'Control of methane production in terrestrial ecosystems', in M. Andreae and D. Schimel (eds) *Exchange of Trace Gases between Terrestrial Ecosystems and the Atmosphere*, Wiley, Chichester, pp39–58

Conrad, R. (1989b) 'Temporal change of gas metabolism by hydrogensyntrophic methanogenic bacterial associations in anoxic paddy soil', *FEMS Microbiology Ecology*, vol 62, pp265–274

Conrad, R. (1993) 'Mechanisms controlling methane emission from wetland rice fields', in R. Oremland (ed) *The Biogeochemistry of Global Change: Radiative Trace Gases*, Chapman and Hall, New York, pp317–335

Conrad, R (1996a) 'Anaerobic hydrogen metabolism in aquatic sediments', in D. D. Adams, S. P. Seitzinger and D. M. Crill (eds) *Cycling of Reduced Gases in the Hydrosphere*, Schweitzerbart'sche Verlagsbuchhandlung, Stuttgart, pp15–24

Conrad, R. (1996b) 'Soil microorganisms as controllers of atmospheric trace gases (H_2, CO, CH_4, OCS, N_2O, and NO)', *Microbiological Reviews*, vol 60, pp609–640

Denier van der Gon, H. A. C. and Neue, H.-U. (1995) 'Influence of organic matter incorporation on the CH_4 emission from a wetland rice field', *Global Biogeochemical Cycles*, vol 9, pp11–22

Ding, A., Willis, C. R., Sass, R. L. and Fisher, F. M. (1999) 'Methane from rice fields: Effect of plant height among several rice cultivars', *Global Biogeochemical Cycles*, vol 13, pp1045–1052

Fetzer, S. and Conrad, R. (1993) 'Effect of redox potential on methanogenesis by *Methano-sarcinabarkeri*', *Archives of Microbiology*, vol 160, pp108–113

Fitzgerald, G. J., Scow, K. M. and Hill, J. E. (2000) 'Fallow season straw and water management effects on methane emissions in California rice', *Global Biogeochemical Cycles*, vol 14, pp767–775

Fung, I., John, J., Lerner, J., Matthews, E., Prather, M., Steele, L. P. and Fraser, P. J. (1991) 'Three-dimensional model synthesis of the global methane cycle', *Journal of Geophysical Research*, vol 96, pp13033–13065

Garcia, J.-L., Jacq, V., Rinaudo, G. and Roger, P. (1974) 'Activités microbiennes dans les sols de rizières du Sénégal: Relations avec les caractéristiques physico-chimiques et influence de la rhizospère', *Revue d'Ecologie et de Biologie du Sol*, vol 11, pp169–185

Gilbert, B. and Frenzel, P. (1995) 'Methanotrophic bacteria in the rhizosphere of rice microcosms and their effect on porewater methane concentration and methane emission', *Biology and Fertility of Soils*, vol 20, pp93–100

Greenland, D. J. (1997) *The Sustainability of Rice Farming*, CAB International, Wallingford, Oxon, in association with IRRI, Los Baños, The Philippines

Happell, J. D., Chanton, J. P. and Showers, W. S. (1994) 'The influence of methane oxidation on the stable isotopic composition of methane emitted from Florida swamp forests', *Geochimica et Cosmochimica Acta*, vol 58, pp4377–4388

Hein, R., Crutzen, P. J. and Heimann, M. (1997) 'An inverse modeling approach to investigate the global atmospheric methane cycle', *Global Biogeochemical Cycles*, vol 11, pp43–76

Huang, Y., Sass, R. L. and Fisher, F. M. (1997) 'Methane emission from Texas rice paddy soils: 1. Quantitative multi-year dependence of CH_4 emission on soil, cultivar and grain yield', *Global Change Biology*, vol 3, pp479–489

Huke, R. H. and Huke, E. H. (1997) *Rice Area by Type of Culture: South, Southeast, and East Asia*, International Rice Research Institute, Los Baños, The Philippines

Husin, Y. A., Murdiyarso, D., Khalil, M. A. K., Rasmussen, R. A., Shearer, M. J., Sabiham, S., Sunar, A. and Adijuwana, H. (1995) 'Methane flux from Indonesian wetland rice: The effects of water management and rice variety', *Chemosphere*, vol 31, pp3153–3180

Inubushi, K., Cheng, W. G., Aonuma, S., Hoque, M. M., Kobayashi, K., Miura, S., Kim, H. Y. and Okada, M. (2003) 'Effects of free-air CO_2 enrichment (FACE) on CH_4 emission from a rice paddy field', *Global Change Biology*, vol 9, pp1458–1464

International Rice Research Institute (2006) *Bringing Hope, Improving Lives: Strategic Plan, 2007–2015*, IRRI, Manila, The Philippines

IPCC (Intergovernmental Panel on Climate Change) (2007a) *2006 IPCC Guidelines for National Greenhouse Gas Inventories*, H. S. Eggelston, L. Buendia, K. Miwa, T. Ngara and K. Tanabe (eds), IGES, Kanagawa, Japan

IPCC (2007b) *Climate Change 2007: The Physical Science Basis. Contribution of Working Group I to the Fourth Assessment Report of the Intergovernmental Panel on Climate Change*, S. Solomon, D. Qin, M. Manning, Z. Chen, M. Marquis, K. B. Averyt, M. Tignor and H. L. Miller (eds), Cambridge University Press, Cambridge

Jain, M. C., Kumar, S., Wassmann, R., Mitra, S., Singh, S. D., Singh, J. P., Singh, R., Yadav, A. K. and Gupta, S. (2000) 'Methane emissions from irrigated rice fields in northern India (New Delhi)', *Nutrient Cycling in Agroecosystems*, vol 58, pp75–83

Kimura, M., Murase, J. and Lu, Y. H. (2004) 'Carbon cycling in rice field ecosystems in the context of input, decomposition and translocation of organic materials and the fates of their end products (CO_2 and CH_4)', *Soil Biology and Biochemistry*, vol 36, pp1399–1416

Krüger, M. and Frenzel, P. (2003) 'Effects of N-fertilisation on CH_4 oxidation and production, and consequences for CH_4 emissions from microcosms and rice fields', *Global Change Biology*, vol 9, pp773–784

Krüger, M., Frenzel, P. and Conrad, R. (2001) 'Microbial processes influencing methane emission from rice fields', *Global Change Biology*, vol 7, pp49–63

Krüger, M., Eller, G., Conrad, R. and Frenzel, P. (2002) 'Seasonal variation in pathways of CH_4 production and in CH_4 oxidation in rice fields determined by stable carbon isotopes and specific inhibitors', *Global Change Biology*, vol 8, pp265–280

Li C. S., Qiu, J. J., Frolking, S., Xiao, X. M., Salas, W., Moore, B., Boles, S., Huang, Y. and Sass, R. (2002) 'Reduced methane emissions from large-scale changes in water management of China's rice paddies during 1980–2000', *Geophysical Research Letters*, vol 29, doi:10.1029/2002GL015370

Lindau, C. W., Bollich, P. K. and Delaune, R. D. (1995) 'Effect of rice variety on methane emissions from Louisiana rice', *Agricultural Ecosystems and Environment*, vol 54, pp109–114

Lou, Y. S., Inubushi, K., Mizuno, T., Hasegawa, T., Lin, Y., Sakai, H., Cheng, W. and Kobayashi, K. (2008) 'CH_4 emission with differences in atmospheric CO_2 enrichment and rice cultivars in a Japanese paddy soil', *Global Change Biology*, vol 14, pp2678–2687

Lovley, D. R., Coates, J. D., Blunt Harris, E. L., Phillips, E. J. P. and Woodward, J. C. (1996) 'Humic substances as electron acceptors for microbial respiration', *Nature*, vol 382, pp445–448

Lu, W. F., Chen, W., Duan, B. W., Guo, W. M., Lu, Y., Lantin, R. S., Wassmann, R. and Neue, H. U. (2000) 'Methane emissions and mitigation options in irrigated rice fields in southeast China', *Nutrient Cycling in Agroecosystems*, vol 58, pp65–73

Marik, T., Fischer, H., Conen, F. and Smith, K. (2002) 'Seasonal variation in stable carbon and hydrogen isotope ratios in methane from rice fields', *Global Biogeochemical Cycles*, vol 16, pp41-1–41-11

Masscheleyn, P. H., DeLaune, R. D. and Patrick, W. H. Jr. (1993) 'Methane and nitrous oxide emissions from laboratory measurements of rice soil suspension – effect of soil oxidation-reduction status', *Chemosphere*, vol 26, pp251–260

Mikaloff Fletcher, S. E., Tans, P. P., Bruhwiler L. M., Miller, J. B. and Heimann, M. (2004) 'CH_4 sources estimated from atmospheric observations of CH_4 and its $^{13}C/^{12}C$ isotopic ratios', *Global Biogeochemical Cycles*, vol 18, article GB4005

Neue, H. U. and Roger, P. A. (1993) 'Rice agriculture: Factors controlling emissions', in M. Khalil (ed) *Atmospheric Methane: Sources, Sinks, and Role in Global Change*, Springer, Berlin, pp254–298

Neue, H. U. and Sass, R. (1994) 'Trace gas emissions from rice fields', in R. Prinn (ed) *Global Atmospheric-Biospheric Chemistry*, Plenum Press, New York, pp119–147

Olivier, J. G. J., Bouwman, A. F., Berdowski, J. J. M., Veldt, C., Bloos, J. P. J., Visschedijk, A. J. H., van der Maas, C. W. M. and Zandveld, P. Y. J. (1999) 'Sectoral emissions inventories of greenhouse gases for 1990 on a per country basis as well as on 1° x 1°', *Environmental Science and Policy*, vol 2, pp241–263

Olivier, J. G. J., Van Aardenne, J. A., Dentener, F. J., Pagliari, V., Ganzeveld, L. N. and Peters, J. A. H. W. (2005) 'Recent trends in global greenhouse emissions:

Regional trends 1970–2000 and spatial distribution of key sources in 2000' *Environmental Sciences*, vol 2, pp81–99

Peters, V. and Conrad, R. (1996) 'Sequential reduction processes and initiation of CH_4 production upon flooding of oxic upland soils', *Soil Biology and Biochemistry*, vol 28, pp371–382

Ponnamperuma, F. N. (1972) 'The chemistry of submerged soils', *Advances in Agronomy*, vol 24, pp29–96

Ponnamperuma, F. N. (1981) 'Some aspects of the physical chemistry of paddy soils', in A. Sinica (ed) *Proceedings of Symposium on Paddy Soil*, Science Press-Springer, Beijing and Berlin, pp59–94

Prather, M., Derwent, R., Ehhalt, D., Fraser, P., Sanhueza, E. and Zhou, X. (1995) 'Other trace gases and atmospheric chemistry', in J. T. Houghton, L. G. Meiro Filho, B. A. Callander, N. Harris, A. Kattenberg and K. Maskell (eds) *Climate Change 1994: Radiative Forcing of Climate Change and an Evaluation of the IPCC IS92 Emission Scenarios*, Cambridge University Press, Cambridge, pp73–126

Ratering, S. and Conrad, R. (1998) 'Effects of short-term drainage and aeration on the production of methane in submerged rice field soil', *Global Change Biology*, vol 4, pp397–407

Sass, R. L. (2002) 'Methods for the mitigation of methane emission to the atmosphere from irrigated rice paddy fields', *Proceedings of the 1st Agricultural GHG Mitigation Experts Meeting, Non-CO_2 Network Project on Agricultural GHG Mitigation*, Washington, DC, December

Sass, R. L., Fisher, F. M. Jr., Turner, F. T. and Jund, M. F. (1991), 'Methane emissions from rice fields as influenced by solar radiation, temperature, and straw incorporation', *Global Biogeochemical Cycles*, vol 5, pp335–350

Sass, R. L., Fisher, F. M. Jr., Wang, Y. B., Turner, F. T. and Jund, M. F. (1992) 'Methane emission from rice fields: The effect of floodwater management', *Global Biogeochemical Cycles*, vol 6, pp249–262

Satpathy, S. N., Mishra, S., Adhya, T. K., Ramakrishnan, B., Rao, V. R. and Sethunathan, N. (1998) 'Cultivar variation in methane efflux from tropical rice', *Plant and Soil*, vol 202, pp223–229

Scheehle, E. A., Irving, W. N. and Kruger, D. (2002) 'Global anthropogenic methane emissions', in J. van Ham, A. P. Baede, R. Guicherit and J. Williams-Jacobse (eds) *Non-CO_2 Greenhouse Gases: Scientific Understanding, Control Options and Policy Aspects*, Millpress, Rotterdam, pp257–262

Schütz, H., Seiler, W. and Conrad, R. (1989) 'Processes involved in formation and emission of methane in rice paddies', *Biogeochemistry*, vol 7, pp33–53

Setyanto, P., Makarim, A. K., Fagi, A. M., Wassmann, R. and Buendia, L. V. (2000) 'Crop management affecting methane emissions from irrigated and rainfed rice in Central Java (Indonesia)', *Nutrient Cycling in Agroecosystems*, vol 58, pp85–93

Sigren, L. K., Byrd, G. T., Fisher, F. M. and Sass, R. L. (1997) 'Comparison of soil acetate concentrations and methane production, transport, and emission in two rice cultivars', *Global Biogeochemical Cycles*, vol 11, pp1–14

Smith, P., Martino, D., Cai, Z., Gwary, D., Janzen, H. H., Kumar, P., McCarl, B., Ogle, S., O'Mara, F., Rice, C., Scholes, R. J., Sirotenko, O., Howden, M., McAllister, T., Pan, G., Romanenkov, V., Rose, S., Schneider, U. and Towprayoon, S. (2007) 'Agriculture', in B. Metz, O. R. Davidson, P. R. Bosch, R. Dave and L. A.

Meyer (eds) *Climate Change 2007: Mitigation. Contribution of Working Group III to the Fourth Assessment Report of the Intergovernmental Panel on Climate Change*, Cambridge University Press, Cambridge

Sugimoto, A. and Wada, E. (1993) 'Carbon isotopic composition of bacterial methane in a soil incubation experiment: Contributions of acetate and CO_2/H_2', *Geochimica et Cosmochimica Acta*, vol 57, pp4015–4027

Tyler, S. C., Brailsford, G. W., Yagi, K., Minami, K. and Cicerone, R. J. (1994) 'Seasonal variations in methane flux and $\delta^{13}CH_4$ values for rice paddies in Japan and their implications', *Global Biogeochemical Cycles*, vol 8, pp1–12

Tyler, S. C., Bilek, R. S., Sass, R. L. and Fisher, F. M. (1997) 'Methane oxidation and pathways of production in a Texas paddy field deduced from measurements of flux, $\delta^{13}C$, and δD of CH_4', *Global Biogeochemical Cycles*, vol 11, pp323–348

Wang, J. S., Logan, J. A., McElroy, M. B., Duncan, B. N., Megretskaia, I. A. and Yantosca, R. M. (2004) 'A 3-D model analysis of the slowdown and interannual variability in the methane growth rate from 1988 to 1997', *Global Biogeochemical Cycles*, vol 18, article GB3011

Wang, Z. P., DeLaune, R. D. and Masscheleyn, P. H. (1993) 'Soil redox and pH effects on methane production in a flooded rice soil', *Soil Science Society of America Journal*, vol 57, pp382–385

Wassmann, R., Neue, H. U., Lantin, R. S., Buendia, L. V. and Rennenberg, H. (2000a) 'Characterization of methane emissions from rice fields in Asia: I. Comparison among field sites in five countries', *Nutrient Cycling in Agroecosystems*, vol 58, pp1–12

Wassmann, R., Neue, H. U., Lantin, R. S., Makarim, K., Chareonsilp, N., Buendia, L. V. and Rennenberg, H. (2000b) 'Characterization of methane emissions from rice fields in Asia: II. Differences among irrigated, rainfed and deepwater rice', *Nutrient Cycling in Agroecosystems*, vol 58, pp13–22

Wassmann, R., Aulakh, M. S., Lantin, R. S., Rennenberg, H. and Aduna, J. B. (2002) 'Methane emission patterns from rice fields planted to several rice cultivars for nine seasons', *Nutrient Cycling in Agroecosystems*, vol 64, pp111–124

Yagi, K. and Minami, K. (1990) 'Effect of organic matter application on methane emission from some Japanese paddy fields', *Soil Science and Plant Nutrition*, vol 36, pp599–610

Yagi, K., Tsuruta, H., Kanda, K. and Minami, K. (1996) 'Effect of water management on methane emissions from a Japanese rice paddy field: Automated methane monitoring', *Global Biogeochemical Cycles*, vol 10, pp255–267

Yagi, K. Tsuruta, H. and Minami, K. (1997). 'Possible options for mitigating methane emission from rice cultivation', *Nutrient Cycling in Agroecosystems*, vol 49, pp213–220

Yan, X. Y., Yagi, K., Akiyama, H. and Akimoto, H. (2005), 'Statistical analysis of the major variables controlling methane emission from rice fields', *Global Change Biology*, vol 11, pp1131–1141

Yan, X. Y., Akiyama, H., Yagi, K. and Akimoto, H. (2009) 'Global estimations of the inventory and mitigation potential of methane emissions from rice cultivation conducted using the 2006 Intergovernmental Panel on Climate Change Guidelines', *Global Biogeochemical Cycles*, vol 23, article number GB2002

Yao, H. and Conrad, R. (1999) 'Thermodynamics of methane production in different rice paddy soils from China, the Philippines, and Italy', *Soil Biology and Biochemistry*, vol 31, pp463–473

Yao, H., Conrad, R., Wassmann, R. and Neue, H. U. (1999) 'Effect of soil characteristics on sequential reduction and methane production in sixteen rice paddy soils from China, the Philippines, and Italy', *Biogeochemistry*, vol 47, pp269–295

Yoshida, S. (1981) *Fundamentals of Rice Crop Science*, International Rice Research Institute, Los Baños, The Philippines

Zehnder, A. J. B. and Stumm, W. (1988) 'Geochemistry and biogeochemistry of anaerobic habitats' in A. Zehnder (ed) *Biology of Anaerobic Microorganisms*, Wiley, New York, pp1–38

Zheng, X., Wang, M., Wang, Y., Shen, R., Li, J., Heyer, J., Koegge, M., Papen, H., Jin, J. and Li, L. (2000) 'Mitigation options for methane, nitrous oxide and nitric oxide emissions from agricultural ecosystems', *Advances in Atmospheric Sciences*, vol 17, pp83–92

9
Ruminants

Francis M. Kelliher and Harry Clark

Introduction

This chapter introduces enteric CH_4 emissions from animals. Some animals are herbivores or plant eaters. A subgroup assists their digestive process by regurgitating their food and chewing it a second time. This is called chewing their 'cud' or ruminating, so these animals have been dubbed ruminants. Domesticated ruminant livestock include sheep, cattle, goats and deer. Ruminants have a four-part stomach with two anterior chambers, including the rumen, forming a relatively large fermentation vat. Ruminants cannot digest the cellulose in plants, but these chambers contain an abundant, diverse microbial community 'subcontracted' for the job. Up to 75 per cent of the ruminant's energy supply comes from the products of microbial metabolism of dietary carbohydrate (Johnson and Ward, 1996). This symbiosis includes the microbial community gaining some of the feed's energy and the ruminant must gather the food and provide the community a stable environment (warm, wet and free of oxygen). The community completes their job by passing the fermentation products into two posterior chambers of the 'true' stomach for digestion, the prime energy source being volatile fatty acids. In principle, the bigger the ruminant, the bigger its energy requirement and fermenting vat, the longer its food can be 'cooked', the lower the feed quality that can be 'tolerated'.

Enteric CH_4 is a by-product of feed fermentation in the rumen and, to a lesser extent, in the large intestine. Typically >80 per cent of the CH_4 is produced in the rumen and the rest in the lower digestive tract (Immig, 1996). In sheep and cattle subjected to measurement, 92–98 per cent of the CH_4 gas was emitted from the mouth, the rest via flatus (Murray et al, 1976; Grainger et al, 2007). For enteric CH_4, gas emission has also been called gas eruction but not in this chapter. The rumen microbial community and 'ecosystem' is complex and its long evolution has led to a disposal mechanism for hydrogen through the reduction of carbon dioxide to CH_4 by methanogens (McAllister and Newbold, 2008). A high partial pressure of hydrogen inhibits microbial growth and digestion in the rumen (Wolin et al, 1997). In summary, with its microbial community, the ruminant consumes carbohydrates to meet its energy requirement and CH_4 emissions 'naturally' rid it of some of the hydrogen, potentially harmful in excess.

Recently, enteric CH_4 emissions have gained unprecedented interest with respect to international public policy related to greenhouse gases. Our involvement has focused on the development of enteric CH_4 emissions measurement methods, inventory calculations and the verification of technologies that have been proposed to mitigate or reduce the emissions. We thus reveal our disposition to a numerate approach, spanning significant temporal and spatial scales. Scaling raises the spectre of uncertainty in determining the strength of this source of CH_4 emissions to the atmosphere. After the reader has been introduced to the determination of the emissions, we present a synopsis of the challenges facing ruminant scientists, policy analysts and farmers to develop effective ways to mitigate enteric CH_4 emissions.

Determining enteric methane emissions

The UNFCCC has led to the annual publication of enteric CH_4 emissions (F_{CH4}; the emissions are reported as a flux in units of CH_4 mass emitted to the atmosphere per year) inventories by country. In the year 2005, global annual F_{CH4} have been estimated as 92Tg according to US EPA (2006). This compilation of calculations was done by Tier 1 and Tier 2 approaches. The Tier 1 approach utilized animal population estimates with an emission factor based on recommendation by the IPCC according to an international data synthesis, or country-specific information if available. The Tier 2 approach will be described here, but for now, it suffices to indicate this approach is more complex and hopefully accurate, reflecting the underlying principles. Using a mixture of Tier 1 and 2 approaches, for the year 2005, global annual enteric F_{CH4} were 84Tg according to Steinfeld et al (2006). Using a Tier 2 approach, for the year 2003, global, annual enteric F_{CH4} were 70Tg according to Clark et al (2005). The global emissions increased by 1.4 per cent yr^{-1} between 2000 and 2005 according to a linear interpolation of the data reported by US EPA (2006), so the Clark et al (2005) estimate was adjusted upwards to 72Tg for the year 2005. Thus, the three global estimates differed by up to 22 per cent and the 'true' target is moving, so for the year 2010, the emissions should be 99Tg according to US EPA (2006).

For the year 2010, according to US EPA (2006), 34, 24 and 15 per cent of the global emissions came from Asia, Latin America and Africa, respectively. By country, the top ten emitters accounted for 58 per cent of the global total, including China (13.9Tg), Brazil (11.7Tg), India (11.2Tg), US (5.5Tg), Australia (3.1Tg), Pakistan (3.0Tg), Argentina (2.9Tg), Russia (2.5Tg), Mexico (2.3Tg) and Ethiopia (1.9Tg). Using their Tier 2 approach, Clark et al estimated 63 per cent of the global emissions in 2003 (70Tg according to them) had come from grassland-derived feed, 35Tg from cattle and 9Tg from other domesticated ruminants including sheep, goats, buffalo (*Bubalus bubalis*) and camelids.

The strength of enteric emissions as a CH_4 source may be estimated by compiling an inventory, as done in accordance with the UNFCCC. The inventory may be represented by an equation:

$$F_{CH4} = n\ R\ (1/e)\ m \qquad (8)$$

where 'n' is the number of animals, 'R' is the mean animal's energy requirement (J per unit time) and 'm' the mean CH_4 yield expressed as a proportion of R. By convention, variable R has been expressed in terms of gross energy (GE) intake. Afterwards, variable R may be expressed in terms of the 'metabolizable' energy (ME). The ME is equal to the GE minus the combined GE of the emitted CH_4 and the excreted urine and faeces. To express F_{CH4} in mass flux units, we require term e, the ME content of the feed dry matter (DM, MJ ME kg^{-1} DM).

The F_{CH4} depends on feed intake as implied by variable R in Equation 8 and this will be analysed below. Expressing F_{CH4} in mass flux units allows us to express variable m as mass of CH_4 emitted per unit feed DM intake (DMI). Feed DMI can be measured for individually contained animals by measuring the DM masses of offered feed and that refused and/or wasted. Given available feed and the time to eat it, the rumen volume sets an upper limit for intake during a feeding 'bout'. Under such halcyon conditions that may be established by a farmer seeking optimum production from a domesticated ruminant, the period between feedings will be determined by the digestion rate.

The variables in Equation 8 are means based on sets of imperfect measurements or judgements. We can assess the uncertainty of each variable, expressing it as the coefficient of variation (CV). Here we distinguish between two sources of uncertainty or variation. First, there is variability within a ruminant population that may be quantified by the standard deviation. Second, there is uncertainty about true population means, typically provided by sampling, so the uncertainty may be quantified by the standard error. We have expressed the CV according to the standard error, the standard deviation of the distribution of the sample means.

To measure F_{CH4}, the ruminant may be contained in a chamber and the emitted gas sampled for measurement according to a calorimeter method (Pinares-Patino et al, 2008). Alternatively, a tracer (sulphur hexafluoride, SF_6) contained in a canister can be inserted into the rumen and an emitted gas collection system worn by the ruminant (Lassey, 2007). In New Zealand, the calorimeter and SF_6 methods were recently compared for contained, individual sheep fed cut and carried fresh forage (Hammond et al, 2009). From this meta-analysis of 357 records, the two methods had virtually indistinguishable mean values of variable m (23.1g versus 23.5g CH_4 kg^{-1} DMI), but the SF_6 method's standard error was twice that of the calorimeter method. This comparison suggested the SF_6 method's larger standard error included 'noise' that has been attributed to the method itself. These data included measurements from 21 experiments that had involved 187 sheep; records thus included repeated measures of the same sheep. Additional analysis has been undertaken by numerical simulation that added accounting of the variation (repeatability) of data from different experiments, each of which had been conducted for a different purpose (Murray H. Smith and Keith Lassey, personal communication). This supported the methods comparison conclusion of Hammond et al (2009) using a conventional, bulk approach to meta-analysis, but suggested Hammond et al had underestimated the two CVs of variable m. Based on numerical simulation for the calorimeter method data, the CV for variable m has been estimated to be 3 per cent.

Assuming each variable in Equation 8 is independent and CVs <10 per cent, we may use a root mean square method to estimate a CV for F_{CH4} that may be written:

$$CV(F_{CH4}) = [CV(n)^2 + CV(R)^2 + CV(e)^2 + CV(m)^2]^{0.5} \qquad (9)$$

As an example, for New Zealand's annual enteric CH_4 emissions inventory of up to 85 million sheep and cattle, the CVs for variables n, R and e were 2, 5 and 5 per cent, respectively, according to Kelliher et al (2007). Determination of these CVs was described in their paper. As stated, we have recommended CV = 3 per cent for variable m. Inserting these values into Equation 9 gives $CV(F_{CH4}) = [CV(2 \text{ per cent})^2 + CV(5 \text{ per cent})^2 + CV(5 \text{ per cent})^2 + CV(3 \text{ per cent})^2]^{0.5} = 8$ per cent. The uncertainty of F_{CH4} may be expressed as a (±)95 per cent confidence interval by multiplying the CV by the t-statistic (= 1.96). Thus, we were 95 per cent certain that New Zealand's inventory's true value was ±16 per cent.

For the year 2005, New Zealand's annual, enteric CH_4 emissions were 1.1Tg (US EPA, 2006). This inventory had been compiled by H. Clark using a Tier 2 Approach. Including the uncertainty estimate, New Zealand's inventory calculations yielded 1.1 ± 0.2Tg yr^{-1}, calculations that cannot be directly verified because the involved temporal and spatial scales are beyond measurement. However, as suggested, this inventory has been based on generalization from measurements made at smaller scales. For example, the SF_6 method has been used to measure daily F_{CH4} from small groups of freely grazing ruminants (one of New Zealand's early studies was reported by Judd et al, 1999). Micrometeorological methods have also been developed to measure the mean F_{CH4} of flocks and herds during field campaigns that have lasted up to a month (Laubach et al, 2008). An integrated horizontal flux method seemed the most promising micrometeorological method, based on atmospheric concentration measurements up- and down-wind of the animals by open-path lasers. During a series of beef and dairy cattle herd measurement campaigns, micrometeorological and SF_6 methods (the latter with samples of up to 58 animals) had statistically indistinguishable mean values of F_{CH4}.

The UNFCCC has a goal to stabilize greenhouse gas emissions. This aspiration has been considered with respect to a base year and a change of emissions from the base year to another thereafter. As an example, for New

Zealand in the year 1990, the enteric CH_4 emissions inventory yielded 992Gg (10^3Gg = 1Tg). Earlier, based on development of the approach that yielded Equation 9, we estimated uncertainty of the inventory calculation, an increase of 88Gg (9 per cent) in the emissions (ΔF_{CH4}) between 1990 and 2003 (Kelliher et al, 2007). We noted the inherent challenge of precisely estimating a relatively small change. The calculated uncertainty ΔF_{CH4} was expressed as a (±)95 per cent confidence interval, ±59 per cent including 95 per cent certainty that ΔF_{CH4} >0 was a true increase of the emissions.

Based on Equation 8, we have emphasized variable m as a key determinant of F_{CH4} and ΔF_{CH4}. We illustrated the mean and associated uncertainty of variable m by data analyses of mass-based values for sheep fed grass in New Zealand. Based on the GE intake expression in Equation 8, the mean value of variable m from published studies had a much wider international range of 2–15 per cent according to Johnson and Ward (1996). However, they argued m >7 per cent occurred only when low-quality diets were fed in quantities restricted by researchers to near or below the maintenance (basal metabolic) level of R, 'an unlikely practice by farmers'. Further, they considered m <5 per cent was restricted to special feeds not commonly used by farmers such as finely ground pellets of forage (to accelerate passage rate), distillery or barley mash and concentrates at a very high level (>90 per cent of dietary intake). When the feed was temperate forages, the range of mean m from published studies was 4–8 per cent according to Clark et al (2005), suggesting an overall mean was 6 per cent. For *Bos indicus* beef cattle fed tropical forages in northern Australia, the mean m was 11 per cent when individuals were enclosed in a calorimeter chamber and measured continuously for 24 hour periods (Kurihara et al, 1999). For *Bos taurus* beef cattle fed a grain-based diet in a feedlot in Alberta, Canada, the mean m was 5 per cent when F_{CH4} had been measured by a micrometeorological method (McGinn et al, 2008). For beef cattle fed different grain-based diets, typical of American feedlots, the mean m was also 5 per cent when estimated by a mechanistic model (Kebreab et al, 2008). These differences in variable m reflected the feed 'quality', so, all else equal, the mean m should follow a feed order of grain <temperate forage <tropical forage.

Feed intake and enteric methane emissions

The feed intake of individual animals grazing in a flock or herd cannot be directly measured. To circumvent this dilemma, methods have been developed to estimate the ME requirement of grazing ruminants, including CSIRO (2007). The ME requirement for maintenance (ME_m) has been determined by the amount of feed ME needed to maintain an animal's weight (Blaxter, 1989), called the live weight (LW), on a daily basis. Living can be expensive. For the ruminant, ME_m sets a lower limit based on the ME required to maintain its 'machinery of life'. Based on energy dissipation via carbohydrate oxidation, measurement has focused on the respiration rate of healthy, awake, inactive, non-reproductive, fasting adults in a moderate temperature environment. Energy loss from a body depends on its surface area. For the homeotherm, this leads to a dimensional argument for the dependence of ME_m on animal size via the square of a linear size variable (Brody, 1945). Extending the argument, ME_m would depend on animal volume raised to a power of two thirds. Consider density and ME_m would depend on the animal's live weight (kg LW), also raised to a power of 0.67. The curvilinear relation between ME_m and kg LW may be portrayed graphically as a straight line if both axes are transformed to logarithmic scales. The classic synthesis of Kleiber (1932) showed the line's slope, the power coefficient, was equal to 0.75 for values of kg LW across four orders of magnitude based on measurements done for animals from rats to steers. This was verified across 27 orders of magnitude (10^{-18} to 10^{10}kg LW) by West and Brown (2005). Alternatively, if the number of cells in an animal was proportional to w, one might have expected a power coefficient = 1.0, so ME_m per unit of kg LW was constant. Recently, compiling the largest database to date (including 3006 species) and averaging data on the basis of life forms, Makarieva et al (2008) calculated ME_m was 0.3–9W kg^{-1} LW (W = J s^{-1}), a 'strikingly' narrow range they suggested was evidence for 'life's metabolic optimum'. This was indeed remarkable, but recalling Equation 8 and $F_{CH4} \propto R$, the F_{CH4} inventory compiler must accurately account for a heavier animal having a lower ME_m per unit of w than a lighter animal.

To estimate a ruminant's ME requirement according to CSIRO (2007), feed ME content and digestibility are required. As an example, we have utilised indicative values of 10MJ ME kg^{-1} DM and 70 per cent, respectively, for beef cattle grazing temperate forage. The ME_m was 15 per cent larger for intact males than castrated males and females, decreasing about 2 per cent per year with age and the net efficiency of use of ME for maintenance was 70 per cent, calculated from the feed ME content. As an example, for 150–700kg LW beef cattle, ME_m was estimated to be 30–96MJ ME d^{-1}. The ruminant's ME requirement may have exceeded ME_m owing to the needs of food gathering and growth and for breeding females, pregnancy and lactation. These requirements have been expressed in the form of ME_m multipliers. For grazing beef cattle in the same LW range as before, the multiplier was proportional to the LW, adding 11–32 per cent to the ME_m. For weight gain of 0.2kg LW d^{-1}, the ME multiplier added a further 20 per cent to the ME_m. For breeding females, the pregnancy and lactation ME multiplier added another 13 and 20 per cent, respectively, to the ME_m. As an example, for grazing, breeding females that weighed 430kg LW, the total ME requirement was 104MJ ME d^{-1}, 66 per cent more than ME_m. As an example for males, a 700kg LW grazing bull had a total ME requirement of 134MJ ME d^{-1}.

While F_{CH4} implicitly depends on feed intake according to variable R in Equation 8 and the animal's ME requirement may be estimated, the relation between feed DMI and F_{CH4} may also be determined by measurement. This was applied earlier to the determination of variable m in Equation 8. Here, we propose a different approach based on the analysis of different data from an

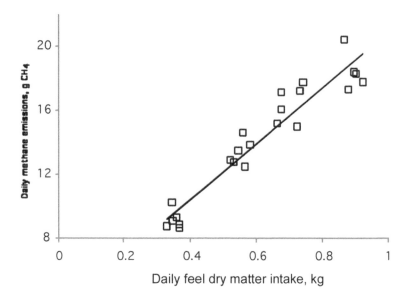

Figure 9.1 The relationship between daily feed DMI and enteric CH_4 emissions for 23 weaned lambs fed cut and carried grass in calorimeter chambers

Note: Each data record was a two-day mean during 3–13 June 2008. Linear regression yielded a slope of 17.4 ± 1.2g CH_4 kg^{-1} DMI, offset of 3.4 ± 0.8g CH_4 and 91 per cent of variability in the CH_4 emissions was associated with feed DMI.

experiment purposely designed as a verification test. For weaned lambs in calorimeter chambers fed cut and carried grass, so DMI and F_{CH4} could be directly measured, the relation that best fitted the data was linear (Figure 9.1). The 23 data records portrayed in Figure 9.1 came from an experiment denoted FLL for feed level lamb, comprising 23 sheep <1 year old that weighed 35–41kg. The linear relation between DMI and F_{CH4} was interpreted to have suggested the CH_4 yield, variable m in Equation 8, was constant with feed intake, indicated by the relation's slope (17.4 ± 1.2g CH_4 kg^{-1} DMI). Although this regression accounted for 91 per cent of the variability, the relation included an offset of 3.4 ± 0.8g CH_4 d^{-1}. For the 23 records, the mean CH_4 yield was 23.6 ± 0.5g CH_4 kg^{-1} DMI.

We also explored the relation between CH_4 yield and feed intake expressed as a proportion of ME_m, independently calculated according to the animal's metabolic weight following CSIRO (2007). While this expression of the independent variable is different to the feed intake, the feed intake remained part of it. Thus, the relation between CH_4 yield and intake as a proportion of ME_m could only be explored, recognizing the limitation of having both the independent and dependent variables determined using the measured feed intake.

Determination of ME_m for weaned lambs in the calorimeter chambers included resolution of an issue about an appropriate value of live weight required for the calculation. This was instructive so, candidly, we shall explain our approach. In the calorimeter chambers, the lambs were fed twice daily. For a grazing lamb, the maximum daily DMI was estimated to have been 3 per cent of the live weight. For a lamb confined to a calorimeter chamber, as an approximation, this proportion was reduced to 2 per cent. Thus, after each completed meal in the chamber, we estimated the DMI could have been up to 1 per cent of the live weight. This quantity of (ingested but undigested) feed was called 'gut fill'. We emphasize the feed intake was actually measured and this estimate has been formed only to explain the gut fill issue. Over time after the meal, digestion reduced the gut fill depending on the feed passage rate. For calculation, we considered a lamb of LW 40kg. For grass, we assumed the DM content was 15 per cent. Thus, the lamb's meal could have included up to 0.4kg DM. Including water contained in the fresh cut and carried grass, the maximum gut fill was estimated to have been 2.7kg. Consequently, the estimated maximum gut fill was equivalent to 7 per cent of the lamb's LW. For determination of the lamb's ME_m based on LW, unknown gut fill represented a potentially significant bias error. For each lamb, a fasting LW was determined as the weight measured 24 hours after feeding and before placement in a calorimeter chamber. Fasting weight was used to determine ME_m, providing a rational basis for the calculation.

As the feed intake, expressed as a proportion of ME_m increased, the CH_4 yield decreased (Figure 9.2). Increasing the proportional expression of feed intake from 1 to 2 corresponded with an 18 per cent reduction in the CH_4 yield

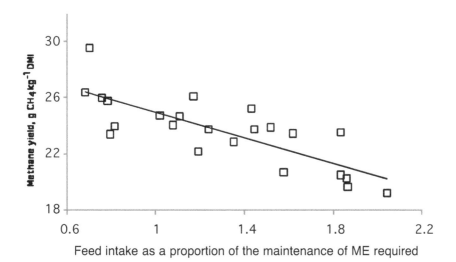

Figure 9.2 The relationship between feed intake as a proportion of the maintenance energy (ME) requirement and CH_4 yield for 23 weaned lambs fed cut and carried grass in calorimeters chambers

Note: Linear regression yielded a slope of -4.5 ± 0.8 g CH_4 kg^{-1} DMI, offset of 29.5 ± 1.0 g CH_4 and 62 per cent of variability in the CH_4 yield was associated with the proportional expression of feed intake.

according to the regression. Though preliminary, this analysis suggested a limitation of the common approach to estimating F_{CH4} according to feed intake and a mean value of the CH_4 yield. To visualize this limitation, consider that increased feed intake must have physically (due to a fixed rumen volume) corresponded with increased passage rate through the lamb, so decreased time available for microbial metabolism of dietary carbohydrate in the rumen. The alternative approach to estimating CH_4 yield suggested by the relation in Figure 9.2 could readily be incorporated calculations including enteric CH_4 emissions inventories. Though not shown here, preliminary analyses have also suggested separating animals into types based on their age as well as physiological state may be necessary. Further research is warranted to verify these suggestions and the merit of an alternative approach for enteric CH_4 emissions inventories in particular. The alternative approach would be more complex than the common, current method but should give a more accurate estimate of the 'true' CH_4 emissions.

Mitigation by reducing enteric methane emissions

It will be challenging for farmers to mitigate or reduce F_{CH4}, although some prospective opportunities have emerged (Table 9.1). Further, even if technical solutions are available, these need to be practical to implement and economically viable if they are to be adopted by farmers. This latter point is highly important since, at present, there are limited incentives for the adoption of mitigation technologies by farmers because of the general lack of a price signal; emission trading schemes are in place in a number of countries but so far none of these schemes include ruminant agriculture. A further point is that reductions in greenhouse emissions need to be viewed holistically. For example, reducing on-farm F_{CH4} by feeding more grain to ruminants brings no net benefit if carbon dioxide and nitrous oxide emissions are increased elsewhere in the food production value chain. This displacement of emissions is sometimes referred to as 'leakage'.

Table 9.1 Methods that have been proposed to reduce enteric CH_4 emissions from ruminant livestock

Short term	Medium term	Long term
Reduce animal numbers	Rumen modifiers	Targeted manipulation of rumen ecosystem
Increase productivity per animal	Select plants that produce lower CH_4 yield by the animals	Breed animals with low CH_4 yield
Manipulate diet		
Rumen modifiers		

Note: The methods have been classified by judgement of the prospect for availability in the short (available now), medium (likely to be commercially available within ten years) and long term (unlikely to be commercially available within ten years).

Short-term opportunities

Reducing animal numbers may seem an obvious way to reduce F_{CH4}. However, equally obvious, this method may not be acceptable to many farmers if it threatened their livelihood. Moreover, while animal numbers have been reducing in some countries (for example European Union and US), the global ruminant population has increased slightly over the past ten years according to FAO (2009). This reflected greater demand for ruminant animal products, milk and meat (Steinfeld et al, 2006). Demand has increased more for meat, which has been increasingly supplied by the production of monogastric animals like pigs and chickens (Galloway et al, 2007). In addition, there are plant-based alternatives for producing human dietary protein. However, utilization of these production systems has not been fully realized, especially for economically advanced and advancing societies, though change might be possible according to Smil's (2002) insightful and constructive analysis. Thus, animal numbers should not be considered in isolation from production.

Farmers have always sought improvement in production efficiency. This may be defined on the basis of feed intake or ME requirement. With this definition, efficiency may be related to a partitioning of feed intake into ME_m and production requirements. The ME_m and associated F_{CH4} may be considered fixed, depending on the ruminant's weight. Thus, the additional ME requirement and F_{CH4} will be determined by the production rate. This means an increase in feed intake and thus production corresponds with decreases in the proportion of F_{CH4} attributable to ME_m and F_{CH4} per unit product. Consequently, for a given quantity of product, the farmer can reduce F_{CH4} by utilizing the most productive ruminants. For a static production level, this argument has been interpreted to suggest an F_{CH4} mitigation strategy based on ruminant selection. However, a farm's production level is unlikely to be static, so reducing F_{CH4} per unit product does not necessarily correspond with reduced F_{CH4}.

To illustrate the relationship between production efficiency and F_{CH4}, we have analysed dairy production for cows fed by grazing fresh pasture. For pasture in New Zealand, the GE content was 18.4MJ GE kg^{-1} DM (Judd et al, 1999). The GE content of CH_4 is 55.6MJ kg^{-1}. Combining these values with a mean value for variable m of 6 per cent and converting the units, the mean m was 20g CH_4 kg^{-1} DM. For context, this calculated value was 14 per cent less than the mean value based on meta-analysis of the sheep (fed grass) measurements discussed earlier. For a grazing, non-lactating cow with constant w = 450kg, daily DMI was 5kg according to Clark et al (2005). Based on this DMI and the mean m, the daily F_{CH4} was 100g. For argument, we attributed this DMI and F_{CH4} to ME_m. When this cow's daily DMI was 10kg and she was lactating, daily milk production was 12kg (Clark et al, 2005). The corresponding daily F_{CH4} was 200g, so half was attributable to production and there was 17g CH_4 kg^{-1} milk. Increasing her daily DMI to 14kg, daily milk production was 24kg (Clark et al, 2005). Now, the corresponding daily F_{CH4}

was 280g, so one third was attributable to production and there was 12g CH_4 kg^{-1} milk. Thus, although doubling the milk production corresponded with a 29 per cent decrease of F_{CH4} per unit product, F_{CH4} itself had increased by 40 per cent.

Diet manipulation has been considered to present other opportunities to reduce F_{CH4} from ruminants. For example, increasing the proportion of grain has been shown to reduce F_{CH4} (Lovett et al, 2006). While increased profitability was also reported, circumstances can be different or change, so this cannot yet be considered a general recommendation. Adding lipids to the diet has also significantly reduced F_{CH4} according to Beauchemin et al (2008). However, 40–55 per cent of the global F_{CH4} from ruminants may be attributed to grazing animals, outdoors for most of their lives and consuming forage diets (Clark et al, 2005). The type of forage grown can be changed to legumes, for example, and an increased concentration of condensed tannins has been shown to reduce F_{CH4} of grazing ruminants (Waghorn and Woodward, 2006). However, alternative species are unlikely to be grown alone, so they must be able to grow with, and not be overwhelmed by, the species currently utilized (for example ryegrass).

Modification of the rumen can involve the administration of chemicals. For example, use of the ionophore, monensin, has reduced F_{CH4} in some circumstances (Beauchemin et al, 2008). Monensin also conveyed productivity and health benefits, so it has attracted widespread scientific attention. However, the rumen microbial community tends to be very adaptive (McAllister and Newbold, 2008), so the reduced F_{CH4} may not be long lasting. These compounds have been classed as antibiotics and their use may be unacceptable to some ruminant product customers and illegal in some production locations. There have been other compounds with claimed efficacy for reducing F_{CH4} that can broadly be classed as rumen modifying agents; these include yeasts, condensed tannin extracts, probiotics and enzyme-based feed additives. Although available commercially, we have found data supporting the claimed efficacy to be sparse and further evidence would be needed before any of these products could be recommended to farmers on this basis.

Medium-term opportunities

Although rumen modifiers are available now, a more realistic appraisal is that they hold promise for the future. To be fair, these products have been developed to increase productivity. The reduction of F_{CH4} has been an emergent 'value' that has suffered in the absence of an agreed price and obligation/market for CH_4. A change from the current, unclear situation to international market agreement could generate clear, tangible co-benefits for the commercial manufacturers of such products.

Plant extracts such as allicin (for example garlic), bacteriocins and improved yeast products have reduced F_{CH4} by a range of mechanisms (McAllister and Newbold, 2008). Breeding for plant traits that have reduced

F_{CH4} has also been considered a medium-term possibility. One example has been the breeding of so-called 'high sugar' grasses (Abberton et al, 2008). On theoretical grounds, at least, ruminants fed by grazing 'high sugar' grass should require less feed to meet their ME requirement, so have reduced F_{CH4}. A major challenge for animal scientists has been to clearly identify the specific plant chemical characteristics that reduced F_{CH4}. This has not been a simple task. For forage-based diets, for example, proximate analysis has been unable to determine which diet components influence F_{CH4} (Hammond et al, 2009).

Long-term opportunities

Targeted manipulation of the rumen ecosystem has long been thought to provide the best hope for reducing F_{CH4}, but also the biggest challenge. In principle, vaccines that stimulate ruminants to produce antibodies against their rumen methanogens can be developed (Wright et al, 2004). However, successful development should be considered a long way off. Technical requirements and challenges for a vaccine include a substantive effect and long-term efficacy, so broad spectrum activity in the rumen. Phages are viruses and phage therapy may be a possible vaccine alternative, though no phages specific to rumen methanogens have been reported (McAllister and Newbold, 2008). A new generation of 'designer' inhibitors may be developed based on knowledge of the genome of rumen methanogens (Attwood and McSweeney, 2008). Breeding animals with reduced F_{CH4} may be possible based on improved feed conversion efficiency (Alford et al, 2006; Hegarty et al, 2007) and lower emissions per unit of feed consumed (Pinares-Patino et al, 2008). The first approach has been considered attractive since the selected animals will combine higher productivity with lower feed intake. This should mean lower F_{CH4} per unit of product, lower F_{CH4} per animal (that may be offset by increased animal stocking rate) and reduced feed cost per animal. Breeding animals with lower F_{CH4} per unit of feed intake should guarantee reduced F_{CH4}. However, for successful adoption, productivity traits must be unaffected. While significantly lower F_{CH4} per unit of feed intake has recently been measured among ruminants, the 'effect' was transitory for the animals measured (Pinares-Patino et al, 2008).

Acknowledgements

We acknowledge the valuable contribution of Murray H. Smith and Keith Lassey whose meta-analysis quantified the uncertainty of CH_4 yield for sheep fed grass. We are grateful to Kirsty Hammond for sharing her draft manuscript and Stefan Muetzel for helping us to compile the calorimeter data.

References

Abberton, M. T., Marshall, A. H., Humphreys, M. W., Macduff, J. H., Collins, R. P. and Marley, C. L. (2008) 'Genetic improvement of forage species to reduce the environmental impact of temperate livestock grazing systems', *Advances in Agronomy*, vol 98, pp311–355

Alford, A. R., Hegarty, R. S., Parnell, P. F., Cacho, O. J., Herd, R. M. and Griffith, G. R. (2006) 'The impact of breeding to reduce residual feed intake on enteric methane emissions from the Australian beef industry', *Australian Journal of Experimental Agriculture*, vol 46, pp813–820

Attwood, G. and McSweeney, C. (2008) 'Methanogen genomics to discover targets for methane mitigation technologies and options for alternative H_2 utilisation in the rumen', *Australian Journal of Experimental Agriculture*, vol 48, pp28–37

Beauchemin, K. A., Kreuzer, M., O'Mara, F. P. and McAllister, T. A. (2008) 'Nutritional management for enteric methane abatement: A review', *Australian Journal of Experimental Agriculture*, vol 48, pp21–27

Blaxter, K. L. (1989) *Energy Metabolism in Animals and Man*, Cambridge University Press, New York

Brody, S. (1945) *Bioenergetics and growth*, Reinhold Publishing Corporation, New York

Clark, H., Pinares-Patino, C. and deKlein, C. A. M. (2005) 'Methane and nitrous oxide emissions from grazed grasslands', in D. A. McGilloway (ed) *Grassland: A Global Resource*, Wageningen Academic Publishers, Netherlands

CSIRO (Commonwealth Scientific and Industrial Research Organization) (2007) *Nutrient Requirements of Domesticated Ruminants*, CSIRO Publishing, Collingwood, Victoria, Australia

FAO (Food and Agriculture Organization) (2009) 'Global livestock and health production atlas', www.fao.org/ag/aga/glipha/index.jsp

Galloway, J. N., Burke, M., Bradford, G. E., Naylor, R., Falcon, W., Chapagain, A. K., Gaskell, J. C., McCullough, E., Mooney, H. A., Oleson, K. L. L., Steinfeld, H., Wassenaar, T. and Smil, V. (2007) 'International trade in meat: The tip of the pork chop', *Ambio*, vol 36, pp622–629

Grainger, C., Clarke, T., McGinn, S. M., Auldist, M. J., Beauchemin, K. A., Hannah, M. C., Waghorn, G. C., Clark, H. and Eckard, R. J. (2007) 'Methane emissions from dairy cows measured using the sulphur hexafluoride (SF_6) tracer and chamber techniques', *Journal of Dairy Science*, vol 90, pp2755–2766

Hammond, K. J., Muetzel, S., Waghorn, G. C., Pinares-Pitano, C. S., Burke, J. L. and Hoskins, S. O. (2009) 'The variation in methane emissions from sheep and cattle is not explained by the chemical compiosition of ryegrass', *New Zealand Society of Animal Production Proceedings*, vol 69, pp174–178

Hegarty, R. S., Goopy, J. P., Herd, R. M. and McCorkell, B. (2007) 'Cattle selected for lower residual feed intake have reduced daily methane production', *Journal of Animal Science*, vol 86, pp1479–1486

Immig, I. (1996) 'The rumen and hindgut as source of ruminant methanogenesis', *Environmental Monitoring and Assessment*, vol 42, pp57–72

Johnson, D. E. and Ward, G. M. (1996) 'Estimates of animal methane emissions', *Environmental Monitoring and Assessment*, vol 42, pp133–141

Judd, M. J., Kelliher, F. M., Ulyatt, M. J., Lassey, K. R., Tate, K. R., Shelton, I. D., Harvey, M. J. and Walker, C. F. (1999) 'Net methane emissions from grazing sheep', *Global Change Biology*, vol 5, pp647–657

Kebreab, E., Johnson, K. A., Archibeque, S. L., Pape, D. and Wirth, T. (2008) 'Model for estimating enteric methane emissions from United States dairy and feedlot cattle', *Journal of Animal Science*, vol 86, pp2738–2748

Kelliher, F. M., Dymond, J. R., Arnold, G. C., Clark, H. and Rys, G. (2007) 'Estimating the uncertainty of methane emissions from New Zealand's ruminant animals', *Agricultural and Forest Meteorology*, vol 143, pp146–150

Kleiber, M. (1932) 'Body size and metabolism', *Hilgardia*, vol 6, pp315–353

Kurihara, M., Magner, T., Hunter, R. A. and McCrabb, G. J. (1999) 'Methane production and energy partition of cattle in the tropics', *British Journal of Nutrition*, vol 81, pp227–234

Lassey, K. R. (2007) 'Livestock methane emission: From the individual grazing animal through national inventories to the global methane cycle', *Agricultural and Forest Meteorology*, vol 142, pp120–132

Laubach, J., Kelliher, F. M., Knight, T., Clark, H., Molano, G. and Cavanagh, A. (2008) 'Methane emissions from beef cattle – A comparison of paddock- and animal-scale measurements', *Australian Journal of Experimental Agriculture*, vol 48, pp132–137

Lovett, D. K., Shaloo, D., Dillon, P. and O'Mara, F. P. (2006) 'A systems approach to quantify greenhouse gas fluxes from pastoral dairy production as affected by management regime', *Agricultural Systems*, vol 88, pp156–179

Makarieva, A. M., Gorsgkov, V. G., Li, B., Chown, S., Reich, P. and Garrilov, V. M. (2008) 'Mean mass-specific metabolic rates are strikingly similar across life's major domains: Evidence for life's metabolic optimum', *Proceedings of the National Academy of Sciences*, vol 105, pp16994–16999

McAlister, T. A. and Newbold, C. J. (2008) 'Redirecting rumen fermentation to reduce methanogenesis', *Australian Journal of Experimental Agriculture*, vol 48, pp7–13

McGinn, S. M., Chen, D., Loh, Z., Hill, J., Beauchemin, K. A. and Denmead, O. T. (2008) 'Methane emissions from feedlot cattle in Australia and Canada', *Australian Journal of Experimental Agriculture*, vol 48, pp183–185

Murray, R. M., Bryant, A. M. and Leng, R. A. (1976) 'Rates of production of methane in the rumen and large intestine of sheep', *British Journal of Nutrition*, vol 36, pp1–14

Pinares-Patino, C. S., Waghorn, G. C., Machmuller, A., Vlaming, B., Molano, G., Cavanagh, A. and Clark, H. (2008) 'Variation in methane emission – effect of feeding and digestive physiology in non-lactating dairy cows', *Canadian Journal of Animal Science*, vol 88, pp309–320

Smil, V. (2002) 'Worldwide transformation of diets, burdens of meat production and opportunities for novel food proteins', *Enzyme and Microbial Technology*, vol 30, pp305–311

Steinfeld, H., Gerber, P., Wassenaar, T., Castel, V., Rosales, M. and de Haan, C. (2006) *Livestock's Long Shadow*, United Nations, FAO, Rome

US EPA (United States Environmental Protection Agency) (2006) *Global Anthropogenic Non-CO_2 Greenhouse Gas Emissions: 1990–2020*, United States Environmental Protection Agency, Washington, DC

Waghorn, C. C. and Woodward, S. L. (2006) 'Ruminant contributions to methane and global warming – A New Zealand perspective', in J. S. Bhatti, R. Lal, M. J. Apps and M. A. Price (eds) *Climate Change and Managed Ecosystems*, Taylor and Francis, Boca Raton, pp233–260

West, G. B. and Brown, J. H. (2005) 'The origin of allometric scaling laws in biology from genomes to ecosystems: Towards a quantitative unifying theory of biological structure and organization', *Journal of Experimental Biology*, vol 208, pp1575–1592

Wolin, M. J., Miller, T. J. and Stewart, C. S. (1997) 'Microbe-microbe interaction', in P. N. Hobson and C. S. Stewart (eds) *The Rumen Microbial Ecosystem*, Blackie Academic and Professional, London, pp467–491

Wright, A. D. G., Kennedy, P., O'Neill, C. J., Toovey, A. F., Popovski, S., Rea, S. M., Pimm, C. L. and Klein, L. (2004) 'Reducing methane emission in sheep by immunization against rumen methanogens', *Vaccine*, vol 22, pp3976–3985

10
Wastewater and Manure

Miriam H. A. van Eekert, Hendrik Jan van Dooren, Marjo Lexmond and Grietje Zeeman

Introduction

The agricultural contribution to global non-CO_2 greenhouse gas emissions is estimated to have been 5.1–6.1Gt CO_2-equivalents (eq) per year in 2005. That is 10–12 per cent of the total global anthropogenic emission of greenhouse gases (Smith et al, 2007). The dominant gases emitted in agriculture are CH_4 and N_2O, contributing to about 47 per cent and 58 per cent of global anthropogenic emissions of CH_4 and N_2O, respectively. According to the IPCC, the main sources of agricultural emissions are soils (38 per cent), enteric fermentation (32 per cent), biomass burning (12 per cent), rice production (11 per cent) and manure handling (7 per cent) (Smith et al, 2007).

The United Nations Food and Agricultural Organization (FAO) reports global anthropogenic emissions of CH_4 and N_2O of 5.9 and 3.4Gt CO_2-eq, respectively. Of these, some 2.2Gt CO_2-eq of both CH_4 and N_2O are emitted by livestock (Steinfeld et al, 2006). These livestock emissions are expected to increase to 2.8Gt CO_2-eq in 2020 (US EPA, 2006). Livestock represents about 80 per cent of agricultural methane emission. There are two main sources of CH_4 from livestock: enteric fermentation and manure. Enteric fermentation represents around 80 per cent of the total emission by livestock, but the emissions are varying in time and between different regions in the world. Emissions from enteric fermentation are calculated by multiplying the number of animals with an emission factor, specific per category of animals and country. This emission factor can be determined at several levels of accuracy differing from country to country (VROM, 2008a). Methane emission is influenced by the milk production, the level of feed intake, the energy consumption, feed composition and rumen conditions (Monteny and Bannink, 2004).

Methods for mitigation of enteric fermentation focus on improving the productivity and efficiency of livestock production and on increasing the digestibility of feedstuff either by modifying feed composition or by interfering with the rumen digestive processes (Steinfeld et al, 2006). Although these strategies may be beneficial to emissions at regional or national level, they may increase the CH_4 emissions per animal (US EPA, 2006). Enteric fermentation as a source of CH_4 emission is discussed in more detail elsewhere in this book (see Chapter 9).

Biological processes are gaining more and more interest as a solution for environmental challenges. Among these processes is anaerobic digestion, which is increasingly applied throughout the world for the treatment of solid organic waste, sludge and wastewater and plays a central role in the treatment of waste and biomass (Figure 10.1). Organic compounds are converted into biogas, which is a mixture of CH_4, CO_2, hydrogen sulphide (H_2S), H_2 and in some cases N_2O. In conventional aerobic wastewater treatment plants, organic matter is oxidized to CO_2 combined with the removal of nitrogen and phosphorus. The production of CH_4 during anaerobic processes is an advantage over other biological processes because of its possible reuse as an energy source. All applications that are fit for the use of natural gas are also able to handle biogas (after pretreatment). The energy content of 1kg chemical oxygen demand (COD) is similar to c. 1kWh (Aiyuk et al, 2006). Today, biogas is mostly converted into electricity and/or heat with electrical conversion efficiencies of up to 25 per cent (for less than 200kW combustion engines) or up to 30–35 per cent for larger (up to 1.5MW) engines. Combining

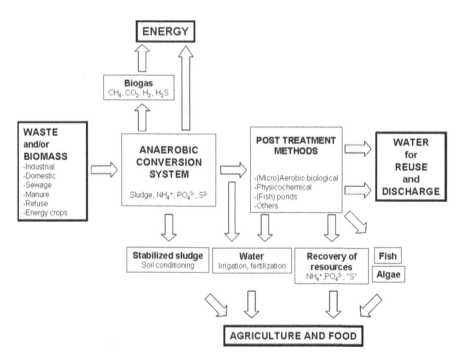

Figure 10.1 Anaerobic processes as a core technology in waste and biomass treatment

electricity generation with hot water recovery (via the exhaust gases of the combustion engine) may yield an efficiency as high as 65 to 85 per cent (IEA, 2001). However, care has to be taken that undesired emissions of greenhouse gases – CH_4 (GWP 21) (US-EPA, 2002) and N_2O (GWP 310) (US EPA, 2002) – are avoided.

The microbiology of the processes involved in this conversion is discussed in more detail elsewhere in this book. In short, the organic matter is converted to biogas in four subsequent reaction steps: hydrolysis, acidogenesis, acetogenesis and methanogenesis. Theoretically $0.35m^3$ (at standard temperature and pressure conditions; STP) CH_4 can be produced per kg of converted COD. Together with the lower sludge yield, the formation of biogas is the most clear advantage of anaerobic treatment over aerobic treatment. By contrast, the effluent quality of aerobic treatment is better than of anaerobic treatment, in which nitrogen (N) and phosphorus (P) are not removed. When reuse of treated wastewater for fertilization and irrigation can be established, the latter becomes an advantage rather than a disadvantage. Methane emissions in aerobic treatment other than the CH_4 that is produced during digestion of primary and secondary sludge appear to be negligible (El-Fadel and Massoud, 2001). Carbon dioxide is evidently emitted in open aerobic wastewater treatment systems, but this is 'short-cycle' CO_2 not originating from fossil fuel. The CO_2 emission related to the energy demand of an aerobic wastewater treatment plant also has to be taken into account. The formation of N_2O seems to be more of a problem especially during nitrification and denitrification when oxygen levels are low, and when low C/N ratios in the wastewater lead to increased nitrite concentrations. The emission of N_2O during nitrification is not related to the activity of nitrifying bacteria but is due to stripping of N_2O formed in the denitrifying stage of the wastewater treatment plant. This is a very slow process, so N_2O emission may even take place after discharge of the effluent to the surface water. Therefore the emission of N_2O from aerobic wastewater treatment plants is small compared to total anthropogenic emissions each year (around 3 per cent), but much higher (26 per cent) when the total water chain is taken into account (Kampschreur et al, 2009). So far, it remains unclear which group of microorganisms is predominantly responsible for N_2O formation in aerobic wastewater treatment plants. Since N_2O is not the main focus of this book its emissions will not be discussed in more detail in this chapter.

Technology

Anaerobic reactor systems

Solid waste and wastewater may be treated on-site or transported to a centralized wastewater treatment plant or digester. Anaerobic treatment may be one of the first steps in the treatment of such waste streams. Technologically, a distinction can be made between 'low-rate' systems with long hydraulic retention times and 'high-rate' systems in which the hydraulic retention time is relatively short (de Mes et al, 2003). Septic tanks are also commonly used for house on-site treatment but these systems cannot be classified as either 'low rate' or 'high rate'. Biogas produced in septic tanks is often not recovered. Slurries and solid wastes such as manure are usually treated in low-rate systems. In those systems the hydraulic retention time and the solid retention time are equal. This is not the case in high-rate systems that are commonly used for the treatment of wastewaters. Those systems are recognized by the extreme uncoupling of solid and hydraulic retention time. This can be achieved via retention of the bacterial sludge biomass in the reactor or by separating the sludge from the effluent. Sedimentation, growth of the biomass on (fixed) carrier material or granulation of the biomass are a few methods used for biomass retention. Batch and accumulation systems, plug flow and continuously stirred tank reactors (CSTRs) are examples of low-rate anaerobic systems. Examples of 'high-rate' systems include the contact process, anaerobic filter, fluidized bed and the upflow anaerobic sludge blanket (UASB) or expanded granular sludge bed (EGSB) reactors (de Mes et al, 2003; Lehr et al, 2005). Septic tank systems are in general operated at low loading rates, though the slow-rate tank (SRT) is commonly much larger than the high-rate tank (HRT).

UASB and EGSB reactors are the most frequently applied systems for the anaerobic treatment of wastewater (IEA, 2001). These systems usually contain granular sludge. These granules are formed through auto-immobilization, vary in size from 0.5 to 3mm diameter and are made up of a consortium of bacteria able to convert the COD in the wastewater to CH_4. They have a high specific methanogenic activity and a high settle ability (sedimentation rate) (up to 6–9m h^{-1}). Furthermore, these granules are very resistant to external shear forces (Lehr et al, 2005).

CSTR units are the most commonly applied systems for manure digestion with and without co-digestion material (Braun, 2007). Reactors have volumes of a few hundred to a few thousand m^3 depending on availability of manure and co-digestion material. Batch reactors are only applied when the substrate is very dry and stackable (>30 per cent dry matter). In that case material is stacked in a gas-tight container, sprinkled with heated inoculate and removed after the desired retention time. The Biocell installation in Lelystad, The Netherlands, is an example of such an installation. Plug flow reactors may be used for the anaerobic treatment of wastes with a dry matter content of between 20 and 30 per cent. The plug flow reactor has the advantage of being able to digest drier material but recirculation of digestate is needed to increase gas production. Units are usually relative small. In all other cases CSTR reactors are applied.

Anaerobic treatment systems may be one-stage and two-stage systems. One-stage systems are most often applied and the different processes needed for the conversion of complex substrates to CH_4 (from hydrolysis to methanogenesis) take place in one reactor without any spatial separation between the different steps. In a two-stage reactor the hydrolysis and acidification usually take place in the first stage and the methanogenesis in the second stage. The design of a system for anaerobic treatment of wastes or wastewater strongly depends on the composition and concentration of the wastewater and the nature of the constituents. Agricultural installations for manure digestion generally are one-stage reactors, though digestion processes will continue at a lower rate in the post-storage of digested manure.

Factors influencing methane production in anaerobic systems

Of course, the amount of CH_4 formed in anaerobic systems largely depends on the nature of the substrates present. Table 10.1 gives a few examples of the biogas or CH_4 production from different substrates determined in laboratory experiments. For the anaerobic conversion of organic compounds, the amount

Table 10.1 Typical biogas or methane production potential at 35°C of different (co)substrates

Substrate	Retention time (days)	Biogas potential (m^3 kg^{-1} VS)	Methane potential (m^3 CH_4 kg^{-1} VS)
Pig manure	20	0.3–0.5	
Cattle manure	20	0.15–0.25	
Chicken manure	30	0.35–0.6	
Carbohydrates (theoretical)		0.747	0.378
Fat (theoretical)		1.25	0.85
Protein (theoretical)		0.7	0.497
Animal fat	33	1.00	
Blood	34	0.65	
Food leftovers	33–35	0.47–1.1	
Edible oils	30	1.104	
Distillery slops (molasses, maize, potato)	10–21	0.4–0.47	
Brewery waste	14	0.3–0.4	
Potato (starch) waste	25–45	0.35–0.898	
Vegetable and fruit processing	14	0.3–0.6	
Carrot			0.31
Winter barley			0.30
Oil seed rape			0.29
Sweet pea			0.37

Source: Braun (2007); Pabon Pereira (2009)

of CH_4 that is produced can also be calculated using the Buswell equation (Buswell and Neave, 1930). In such a case, the assumption is made that an organic compound is completely biodegradable and would be completely converted by the anaerobic microorganisms (sludge yield is assumed to be zero) into CH_4, CO_2 and NH_3.

The actual amount of CH_4 formed depends on the nature of the substrate as a whole (biodegradability, toxicity) and existing physicochemical and process conditions (such as retention time, temperature and pH) (Lehr et al, 2005).

The effect of pH on microbiological processes is well known (Prescott et al, 2002). Every group of microorganisms has its own pH optimum. The pH is of importance for methanogenic processes because of the limited pH range in which the methanogenic microorganisms are active (pH 6.5 to 8.0). The same is true for the effect of temperature on microbial activity. Again, there are several groups of microorganisms, each with its own temperature range for optimal growth. For anaerobic treatment three temperature ranges are distinguished: psychrophilic microorganisms are active below 20°C, mesophilic bacteria function optimally between 20 and 40°C, and thermophilic bacteria at temperatures above 40°C, although these limits are not absolute. Different microbial populations may be active in the different temperature ranges although they may transfer the same substrate to CH_4. Transfer of a mesophilically operated reactor to operation in the thermophilic temperature range (for example 55°C) cannot be expected to proceed smoothly and instantly. A different microbial population will have to develop, which may take time. Besides its effect on the activity of the microorganisms, there is also the effect of temperature on the solubility of CH_4 in water. In the psychrophilic temperature range the maximum solubility of CH_4 is substantially higher than at higher temperatures (Figure 10.2), and as a result a substantial amount of the COD present in the effluent of an anaerobic reactor may be present as CH_4. Without additional measures, dissolved CH_4 may escape to the atmosphere when discharged from the reactor system.

One other important factor influencing the CH_4 production in anaerobic reactor systems is the (organic) load that is applied to the reactor and the variation in the load. Sudden overloads of the system may lead to problems resulting in a vicious circle (Figure 10.3). The non-stabilized effluents that result may lead to CH_4 emissions in post-storage systems or receiving surface waters.

Emissions from manure

There are several ways to treat manure that is produced on a farm (Figure 10.4). Emissions from manure related to storage, treatment and handling are also a result of anaerobic fermentation processes, which partially take place due to the presence of enteric bacteria that are excreted into the manure by the animals. On-site or central digestion is a recognized pathway for the treatment

WASTEWATER AND MANURE | 157

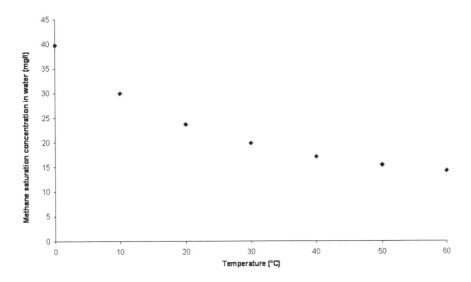

Figure 10.2 Maximum solubility of methane at different temperatures
Source: Data for Henry's constant taken from Metcalf & Eddy Inc. (1991)

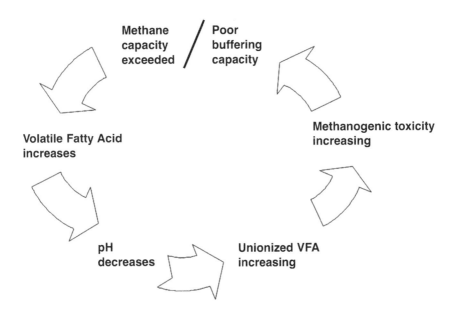

Figure 10.3 Problems related to overloading of anaerobic treatment systems
Source: Redrawn from van Lier et al (2008)

of manure for producing biogas and reducing CH_4 emissions from storage. The biogas production in digesters depends on the nature of the manure (Table 10.1) and on process parameters that have been discussed elsewhere in this chapter. The digestion of manure and factors related to its success have been reviewed extensively (see van Velsen, 1981; Zeeman, 1991; El-Mashad, 2003). Similar to other anaerobic processes, the biogas that is produced contains CH_4 and CO_2. The CH_4 emission from manure is calculated by multiplying the amount of manure per animal category and/or country or region with an emission factor. This emission factor depends on the fraction of organic matter, the gross CH_4 production potential and a conversion factor representing the percentage of the CH_4 potential that is actually realized in a specific manure handling system (VROM, 2008b).

Enhancing the production of CH_4 and capturing the biogas is an abatement strategy, generally referred to as anaerobic digestion. The captured gas can then be used as fuel for heating, lighting, transport or production of electricity or can be burned without beneficial use (flaring). The positive greenhouse gas emission effect is based on the conversion of CH_4 (GWP of 21–25) to CO_2 and the replacement of electricity from fossil fuels by renewable energy in the form of CH_4.

The emissions reduction potential of anaerobic digestion in manure treatment and storage systems depends on the emissions from the reference system, and ranges for liquid manure systems from 50 per cent in cool climates to 75 per cent in hot climates (Steinfeld et al, 2006). An emission reduction of 85 per cent for large-scale CSTR and plug-flow reactors has been achieved in developed countries, with a reduction of around 50 per cent reported for small-scale reactors in developing countries (US EPA, 2006).

Methane production is often increased by so-called co-digestion of manure with other biomass. This can be dedicated crops, crop residues, residues and leftovers from the food industry, or biomass from nature reserve areas (refer to Table 10.1 for examples). Digestion of manure by co-digestion is considered to shorten the length of time in pre-storage at farms. However, because of the relatively low biochemical methane potential (BMP) of manure (Table 10.1) compared to, for example, maize (around $0.3m^3$ CH_4 tonne^{-1}), and the lower organic matter content of manure compared to maize (Amon et al, 2007) and other co-substrates, the effect of shorter storage times may be offset. In The Netherlands, the amount of co-substrate that is permitted while still treating the digestate as 'fertilizer of animal origin' is 50 per cent on a volumetric basis.

Total greenhouse gas emission reduction for the whole biogas production chain, including co-digestion of biomass, is difficult to establish as emissions of alternative processes and the attribution of emissions in proceeding processes are debatable.

Post-storage of digestate of manure is gaining increasing interest as a possible source of unwanted CH_4 emissions. An inventory of 15 different digestates showed an average residual gas formation of $5m^3$ CH_4 tonne^{-1} (range $1–10m^3$ CH_4 tonne^{-1} digestate), for central manure systems, which may be 8

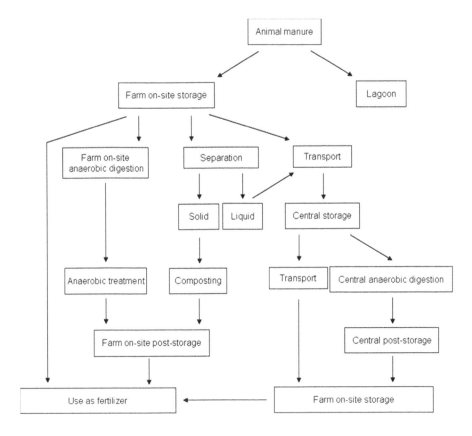

Figure 10.4 Different ways to handle manure

per cent of the total CH_4 mitigation potential (I. Bisschops, personal communication, 2009). Post-digestion in, for example, gas-tight covered storage is therefore promoted to optimize CH_4 production and to avoid unwanted CH_4 emissions. The BMP is often not achieved in digestors due to insufficient hydrolysis. Angelidaki et al (2005, 2006) analysed the potential CH_4 production/emission of digestates of central manure digestion systems. In the case of central manure digestion, the central storage time of digestates is generally short due to high land prices, while for on-farm storage sites, incentives for biogas collection are often lacking (Angelidaki et al, 2005).

Existing technology offers several possibilities to mitigate unwanted CH_4 emissions:

- The storage *temperature* of the slurry can significantly influence the emissions. Storage outside at lower ambient temperatures, depending on the climate conditions, can reduce the emission compared to those from animal housing. With active (deep) cooling, the emission of CH_4 can be

further reduced but only with high energy costs and a risk of increased CO_2 emissions associated with electricity production (Sommer et al, 2004).
- *Frequent removal* of the slurry from animal housing has the advantage of reducing the emitting surface and concentrating the slurry. It can also make other abatement options – such as covering, filtering, cooling or air treatment – more effective. The storage areas should be completely emptied to avoid inoculation of fresh manure with active microorganisms (Zeeman, 1994).
- *Gas-tight covering of storage* is a relative easy and cost-effective method to reduce emissions of both CH_4 and ammonia.
- *Solid separation* leads to removal of much of the organic matter required for CH_4 producing processes, and thus lowers the emissions of CH_4 from the remaining liquid part.
- *Aerobic treatment* of stored slurry can reduce CH_4 production by inhibiting microbial methanogenesis, but is associated with high energy consumption and a risk of elevated N_2O emissions.
- *Composting* of slurries together with other organic solids, or composting of solid farmyard manure, can reduce the emission of CH_4 but may also increase emissions of N_2O.
- Possibilities of *treatment of air* from animal houses or manure storages with biofilters (Melse, 2003). A CH_4 removal efficiency of 85 per cent was achieved in a test system but, due to low CH_4 concentrations in exhaust air of housing and storage, the required size of the biofilter to treat the air from a 1000m³ storage area would be 20–80m³. The costs of such a strategy were calculated at EU€100–500 per tonne CO_2-eq emission reduction.

Emissions from wastewater

Wastewater may be of domestic or industrial origin. The degree of treatment varies. Wastewater may be discharged directly in (coastal) surface water or treated prior to discharge (Figure 10.5). Wastewater in developed countries is usually treated in conventional centralized aerobic treatment plants. Industrial wastewater may be treated (in-plant) for process water recirculation and prior to discharge. Depending on the nature of the industry this treatment may also be anaerobic. Wastewater from the food industry is often easily treated with anaerobic methods, and even wastewaters from chemical industries can often be treated with anaerobic technologies, despite the presence of toxic compounds (Lettinga, 1995). Special measures have to be taken to enable anaerobic treatment in the latter case, i.e. application of special reactor designs with sludge retention or inoculation of the anaerobic sludge with specialized bacteria. In warmer climates, with higher ambient temperatures, anaerobic treatment is also used for domestic wastewater treatment.

One of the first factors that must be taken into account when considering CH_4 emissions from domestic wastewater is the collection and transport

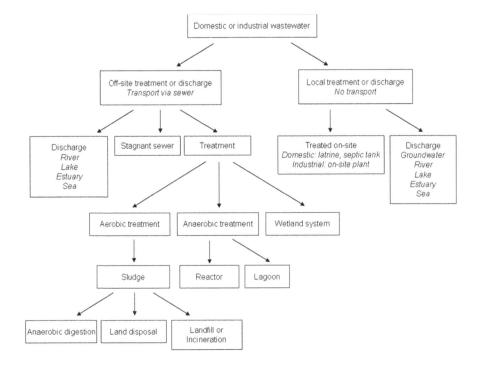

Figure 10.5 Discharge pathways of wastewater

system. Open sewers, networks of open canals, gutters and ditches, may all act as a source of CH_4. Closed sewers are not considered to be a source of CH_4 emission (IPCC, 2006d). The amount of CH_4 formed in open sewers will be directly related to the amount of biodegradable COD present in the wastewater and local conditions, particularly temperature. This is also the case for direct discharge of wastewater to the land surface and the treatment of sludge that is formed during primary, secondary and tertiary treatment. Therefore, the amount of CH_4 formed is very difficult to assess and will not be discussed in more detail in this chapter.

Domestic wastewater

Methane emissions from domestic wastewater for each treatment system or discharge step can be calculated using a default maximum CH_4 producing capacity (B_0) that is corrected by multiplication with the methane correction factor (MCF) of the wastewater treatment system used. From the MCFs in Table 10.2 it becomes clear that recovery of CH_4 in such systems is very important. The total amount of degradable organics in the wastewater (TOW) is corrected for organics removed as sludge (S) and the amount of CH_4

recovered (R) and subsequently adjusted with the emission factor EF_j after correction for the influence of different income groups (U_i) in the wastewater-producing population and the degree $T_{i,j}$ with which the treatment step (j) is used by each income group i (IPCC, 2006d) according to the following equation:

$$CH_4 \text{ emission, domestic} = \left[\sum_{i,j}(U_i \times T_{i,j} \times EF_j)\right](TOW - S) - R \quad (10)$$

with:

$$TOW = P \times \frac{BOD}{1000} \times I \times 365 \text{ (kg BOD year}^{-1}) \quad (11)$$

where P = population in inventory year; BOD = biological oxygen demand, country specific per capita BOD; and I = correction factor for additional industrial discharge.

The accuracy of the calculation is largely determined by the uncertainty in the data used. The uncertainty in the actual values of B_0 may be as high as 30 per cent and up to 50 per cent for the MCF. In addition, the per capita BOD varies largely per region and country, as do the urbanization constant and degree of utilization. Nevertheless, using the equations above, it is possible to estimate the amount of CH_4 emitted (IPCC, 2006d).

Anaerobic treatment of domestic wastewater in combination with CH_4 recovery is able to limit CH_4 emissions (Table 10.2) while at the same time reducing CO_2 emissions by displacing fossil fuel use (Aiyuk et al, 2006). The first pilot plant for anaerobic treatment of domestic sewage was built in Colombia (Cali) to treat a rather dilute sewage source (Schellinkhout et al, 1985). At an average temperature of 25°C, COD and BOD removal efficiencies higher than 75 per cent were obtained (Lettinga et al, 1987). Full-scale applications followed in India (Kanpur and Mirzapur), Colombia (Bucaramanga), Brazil, Portugal, Mexico and others (Maaskant et al, 1991; Draaijer et al, 1992; Schellinkhout and Collazos, 1992; Haskoning, 1996; Monroy et al, 2000). Trickling filters or polishing ponds have also been used for effluent polishing.

Low-temperature domestic sewage treatment has been researched extensively, but so far without full-scale application (Seghezzo et al, 1998). One of the disadvantages of this strategy is the large fraction of total methane production CH_4 that is found dissolved in water at low temperatures (Figure 10.2).

Based on the COD produced per person per day with grey water, faeces and urine (Table 10.3) (Kujawa-Roeleveld and Zeeman, 2006) and considering an anaerobic biodegradability of domestic sewage of 70 per cent (Elmitwally et al, 2001) the total amount of CH_4 that can theoretically be produced from a human population of 6.79 billion people amounts to 70Tg yr^{-1}. Likewise, controlled anaerobic treatment of this human waste would result in 98 × 10^{9}m^3 CH_4 being available for energy production.

Table 10.2 Methane emissions related to wastewater treatment and discharge

Treatment		MCF[a]	Methane emission
Untreated wastewater	Discharge in surface water	0–0.2	Depending on COD of wastewater and local conditions with maximum production: $CH_{4,produced,max}$
	Closed sewers		No
	Open sewers	0–0.8	Depending on COD of wastewater and local conditions with maximal production: $CH_{4,produced,max}$
Aerobic centralized treatment of wastewater	Wastewater	0.0–0.2 (good management)	Proper design and management: no or minimal CH_4 emission
		0.2–0.4 (poor management)	Poor design or management: CH_4 emission that strongly depends on conditions in the plant
	Sludge	0.8–1.0 (in the absence of recovery)	In the case of anaerobic digestion of primary and secondary sludge, CH_4 may be emitted in the absence of proper recovery
Aerobic shallow ponds		0–0.3	Proper design and management: no or minimal CH_4 emission
			Poor design or management: CH_4 emission that strongly depends on conditions in the plant
Anaerobic deep lagoons		0.8–1.0	Methane emission possible with maximal production: $CH_{4,produced,max}$
Anaerobic reactors		0.8–1.0 (in the absence of recovery)	Methane may be emitted in the absence of proper recovery
Septic tanks		0.5	Methane produced. Amount is dependent on the management of the system
Open pits/latrines		0.05–1.0	Methane production is likely depending on local conditions (temperature) and management (retention time)

Note: Methane correction factor (MCF) based on expert judgement of lead authors in IPCC (2006d)
Source: adapted from IPCC (2006d)

Industrial wastewater

The use of the anaerobic digestion (AD) process to recover CH_4 has considerable potential for industries with organic waste streams. Many of these industries use AD as a pretreatment step to lower sludge disposal costs,

Table 10.3 Volume and composition of separated domestic wastewater streams

Parameter	Unit	Urine	Faeces	Grey water	Kitchen refuse
Volume	g or L p^{-1} day^{-1}	1.25–1.5	0.07–0.17	91.3	0.2
Nitrogen	g N p^{-1} day^{-1}	7–11	1.5–2	1.0–1.4	1.5–1.9
Phosphorus	g P p^{-1} day^{-1}	0.6–1.0	0.3–0.7	0.3–0.5	0.13–0.28
Potassium	g K p^{-1} day^{-1}	2.2–3.3	0.8–1.0	0.5–1	0.22
Calcium	g Ca p^{-1} day^{-1}	0.2	0.53		
Magnesium	g Mg p^{-1} day^{-1}	0.2	0.18		
BOD	g O_2 p^{-1} day^{-1}	5–6	14–33.5	26–28	
COD	g O_2 p^{-1} day^{-1}	10–12	45.7–54.4	52	59
Dry matter	g p^{-1} day^{-1}	20–69	30	54.8	75

Source: Kujawa-Roeleveld and Zeeman (2006)

control odours and to reduce the costs of final treatment at a municipal wastewater treatment facility. From the perspective of the municipal facility, pretreatment effectively expands existing treatment capacity (IEA, 2001).

The following industries have been identified as sectors with high potential CH_4 emissions (IPCC, 2006d) or production (IEA, 2001):

- pulp and paper industry;
- meat and poultry processing;
- alcohol, beer and starch (and other food processing industries such as dairy, edible oil, fruit, cannery, juice making and others);
- organic chemicals production.

If industrial wastewater is treated via the same systems as domestic wastewater, it should be included in the calculations for domestic wastewater. However, in many cases industrial wastewater is (pre)treated on-site or in-plant prior to discharge to sewer systems, surface waters or in-plant recycling. These industries all produce significant quantities of wastewater with moderate to high COD concentrations that are attractive for anaerobic treatment and subsequent CH_4 recovery. Typical values for the generated wastewater quantity (m^3 $tonne^{-1}$ product) and COD concentration are given in Table 10.4.

Similar to domestic wastewater, the CH_4 emission from industrial wastewater can be calculated as follows:

$$CH_4 \text{ emission, domestic} = \sum_i \left[(TOW_i - S_i)EF_i - R_i \right] \quad (12)$$

with:

$$TOW_i = P_i \times W_i \times COD_i \text{ (kg COD } yr^{-1}) \quad (13)$$

where P = total industrial product for industrial sector 1 (tonne yr^{-1}); and W = wastewater generated (m^3 $tonne^{-1}$ product) (IPCC, 2006d).

Table 10.4 Typical data for industrial wastewater production

Industry	Wastewater usage		COD		B_0	CH_4 production
	Typical (m^3 tonne^{-1})	Range (m^3 tonne^{-1})	Typical (kg m^{-3})	Range (kg m^{-3})	(m^3 CH_4 kg^{-1} COD) (STP)	calculated[a] (m^3 tonne^{-1} product)
Beer and malt	6.3	5.0–9.0	2.9	2–7	0.33	6.0
Dairy products	7	3–10	2.7	1.5–5.2	0.35	6.6
Organic chemicals	67	0–400	3	0.8–5	Variable	Not determined
Pulp and paper	162	85–240	9	1–15	0.27	393.7[b]
Starch production	9	4–18	10	1.5–42	0.33	29.7
Sugar refinery	Not available	4–18	3.2	1–6	0.33	4.2

Note: [a] Use of data indicated in Table 10.1. [b] Today, the pulp and paper industry often uses less water and the wastewater contains lower COD concentrations due to the internal closing of water cycles (Pokhrel and Viraraghavan, 2004).
Source: Data from IPCC (2006d); Lexmond and Zeeman (1995)

Based on the typical values for wastewater production (W) and COD concentrations as depicted in Table 10.4, the potential CH_4 emission/production per tonne product for the different distinguished industries are calculated. However, considering the large fluctuations in industrial production – especially in the current economic climate – worldwide potential CH_4 emissions/productions are not calculated.

Domestic sludge treatment

Sludge and other biodegradable organic wastes hold enormous potential for energy generation. The amount of energy that is released upon oxidation of organic matter is around 14.5MJ (kg O_2)$^{-1}$ (Blackburn and Cheng, 2005). The amount of energy released depends on the biodegradability of the organic matter under certain conditions. The most commonly used methods for the treatment of biological wastes will be discussed below.

Landfills

Disposal of waste at landfill sites should be carried out in a proper way to avoid unwanted CH_4 emission due to decomposition (see Chapter 11). The recovery of landfill gases is now a common way to reduce unwanted emissions (IPCC, 2006b). It is difficult to assess the effect of sludge disposal on CH_4 emissions. A generally accepted way is to use the first-order decay method that assumes a slow decay of organic matter. Depending on the conditions (especially temperature), CH_4 generation rate constants for the decay of sewage sludge in landfills range from 0.05 to 0.1 year^{-1} in dry conditions, and from 0.1 to 0.7 year^{-1} in moist and wet conditions (IPCC, 2006b).

Digestion

The emission due to digestion of waste sludge can be calculated with the following equation:

$$CH_4 \text{ emission, domestic} = \sum_i (M_i \times EF_i) \times 10^{-3} - R \qquad (14)$$

where M_i = mass of organic waste treated by biological treatment type i; EF is emission factor for treatment i; i refers to composting or digestion; and R = total amount of CH_4 recovered (IPCC, 2006a).

The CH_4 emission factor is the main determining factor for the CH_4 emission and this is highly variable. It is determined by the nature of the waste and supporting material (if any), the temperature and moisture content, and the energy needed for aeration (in the case of composting) (Table 10.5) (IPCC, 2006d). Emissions from anaerobic digestors due to unintentional leakages of CH_4 are estimated to be between 0 and 10 per cent (5 per cent is the default value).

Table 10.5 The methane emission factors (EF) of anaerobic digestion at a biogas facility and during composting

	Dry weight basis	Wet weight basis (moisture content 60%)
Anaerobic digestion	2 (0–20)	1 (0–8)
Composting	10 (0.08–20)	4 (0.03–8)

Note: There is an assumption of 25–50 per cent dissolved organic carbon (DOC) in dry matter, and 60 per cent moisture in wet waste.
Source: Data taken from IPCC (2006a)

Typical biochemical CH_4 potential data are given in Table 10.6. As expected, the nature of the sludge largely determines the BMP. In some cases the CH_4 production can be further increased by thermal, chemical, physical and/or thermo-chemical pretreatment. In this way more CH_4 may be generated for further use as an energy source, with the concurrent effect that undesired CH_4 emissions after the digestion process are mitigated. An example of the effect of thermo-chemical pretreatment on CH_4 production is given by Kim et al (2003). They reported a 34 per cent increase in CH_4 production during digestion of waste activated sludge after thermo-chemical treatment at 121°C for 30 minutes at pH 12. Likewise, Tanaka et al (1997) observed a 27 per cent increase in CH_4 production after thermo-chemical pretreatment of the sludge.

Composting

The CH_4 emissions during composting of waste sludge can be calculated with the equation given above for digestion. Although composting is an aerobic

Table 10.6 Typical BMPs for different sludge types and other biowaste

Sludge type	m³ CH$_4$ tonne⁻¹ OS	Reference
Primary sludge (domestic)	600	Han and Dague (1997)
Primary sludge (industrial)	300	Braun (2007)
Domestic sludge	172	Tanaka et al (1997)
Secondary sludge (domestic)	200–350	Braun (2007)
Flotation sludge	690	Braun (2007)
Source-separated municipal biowaste	400	Braun (2007)

Note: OS = organic sludge

process, CH$_4$ may be formed in the anaerobic zones of the heap. Significant methane emissions are very unlikely to occur during composting because any CH$_4$ formed will be oxidized in the aerobic zones of the compost by methanotrophic bacteria. Total CH$_4$ emissions from composting will be limited to less than 1 per cent of the initial carbon content of the sludge (or any other material to be composted) (IPCC, 2006c).

Other methods
Incineration of solid wastes – sludge remains after composting for example – is generally carried out in controlled facilities. Emission factors are difficult to assign. Also, the effect of land application on emissions is difficult to assess.

Methane combustion and flaring

Ideally, any CH$_4$ generated during wastewater treatment and the digestion of solid wastes like manure and domestic sewage sludge should be recovered and used as an energy source. There are incentives from governments to promote the use of biologically generated CH$_4$ as a replacement for natural gas. However, in most cases the biogas generated in digesters is used for electricity generation with combined heat and power (CHP) generators. In most European countries electricity from biogas is considered as renewable and is subsidized. The heat generated through biogas combustion in CHP systems is often used to warm the input of the digester and digester content, and the surplus can be used for drying the solid fraction after separation of the digested manure. It may also be used to heat commercial buildings or residential areas. In the latter case it can be more profitable to install the CHP near the heat-demanding area and transport the biogas instead of transporting the heat.

Direct use of the biogas in the national natural gas grid, or use as transport fuel in cars, buses and trains, is becoming more and more popular. For direct use of the gas, it must first be purified to (more or less) natural gas properties (FNR, 2005). Flaring of CH$_4$ has to be applied if recovery is not possible. The emissions mitigation advantage of flaring – the conversion of CH_4 (GWP 21–25) to CO_2 and water vapour – is that total emissions in terms of CO_2-eq decrease considerably.

Conclusion and recommendations

In this chapter the potential CH_4 emissions from animal manure, domestic and industrial wastewaters and sludges were presented. For optimal CH_4 production, recovery and use of CH_4 emissions worldwide the following can be considered.

Animal manure

As early as the 1980s, in The Netherlands a reduction in CH_4 production in animal manure digesters was observed as a result of digestion of the manure in pre-storage (Zeeman et al, 1985). It is difficult to prevent such CH_4 emissions from taking place in many manure storage systems, especially where large storage capacities, able to hold the manure production of more than 100 days, are compulsory. Storage temperature, storage time and the presence of an inoculum (i.e. manure leftovers after emptying the storage) are three crucial factors that determine the amount of CH_4 emitted from a manure storage (Zeeman, 1994). Manure has to be entirely fresh to ensure the highest possible biogas production when it is digested as the sole substrate. Today, co-digestion is more often used and the co-products may be responsible for the largest part of the produced biogas, reducing the attention on fresh manure input. Aside from emissions in pre-storage, CH_4 emissions in post-storage may also occur. After digestion of fresh pig manure in a CSTR at 30°C and 20 days detention time, the additional CH_4 gas production in the post-storage at 10–15°C can amount to 23 per cent of total production (Zeeman, 1994). Co-digestion may result in a variation in composition and concentration of input material, resulting in variations in the organic load to the anaerobic digester. These variations could negatively influence the stability of the digestate and so lead to increased CH_4 emissions during post-storage (depending on the hydraulic retention time (HRT) applied in the digester).

Post-storage CH_4 emissions can be avoided with the application of gas-tight covers on digestate storage tanks. Lately, many biogas plants have installed gas collection systems in after-storage tanks (Angelidaki et al, 2005). Extensive research in the 1980s in The Netherlands and Switzerland into low (ambient) temperature digestion using combined storage/digestion (accumulation) systems could help the optimization of such systems for (co-)digestion elsewhere, and solve emission problems from both pre- and after-storage (Wellinger and Kaufmann, 1982; Zeeman, 1991).

The use of post-digestion for substrate mixtures is reported to improve biogas production (Boea et al, 2009). Applying a mixture of cow and pig manure and industrial wastes during post-digestion resulted in an increase in the biogas production of 11.7 per cent, 8.4 per cent and 1.2 per cent at a temperature of 55°C, 37°C and 15°C, respectively, and an HRT of 5.3 days. The main reactor was operated at 55°C with an HRT of 15 days. For on-site digestion, covered post-storages might be more efficient than post-digestion, as long-term storage is frequently compulsory. For central digestion, as applied in Denmark, short-term post-storage may be applied as land prices are often high. To prevent CH_4 emissions during subsequent long-term on-farm storage (Figure 10.4) and to increase the CH_4 recovery, post-digestion at an elevated temperature – as suggested by Boea et al (2009) – may be an efficient option.

Wastewater

Domestic wastewater has a potential CH_4 production/emission of c. 14m^3 per person per year. The collection, transport and treatment strategy that is applied will determine to what extent CH_4 can be recovered and used for energy production, whether CH_4 will be partially emitted, and if CO_2 instead of CH_4 will be produced. Controlled anaerobic treatment is a challenging but potentially important technology for reducing energy use and increasing energy production, while reducing CO_2-eq emissions (Aiyuk et al, 2006).

Anaerobic treatment of domestic sewage is commonly applied in tropical latitudes, while at higher latitudes it is used more rarely. Low nutrient removal efficiencies and high dissolved CH_4 concentrations are two disadvantages of low-temperature anaerobic treatment. Newly developed nutrient removal techniques, such as anammox and denitrification with CH_4 (Strous and Jetten, 2004) may enable future application of low-temperature domestic sewage treatment. Recently, it was shown that the direct anaerobic oxidation of CH_4 coupled to denitrification of nitrate is possible (Raghoebarsing et al, 2006). Future application of this microbial conversion process could solve two problems in one, viz. removing high dissolved CH_4 and nitrogen contents in low-temperature effluent of anaerobic treatment systems for domestic sewage.

Another possibility is the separation of domestic waste(water) streams at the source, with subsequent community-based on-site anaerobic treatment of the concentrated black water and kitchen waste at an elevated temperature for the production of energy (grey water is separately treated). Successful demonstrations have already been made in Germany and The Netherlands (Otterpohl et al, 1999; Zeeman et al, 2008). Worldwide, domestic on-site septic tanks are frequently applied for domestic sewage treatment, especially in rural areas. The CH_4 that is produced is generally not recovered or used as an energy source. A greater number of septic tank systems could be improved and operated as UASB-septic tanks for increased efficiency and biogas production (Lettinga et al, 1993). Indeed, by feeding black water and kitchen waste to a UASB-septic tank, ~60 per cent of the energy for cooking can be provided by the produced biogas (Zeeman and Kujawa, unpublished results).

High-rate anaerobic treatment for industrial wastewater was first applied on a commercial scale in the sugar industry in the mid-1970s in The

Netherlands. Since that time the technology has developed into a standard method of wastewater treatment for a wide variety of industries (Frankin, 2001). For industrial wastewater, the application of anammox and other nutrient removal processes may further improve the efficiency of wastewater treatment processes. In addition to producing energy by application of anaerobic treatment, many industries are becoming interested in closing water and resource cycles. Rearranging the conventionally applied water cycles in industries offers many more advantages than solely 'increasing' the water resources. The main drivers in Dutch industries to optimize water use are related to (van Lier and Zeeman, 2007):

- optimized usage of raw materials (less wastage);
- optimized energy efficiency; as heating surface water and groundwater to production temperatures requires ~4.2MJ $°C^{-1}$ m^{-3}, a substantial energy benefit can be gained if treated process water is reused in the production process;
- reduced costs related to water intake taxes and costs for drinking water/industrial water;
- reduced costs for wastewater conveyance and treatment.

This chapter has discussed many different issues regarding the emission of CH_4 from wastewater and manure. It is clear that different measures can be taken to either promote controlled CH_4 production or mitigate undesired CH_4 emissions. The recovery of CH_4 is important both from an energetic standpoint and from the perspective of greenhouse gas emissions. Therefore it is important that any CH_4 emitted is at least flared, but preferably used for energy generation. Finally, a matter of immediate concern is the fact that treatment of both industrial and urban wastewater is still comparatively rare in developing countries: for example Asia at around 35 per cent and Latin America at about 14 per cent (WHO/UNICEF, 2000). As such, already-scarce water resources for many millions of people are even more threatened (van Lier and Zeeman, 2007).

References

Aiyuk, S., Forrez, I., De Kempeneer, L., van Haandel, A. and Verstraete, W. (2006) 'Anaerobic and complementary treatment of domestic sewage in regions with hot climates – A review', *Bioresource Technology*, vol 97, pp2225–2241

Amon, T., Amon, B., Kryvoruchko, V., Zollitsch, W., Mayer, K. and Gruber, L. (2007) 'Biogas production from maize and dairy cattle manure: Influence of biomass composition on the methane yield', *Agriculture, Ecosystems & Environment*, vol 118, pp173–182

Angelidaki, I., Boe, K. and Ellegaard, L. (2005) 'Effect of operating conditions and reactor configuration on efficiency of full-scale biogas plants', *Water Science & Technology*, vol 52, pp189–194

Angelidaki, I., Heinfelt, A. and Ellegaard, L. (2006) 'Enhanced biogas recovery by applying post-digestion in large-scale centralized biogas plants', *Water Science & Technology*, vol 54, pp237–244

Blackburn, J. W. and Cheng, J. (2005) 'Heat production profiles from batch aerobic thermophilic processing of high strength swine waste', *Environmental Progress*, vol 24, pp323–333

Boea, K., Karakasheva, D., Trablyb, E. and Angelidaki, I. (2009) 'Effect of post-digestion temperature on serial CSTR biogas reactor performance', *Water Research*, vol 43, pp669–676

Braun, R. (2007) 'Anaerobic digestion: A multi-faceted process for energy, environmental management and rural development', in P. Ranalli (ed) *Improvement of Crop Plants for Industrial End Uses*, Springer Verlag, pp335–416

Buswell, A. M. and Neave, S. L. (1930) *Laboratory Studies of Sludge Digestion*, Bulletin no. 30., Division of the State Water Survey, Urbana, Illinois

de Mes, T. D. Z., Stams, A. J. M., Reith, J. H. and Zeeman, G. (2003) 'Methane production by anaerobic digestion of wastewater and solid wastes', in J. H. Reith, R. H. Wijffels and H. Barten (eds) *Bio-methane and Bio-hydrogen: Status and Perspectives of Biological Methane and Hydrogen Production*, Dutch Biological Hydrogen Foundation, Energy Research Centre of The Netherlands, pp58–102

Draaijer, H., Maas, J. A. W., Schaapman, J. E. and Khan, A. (1992) 'Performance of the 5 MLD UASB reactor for sewage treatment at Kanpur, India', *Water Science & Technology*, vol 25, pp123–133

El-Fadel, M. and Massoud, M. (2001) 'Methane emissions from wastewater management', *Environmental Pollution*, vol 114, pp177–185

El-Mashad, H. E. H. (2003) 'Solar Thermophilic Anaerobic Reactor (STAR) for renewable energy production', PhD thesis, Wageningen University, Wageningen, The Netherlands

Elmitwally, T. A., Soellner, J., de Keizer, A., Bruning, H., Zeeman, G. and Lettinga, G. (2001) 'Biodegradability and change of physical characteristics of particles during anaerobic digestion of domestic sewage', *Water Research*, vol 35, pp1311–1317

FNR (Fachagentur Nachwachsende Rohstoffe) (2005) *Handreichung biogasgewinnung und nutzung*, Fachagentur Nachwachsende Rohstoffe, Gülzow

Frankin, R. (2001) 'Full-scale experiences with anaerobic treatment of industrial wastewater', *Water Science and Technology*, vol 44, no 8, pp1–6

Han, Y. and Dague, R. R. (1997) 'Laboratory studies on the temperature-phased anaerobic digestion of domestic primary sludge', *Water Environment Research*, vol 69, pp1139–1143

Haskoning (1996) *MLD UASB Treatment Plant in Mirzapur, India*, Evaluation report on process performance, Haskoning Consulting Engineers and Architects, Nijmegen, The Netherlands

IEA (International Energy Agency) (2001) *Biogas and More! Systems and Markets Overview of Anaerobic digestion*, IEA, Paris

IPCC (Intergovernmental Panel on Climate Change) (2006a) 'Waste generation, composition and management data', in J. Wagener Silva Alves, Q. Gao, S. G. H. Guendehou, M. Koch, C. Lopez Cabrera, K. Mareckova, H. Oonk, E. Scheehle, A. Smith, P. Svardal and S. M. M. Vieira (eds) *2006 IPCC Guidelines for National Greenhouse Gas Inventories*, Institute for Global Environmental Strategies, Kanagawa, Japan, pp2.1–2.24

IPCC (2006b) 'Solid waste disposal', in J. Wagener Silva Alves, Q. Gao, C. Lopez Cabrera, K. Mareckova, H. Oonk, E. Scheehle, C. Sharma, A. Smith and M. Yamada (eds) *2006 IPCC Guidelines for National Greenhouse Gas Inventories*, Institute for Global Environmental Strategies, Kanagawa, Japan, pp3.1–3.40

IPCC (2006c) 'Biological treatment of solid waste', in J. Wagener Silva Alves, Q. Gao, C. Lopez Cabrera, K. Mareckova, H. Oonk, E. Scheehle, C. Sharma, A. Smith, P. Svardal and M. Yamada (eds) *2006 IPCC Guidelines for National Greenhouse Gas Inventories*, Institute for Global Environmental Strategies, Kanagawa, Japan, pp4.1–4.8

IPCC (2006d) 'Wastewater treatment and discharge', in M. R. J. Doorn, S. Towprayoon, S. M. Manso, W. Irving, C. Palmer, R. Pipatti and C. Wang (eds) *2006 IPCC Guidelines for National Greenhouse Gas Inventories*, Institute for Global Environmental Strategies, Kanagawa, Japan, pp6.1–6.28

Kampschreur, M. J., Temmink, H., Kleerebezem, R., Jetten, M. S. and Loosdrecht, M. C. M. v. (2009) 'Nitrous oxide emission during wastewater treatment', *Water Research*, doi:10.1016/j.watres.2009.03.001

Kim, J., Park, C., Kim, T.-H., Lee, M., Kim, S., Kim, S.-W. and Lee, J. (2003) 'Effects of various pretreatments for enhanced anaerobic digestion with waste activated sludge', *Journal of Bioscience and Bioengineering*, vol 95, pp271–275

Kujawa-Roeleveld, K. and Zeeman, G. (2006) 'Anaerobic treatment in decentralised and source-separation-based sanitation concepts', *Reviews in Environmental Science and Bio/Technology*, vol 5, pp115–139

Lehr, J. H., Keeley, J. and Lehr, J. (2005) *Water Encyclopedia: Domestic, Municipal, and Industrial Water Supply and Waste Disposal*, J. Wiley and Sons, New York

Lettinga, G. (1995) 'Anaerobic digestion and wastewater treatment systems', *Antonie van Leeuwenhoek*, vol 67, pp3–28

Lettinga, G., de Man, A., Grin, P. and Hulshof Pol, L. (1987) 'Anaerobic wastewater treatment as an appropriate technology for developing countries', *Tribune Cebedeau*, vol 40, pp21–32

Lettinga, G., de Man, A., van der Last, A. R. M., Wiegant, W., Knippenburg, K., Frijns, J. and van Buuren, J. C. L. (1993) 'Anaerobic treatment of domestic sewage and wastewater', *Water Science & Technology*, vol 27, pp67–73

Lexmond, M. J. and Zeeman, G. (1995) *Potential of Controlled Anaerobic Wastewater Treatment in Order to Reduce the Global Emissions of the Greenhouse Gases Methane and Carbon Dioxide*, NRP, Bilthoven

Maaskant, W., Magelhaes, C., Maas, J. and Onstwedder, H. (1991) 'The upflow anaerobic sludge blanket (UASB) process for the treatment of sewage', *Environmental Pollution*, vol 1, pp647–653

Melse, R. W. (2003) 'Methane degradation in a pilot-scale biofilter for treatment of air from animal houses and manure storages', *Agrotechnology and Food Innovations*, report no 2003–16, 193pp

Metcalf & Eddy Inc. (1991) *Wastewater Engineering: Treatment, Disposal, and Reuse*, McGraw-Hill International Editions, New York

Monroy, O., Fama, G., Meraz, M., Montoya, L. and Macarie, H. (2000) 'Anaerobic digestion for wastewater treatment in Mexico: State of technology', *Water Research*, vol 34, pp1803–1816

Monteny, G. J. and Bannink, A. (2004) 'Main principles for greenhouse gas abatement strategies for animal houses, manure storage and manure treatment', *Proceedings of the International Conference on GHG Emissions from Agriculture: Mitigation Options and Strategies*, Leipzig, Germany, February 2–4, pp38–42

Otterpohl, R., Albold, A. and Oldenburg, M. (1999) 'Source control in urban sanitation and waste management: Ten systems with reuse of resources', *Water Science & Technology*, vol 39, pp53–60

Pabon Pereira, C. P. (2009) 'Anaerobic digestion in sustainable biomass chains', PhD thesis, Wageningen University, Wageningen, The Netherlands

Pokhrel, D. and Viraraghavan, T. (2004) 'Treatment of pulp and paper mill wastewater – a review', *Science of the Total Environment*, vol 333, pp37–58

Prescott, L. M., Harley, J. P. and Klein, D. A. (2002) *Microbiology*, McGraw Hill, Boston, USA

Raghoebarsing, A. A., Pol, A., van de Pas-Schoonen, K. T., Smolders, A. J., Ettwig, K. F., Rijpstra, W. I., Schouten, S., Sinnighe-Damste, J., Op den Camp, H. J., Jetten, M. S. and Strous, M. (2006) 'A microbial consortium couples anaerobic methane oxidation to denitrification', *Nature*, vol 440, pp918–921

Schellinkhout, A. and Collazos, C. J. (1992) 'Full scale application of the UASB technology for sewage treatment', *Water Science & Technology*, vol 25, pp159–166

Schellinkhout, A., Lettinga, G., van Velsen, L. and Louwe Kooijmans, J. (1985) 'The application of UASB reactor for the direct treatment of domestic wastewater under tropical conditions', *Proceedings of the Seminar-Workshop on Anaerobic Treatment of Sewage*, Amherst, USA, pp259–276

Seghezzo, L., Zeeman, G., van Lier, J. B., Hamelers, H. V. M. and Lettinga, G. (1998) 'A review: The anaerobic treatment of sewage in UASB and EGSB reactors', *Bioresource Technology*, vol 65, pp175–190

Smith, P., Martino, D., Cai, Z., Gwary, D., Janzen, H., Kumar, P., McCarl, B., Ogle, S., O'Mara, F., Rice, C., Scholes, B. and Sirotenko, O. (2007) 'Agriculture', in B. Metz, O. R. Davidson, P. R. Bosch, R. Dave and L. A. Meyer (eds) *Climate Change 2007: Mitigation. Contribution of Working Group III to the Fourth Assessment Report of the Intergovernmental Panel on Climate Change*, Cambridge University Press, Cambridge and New York

Sommer, S. G., Pederson, S. O. and Møller, H. B. (2004) 'Algorithms for calculating methane and nitrous oxide emissions from manure management', *Nutrient Cycling in Agroecosystems*, vol 69, pp143–154

Steinfeld, H., Gerber, P., Wassenaar, T., Castel, V., Rosales, M. and de Haan, C. (2006) 'Livestock's long shadow', *Environmental Issues and Options*, Food and Agriculture Organization of the United Nations, Rome

Strous, M. and Jetten, M. S. M. (2004) 'Anaerobic oxidation of methane and ammonium', *Annual Reviews in Microbiology*, vol 58, pp99–117

Tanaka, S., Kobayashi, T., Kamiyama, K. and Signey Bildan, M. L. N. (1997) 'Effect of thermochemical pretreatment on the anaerobic digestion of waste activated sludge', *Water Science & Technology*, vol 35, pp209–215

US EPA (US Environmental Protection Agency) (2002) *Greenhouse Gases and Global Warming Potential Values*, US Greenhouse Gas Inventory Program Office of Atmospheric Programs US Environmental Protection Agency, Washington, DC

US EPA (2006) *Global Mitigation of Non-CO$_2$ Greenhouse Gas*, United States Environmental Protection Agency, Washington DC

van Lier, J. B. and Zeeman, G. (2007) 'Water and the sustainable use of an "abundant" resource', in *Environmental Technology, Changing Challenges in a Changing World*, Farewell Symposium of Wim H. Rulkens, Wageningen, The Netherlands

van Lier, J. B., Mahmoud, N. and Zeeman, G. (2008) 'Anaerobic wastewater treatment', in M. Henze, M. C. M. van Loosdrecht, G. A. Ekama, and D. Brdjanovic (eds) *Biological Wastewater Treatment Principles: Modelling and Design*, IWA Publishing, London, pp415–457

van Velsen, A. F. M. (1981) 'Anaerobic digestion of piggery waste', PhD thesis, Wageningen University, Wageningen, The Netherlands

VROM (Ministerie van Volkshuisvesting, Ruimtelijke Ordening en Milieu; Ministry of Housing, Spatial Planning and the Environment) (2008a) 'Protocol 8127 Pensfermentatie rundvee t.b.v. NIR 2008. 4A: CH$_4$ ten gevolge van pens- en darmfermentatie', VROM, Den Haag

VROM (2008b) 'Protocol 8130 Mest CH$_4$ t.b.v. NIR 2008. 4B CH$_4$ uit mest', VROM, Den Haag

Wellinger, A. and Kaufmann, R. (1982) 'Psychrophilic methane generation from pig manure', *Process Biochemistry*, vol 17, pp26–30

WHO/UNICEF (World Health Organization/ United Nations Children's Fund) (2000) *Global Water Supply and Sanitation Assessment*, WHO and UNICEF, Geneva

Zeeman, G. (1991) 'Mesophilic and psychrophilic digestion of liquid manure', PhD thesis, Agricultural University, Wageningen, The Netherlands

Zeeman, G. (1994) 'Methane production/emission in storages for animal manure', *Fertilizer Research*, vol 37, pp207–211

Zeeman, G., Treffers, M. E. and Halm, H. D. (1985) 'Laboratory and farmscale anaerobic digestion in The Netherlands', in B. F. Pain and R. Hepherd (eds) *Anaerobic Digestion of Farm Wastes*, Technical Bulletin 7, pp135–140

Zeeman, G., Kujawa, K., de Mes, T., Hernandez, L., de Graaff, M., Abu-Ghunmi, L., Mels, A., Meulman, B., Temmink, H., Buisman, C., van Lier, J. and Lettinga, G. (2008) 'Anaerobic treatment as a core technology for energy, nutrients and water recovery from source-separated domestic waste(water)', *Water Science & Technology*, vol 57, pp1207–1212

11
Landfills

Jean E. Bogner and Kurt Spokas

Introduction and background

Landfill CH_4 accounts for approximately 1.3 per cent (0.6Gt CO_2-eq yr^{-1}) of global anthropogenic greenhouse gas emissions relative to total emissions from all sectors of about 49Gt (Monni et al, 2006; US EPA, 2006; Bogner et al, 2007; Rogner et al, 2007). For countries with a history of controlled landfilling, landfills can be one of the larger national sources of anthropogenic CH_4; for example, US landfills are currently the second largest source of anthropogenic CH_4 after ruminant animals (US EPA, 2008). In general, landfill CH_4 emissions are decreasing in developed countries (Deuber et al, 2005; US EPA, 2008). Because CH_4 has both a relatively high GWP and a relatively short atmospheric lifetime of about 12 years (Forster et al, 2007), many countries have targeted reductions in landfill CH_4 emissions as part of a strategy to stabilize and reduce atmospheric CH_4 concentrations. Historically, landfill CH_4 has been commercially recovered and utilized since 1975 in many countries to provide a local source of renewable energy (for example Themelis and Ulloa, 2007). In many developed and developing countries, landfilling and other waste management activities are highly regulated; thus, their associated greenhouse gas emissions have also come under public scrutiny with many existing and evolving regulatory, planning, energy-related and financial mechanisms that impact their activities at the local, regional, national and international levels.

Potential mechanisms for landfill CH_4 transport and emissions include diffusion (gaseous flux driven by concentration gradients), advection (flux due to pressure gradients), ebullition (bubbling flux through liquid phase), and flux through plant vascular systems (plant-mediated transport). Similar to other soil and wetland ecosystems, diffusion is a major mechanism for landfill CH_4 emissions with periodic influence of advective processes, wind-driven advection and ebullition. Previous studies have addressed advection induced by wind (Poulsen, 2005) and changing barometric pressure (Latham and Young, 1993; Kjeldsen and Fischer, 1995; Nastev et al, 2001; Christophersen and Kjeldsen, 2001; Christophersen et al, 2001; McBain et al, 2005; Gebert and Gröngröft, 2006). In a combined study of several sites in the northeastern US, Czepiel et al (2003) found a robust inverse linear relationship between landfill CH_4 emissions and barometric pressure that has not generally been replicated by subsequent studies. In many studies, these correlations are weak ($r^2<0.50$, as in the study of McBain et al, 2005) indicating that pressure changes are not the primary factor for emissions. It is generally considered that normal barometric pressure variations would tend to have a net zero effect over the course of a day. Gebert and Gröngröft (2006) observed a reciprocal relationship between the change in pressure and gas release through a passive landfill biofilter system where rising atmospheric pressure could result in a flux reversal with atmospheric air penetrating into the landfill. Some specialized circumstances in landfill settings where pressure gradients can develop are below saturated cover soils (Bogner et al, 1987) or under exposed geomembrane composite covers. Under saturated surface conditions, it may also be important to consider ebullition (bubbling) flux mechanisms, especially at the edge of the landfill footprint. Plant-mediated transport mechanisms may also affect observed fluxes (Chanton, 2005), but these have not been extensively studied in landfill settings.

The single most important strategy to reduce landfill CH_4 emissions is the installation of engineered gas extraction systems using vertical wells or horizontal collectors. The collected landfill gas (40–60 per cent CH_4) can be an important local source of renewable energy for industrial or commercial boilers to provide process heating, for on-site electrical generation using internal combustion engines or gas turbines, or for upgrading (by removal of CO_2 and trace components) to a substitute natural gas. In developing countries, where landfills may have been developed without engineered cells, daily and final cover materials, and engineered control systems for liquids and gases, it can be a substantial challenge to construct efficient landfill gas recovery systems – in many cases, regrading and additional cover material can be required.

In addition to landfill gas recovery, CH_4 emissions are also reduced through the combined effects of the thickness, composition, moisture content and methanotrophic activity of the cover materials. Thus both physical processes (retardation of CH_4 transport rates; increased residence time) and biochemical processes (aerobic CH_4 oxidation rates) act in tandem to reduce emissions. Oxidation rates are dependent on the gross CH_4 flux rate to the base of the cover (largely determined by the gas recovery system), CH_4 residence time in the cover, O_2 diffusion from the atmosphere, as well as soil moisture, temperature and other interrelated variables that affect microbial processes in soils. Both CH_4 emissions and the emissions of other hydrocarbon species such as aromatics and lower chlorinated compounds can be mitigated by optimizing the activity of methanotrophic microorganisms in landfill cover soils (Scheutz et al, 2003, 2008; Barlaz et al, 2004; Bogner et al, 2010).

This review will update and complement previous reviews (Barlaz et al, 1990; Christensen et al, 1996; Bogner et al, 1997b; Scheutz et al, 2009) with emphasis on field measurement of landfill CH_4 emissions and oxidation, modelling of emissions and oxidation, and current trends. We will not extensively review biogenic CH_4 production in landfills as this process is similar to other open ecological settings with respect to substrates, redox, nutrients, pH, toxins and microbial pathways (see Chapter 2). One difference with respect to landfills is the complexity of the organic carbon substrates, encompassing whatever is permitted by regulation and local waste management practices. For detailed information on rates and constraints for landfill methanogenesis, please refer to Halvadakis et al (1983) and Barlaz et al (1989a, 1989b, 1990). In all cases, anaerobic decomposition results in biogenic CH_4 production as the terminal step of a complex series of reactions mediated by hydrolytic, fermentative, acidogenic and acetogenic microorganisms. The stable carbon and hydrogen isotopic signatures of landfill CH_4 (Bogner et al, 1996) indicate that both of the biogenic CH_4 pathways (fermentation/acetate cleavage and CO_2 reduction with H_2) can be important in landfill settings (Harris et al, 2007). In a few cases, field and laboratory data have demonstrated that high temperatures and competitive reactions can suppress methanogenesis, resulting in landfill gas enriched in H_2 and CO_2 relative to CH_4. Moreover, the presence of large quantities of sulphate (especially fine-grained gypsum board in construction and demolition wastes) results in H_2S production via sulphate reduction (for example Fairweather and Barlaz, 1998; Lee et al, 2006).

In general, although there are a wide range of landfill design and operational practices in developed and developing countries, landfill CH_4 processes can be placed in an intermediate position between two end points (Bogner and Lagerkvist, 1997): (1) open methanogenic systems in wetlands and soils (for example Segers, 1998); and (2) highly controlled 'in-vessel' optimized systems for anaerobic digestion (for example D'Addario et al, 1993). In contrast to the uncontrolled anaerobic burial of organic matter in ecosystems and geologic settings, anaerobic pathways in landfills are quickly initiated following large daily additions (tens to thousands of tonnes) of degradable organic carbon substrates. Because both CH_4 emission and oxidation rates in landfill cover soils can span six to seven orders of magnitude as discussed below, the landfill literature extends the dynamic ranges of published rates for soils, wetlands and other ecosystems to the more extreme condition of an 'in-ground' engineered methanogenic system. Moreover, because of the high spatial and temporal variability of both oxidation rates and 'net' emissions to the atmosphere, it is a significant challenge to measure and model these dynamic processes in landfill settings. Alternatively, compared to optimized 'in-vessel' systems for anaerobic digestion, landfills function as relatively inefficient digesters under less-controlled conditions (for example Pohland and Al-Yousfi, 1994; Vieitez and Ghosh, 1999).

For each mole of CH_4 produced in a landfill cell, there are several possible pathways. These pathways can be summarized in a mass balance framework as follows (Bogner and Spokas, 1993) (in which all units = mass time^{-1}):

CH₄ production = CH₄ recovered + CH₄ emitted +
lateral CH₄ migration + CH₄ oxidized + ΔCH₄ storage   (15)

Thus, the relevant pathways for generated CH_4 in a landfill cell include capture by engineered systems for flaring or use as a renewable energy resource; oxidation by aerobic methanotrophic microorganisms in cover soils before emission to the atmosphere; 'net' emissions to the atmosphere; lateral subsurface migration (especially at unlined sites); and temporary internal storage within the landfill volume. As will be discussed below, our understanding of 'net' emissions inclusive of other pathways has improved in recent years as the result of integrated field and laboratory measurement programmes using multiple techniques.

Field measurement of landfill methane emissions and laboratory/field measurements of methane oxidation

Unlike for CH_4 emissions and oxidation in wetlands, soils and rice production systems during the last two decades, there have not been comprehensive regional field campaigns addressing landfill CH_4 emissions. In large part, because landfills are discrete sites dispersed across the landscape, field campaigns to date have focused on specific sites and have rarely included multiple seasons or years. Previous summaries, mostly for small-scale (chamber) measurements, have reported positive emission rates ranging over seven orders of magnitude from $0.0004g$ CH_4 m^{-2} d^{-1} to more than $>4000g$ CH_4 m^{-2} d^{-1} (Bogner et al, 1997b, and references cited therein); Scheutz et al (2009) reported maximum rates of $1755g$ CH_4 m^{-2} d^{-1}. Landfill field studies to date have also reported negative fluxes (uptake of atmospheric CH_4) at rates ranging over six orders of magnitude from -0.000025 to $-16g$ m^{-2} d^{-1} (Bogner et al, 1997b, Scheutz et al, 2009 and references cited therein). Techniques that have been used for field measurement of landfill CH_4 emissions include: (1) surface techniques (static and dynamic chambers); (2) above-ground micro-meteorological (eddy correlation, mass balance), remote sensing (lidar, tunable diode laser (TDL), and static/dynamic tracer techniques; and (3) below-ground gradient techniques (concentration profiles). The advantages and disadvantages of the various techniques for landfill applications have been previously summarized in Bogner et al (1997b), Scheutz et al (2009) and references cited therein. In general, it is important to consider site-specific constraints with respect to application of specific techniques, including variable emission signals from adjacent cells with different cover materials (daily, intermediate, final), complex topography, variable slopes and localized meteorology. Often the use of two or more techniques in combination is recommended, such as the use of an above-ground technique (for whole-cell emissions) with static chambers (to characterize the variability of emissions across a cell).

Typically, published data to date indicate that rates at individual sites can vary spatially over two to four orders of magnitude over short distances (m).

Both 'hot spots' with elevated emissions and points of negative emissions are common at landfills over small spatial scales (Figure 11.1); thus, geostatistical methods must be applied to small-scale chamber results to determine whole-landfill fluxes (for example Graff et al, 2002; Spokas et al, 2003; Abichou et al, 2006a). Often the landfill surface emissions are characterized by minimal to no spatial structure as indicated through semi-variogram models (Graff et al, 2002; Spokas et al, 2003). In most cases, inverse distance weighting (IDW) is the recommended interpolation method for deriving area emissions from chamber results due to the lack of spatial structure that is a prerequisite for other kriging methods (Abichou et al, 2006a; Spokas et al, 2003). However, a large number of individual chamber measurements are required to adequately

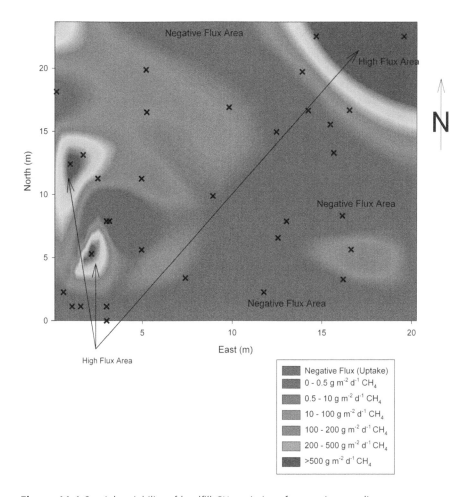

Figure 11.1 Spatial variability of landfill CH_4 emissions from an intermediate cover area at a southern California landfill using static chambers

Note: Note the high positive CH_4 fluxes in close proximity (<5m) to negative fluxes (uptake of atmospheric CH_4)
Source: Unpublished authors' data

address the spatial variability (Webster and Oliver, 1992), which presents challenges for whole-site analyses. The use of a surrogate variable (one that is easier to measure than surface chamber emissions itself) could provide a mechanism through co-kriging to arrive at improved emission estimates if reliable and robust relationships can be drawn between the surrogate variable(s) and the corresponding emissions at a site (for example Ishigaki et al, 2005). Currently, co-kriging methods require further validation.

Explanations for the 'hot spots' (Figure 11.1) include localized areas of thinner cover materials, leakages at interfaces (edge of landfill footprint, localized emissions around gas piping systems requiring maintenance), macropore formation (for example animal burrows or desiccation cracking), and differential settlement from waste inhomogeneities, slope configurations, infiltration characteristics, and variable decomposition rates. Typically at US sites, for example, locations of such sporadic emissions are identified during routine surface monitoring for CH_4 concentrations and then mitigated by follow-up maintenance activities. Imaging of depressions using remote sensing (thermal imagery) has met with mixed success (for example Zilioli et al, 1992; Lewis et al, 2003). Whole-landfill CH_4 emissions measurements reported from Europe, the US and South Africa, also relying predominately on chamber methods, ranged within about one order of magnitude, from approximately 0.1 to 1.0t CH_4 ha^{-1} d^{-1} (Nozhevnikova et al, 1993; Hovde et al, 1995; Czepiel et al, 1996a; Börjesson, 1997; Mosher et al, 1999; Trégourès et al, 1999; Galle et al, 2001; Morris, 2001).

During the last five years, more comprehensive campaigns using multiple techniques at the same site and comparing those techniques across several sites have recently increased our understanding of landfill CH_4 emission rates. Moreover, as discussed below, the deployment of several 'above-ground' techniques in tandem in recent studies have also highlighted the limitations of techniques that are designed for uniform terrain, relatively constant meteorology, and emission sources with less extreme temporal and spatial variability than landfills.

Several years of studies at three French sites involved field-scale measurements of the complete CH_4 mass balance (Equation 15) for nine different landfill cells with varying cover materials and management practices (Spokas et al, 2006, and references cited therein). Figure 11.2 compares modelled landfill gas generation from the nine sites to both measured gas recovery and measured emissions (using both a tracer plume method and dynamic chambers). Note that there is a significant linear relationship between generation and recovery (for the eight sites that had active gas extraction) but no relationship between generation and emissions, with the residual emissions varying over several orders of magnitude. Through a combination of intensive field measurements, supporting laboratory studies, and modelling, high rates of CH_4 recovery were documented at the French sites. For example, only about 1–2 per cent of the CH_4 production is being emitted and about 97 per cent is being recovered with an active gas extraction system at Montreuil-sur-Barse in eastern France (near Troyes). At Lapouyade (near Bordeaux, southwestern France), a minimum of 94 per cent of the CH_4 production was being recovered at two cells with engineered gas recovery, but for a cell without recovery, 92 per cent of the CH_4 production was being emitted (Spokas et al, 2006). Thus, through a combination of intensive field measurements, supporting laboratory studies, and modelling, high rates of CH_4 recovery can be documented at field

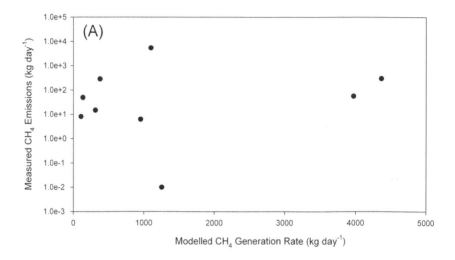

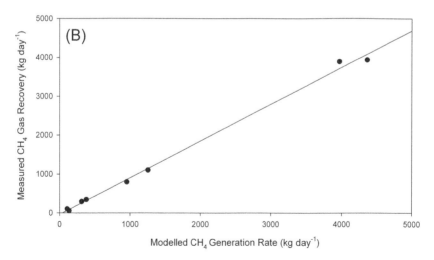

Figure 11.2 Comparison of (A) measured CH_4 emissions and modelled CH_4 generation; and (B) measured CH_4 recovery and modelled CH_4 generation for nine whole-landfill cells in northeastern, western and southwestern France

Note: There is a lack of agreement between measured emissions and modelled generation as opposed to the good agreement between measured recovery and modelled generation.
Source: Based on data from Spokas et al (2006)

scale. Additional mass balance studies are needed to expand our understanding of field-scale processes in various climatic regimes under different management strategies.

A direct field comparison of five above-ground techniques was also completed at the Lapouyade landfill in southwestern France in October 2007 (Babillotte et al, 2008). In addition, the ability of the techniques to quantify flux rates from a blind CH_4 release test on a non-landfill area was also evaluated during the same field campaign. The techniques were (1) vertical radial plume mapping (VRPM) and horizontal radial plume mapping (HRPM) by ARCADIS (US); (2) differential absorption lidar (DIAL) by the National Physics Laboratory (NPL; UK); (3) mobile and static plume methods by ECN (The Netherlands); (4) inverse modelling from vehicle-based spectroscopic measurements by INERIS (France); and (5) airborne laser methane assessment (ALMA), a helicopter-based spectroscopy method by PERGAM (Switzerland). The ALMA/PERGAM method does not, however, yield quantitative measurements of fluxes. In general, the NPL and ARCADIS methods measured emissions from the four cells individually, yielding global site emissions of 12.8 ± 2.9, and 25.2 ± 2.4g sec^{-1}, respectively. The ECN CH_4 reported whole-site dynamic and static plume results of 41 ± 17 and 83 ± 36g sec^{-1} respectively, while the INERIS method reported a whole-site flux of 167.2g sec^{-1}. In addition, a blind test was conducted off the landfill footprint with a CH_4 release of 0.5g sec^{-1} over an area of 100m^2. The comparison among the techniques (in terms of per cent of actual release rate detected by each method) is as follows: NPL (–54 per cent), Arcadis (+78 per cent), ECN (+240 per cent to +360 per cent) and INERIS (+200 per cent to +300 per cent). Both of the Gaussian dispersion methods (ECN and INERIS) significantly overestimated emissions. These results indicated, first, the complexities of landfill emissions measurements (adjacent cells with variable emissions) and some systematic differences among the four measurement methods. In general, there is, as yet, no single method that is appropriate for universal and unambiguous deployment for field measurement of landfill CH_4 emissions.

In late September and early October 2008, Veolia Environmental Services (VES) and Waste Management, Inc. (WMX) collaborated on a comprehensive field comparison campaign at two adjacent landfill sites in southeastern, Wisconsin, US: the WMX Metro site and the VES Emerald Park site (Babillotte et al, 2009). Methane emissions from multiple areas of both sites were measured using multiple techniques, including (1) static chambers by WMX, Florida State University, and Landfills +, Inc.; (2) a micrometeorological method (eddy covariance) by the Finnish Meteorological Institute; (3) VRPM by WMX (in collaboration with ARCADIS, US); (4) DIAL by the National Physics Laboratory (UK); and (5) a mobile plume method using portable Fourier transform infrared (FTIR) spectroscopy by the Swedish company FLUXSENSE, which collaborates with Chalmers University. In addition, the comparative performance of (3)–(5) were evaluated using a blind test with a series of controlled CH_4 release rates on a non-landfill area adjacent to the two sites. The release area encompassed 40 × 40m² within an overall area of 300 × 500m. This area had favourable terrain (flat, uniform) and meteorology (steady wind speed with no contribution from landfill sources). As of the writing of this chapter, only the controlled release results have been published (Babilotte et al, 2009). Four separate trials, each with constant release rates from one to three locations on the test area, were conducted during midday hours (9.00–14.00). All groups were encouraged to optimize their own data collection through technique-specific experimental designs and instrument placement. The release rates (normalized on an area basis) and the results are summarized in Table 11.1. In general, for this trial, the FLUXSENSE mobile plume/portable FTIR method had the most unambiguous results (within 20 per cent of the actual release rate with low standard deviations) by measuring a well-mixed plume about 450m downwind from the release area. However, all three methodologies reported fluxes within about 20–30 per cent of the actual release rate. A major issue associated with the VRPM, which systematically underestimated fluxes, was that the magnitude of the underestimation was a function of the distance between the source and the vertical measurement plane, suggesting that existing models for the area contributing to flux require further refinement. The DIAL method enabled definition of plume and source area but had difficulties with quantification at larger distances, possibly due to measurement interferences. Thus, at the current time, there is no single recommended above-ground technique for field measurement of whole-landfill CH_4 emissions. All of these techniques are being refined by the various research groups for their application to complex area sources such as landfills. In general, because static chambers can quantify the variability of emissions across a particular type of cover material (including locations where negative fluxes, or uptake of atmospheric CH_4, is occurring), their use in parallel with above-ground methods such as those described above is highly recommended.

Methane oxidation for landfill settings has been studied at scales ranging from laboratory batch studies to field-scale measurements. In the field (Figure 11.3), oxidation tends to be optimized at specific depths that vary seasonally and can be observed in soil gas profiles. For small-scale laboratory batch studies, maximum CH_4 oxidation rates have ranged from 0.01 to 117μg CH_4 g^{-1} h^{-1} in landfill cover soils with organic carbon contents of 1.2 to 30 per cent (w/dry w). Q_{10} values ranged from 1.9 to 5.2 over temperature ranges of 2–30°C. Optimum moisture contents were generally below 25 per cent (w/w). For laboratory column experiments simulating landfill cover environments, steady-state CH_4 oxidation rates (aerial basis) ranged from 22 to 210g CH_4 m^{-2} d^{-1} with fractional CH_4 oxidation ranging from 15 to 97 per cent for studies conducted over 30 to >300 days. The use of compost or other highly organic soils in similar column experiments over 35–369 days tended to increase the fractional CH_4 oxidation to a higher range of approximately 70–100 per cent. (Scheutz et al, 2009, and references cited therein). In recent work elucidating the limits and dynamics of CH_4 oxidation in landfill cover soils using batch studies, soils that were pre-incubated with CH_4 for 60 days and with soil moisture potential adjusted to 33kPa (field-holding capacity) exhibited consistently high oxidation rates of 112.1 to 644µg CH_4 g_{soil}^{-1} d^{-1} across all soil types studied (Spokas and Bogner, 2010). These results contrasted with parallel studies of the same soils without pre-incubation and moisture adjustment which exhibited oxidation rates ranging over four orders of magnitude, from 0.9 to 277µg CH_4 g^{-1} day^{-1}. In the same study, the minimum soil moisture threshold for oxidation activity was estimated at approximately 1500kPa. Furthermore, indicating the coupled interaction of soil moisture and temperature, the threshold soil moisture potential for CH_4 oxidation activity shifted to lower soil moisture potentials (higher moisture contents) at extreme temperatures. At <5°C, the minimum soil moisture threshold is approximately 300kPa. At the upper temperature limits of CH_4 oxidation activity (>40°C) the minimum soil moisture threshold is approximately 50kPa.

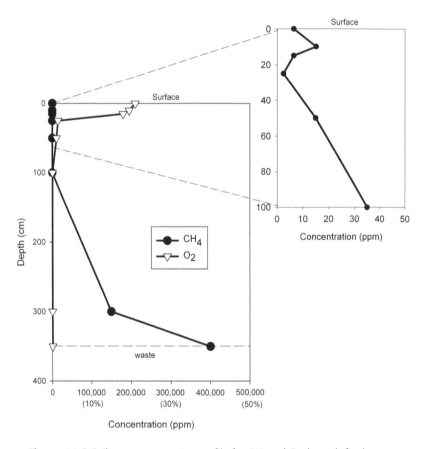

Figure 11.3 Soil gas concentration profile for CH_4 and O_2 through final cover materials at a southern California landfill

Note: Inset graph shows detail for the 0–100cm depth for CH_4 (note decrease at 25cm).
Source: Unpublished authors' data

At field scale, stable carbon isotopic methods, which rely on the difference between the $\delta^{13}C$ of emitted CH_4 and the $\delta^{13}C$ of unoxidized CH_4 in the anaerobic zone, provide the most robust approach to date for the quantification of fractional CH_4 oxidation, that is, the percentage of CH_4 that is oxidized during transport through the landfill cover materials. Isotopic methods have been developed over the last decade and rely on the preference of methanotrophs for the isotope of smaller mass, ^{12}C rather than ^{13}C, according to one or more fractionation factor(s) dependent on soil properties and gaseous transport considerations (Liptay et al, 1998; Chanton et al, 1999; Chanton and Liptay, 2000). Methanotrophic bacteria will oxidize $^{12}CH_4$ at a slightly more rapid rate than $^{13}CH_4$. In general, the isotopic methods can be applied as follows: (1) at ground level, using static chambers and comparing the $\delta^{13}C$ of CH_4 in the refuse (by sampling at gas recovery wells, gas collection headers, or deep gas probes) to the emitted CH_4 collected in the chamber; (2) in the lower atmosphere, relying on a comparison between the $\delta^{13}C$ of atmospheric CH_4 in an upwind transect and a downwind transect; and (3) below ground level, relying on soil gas profiles for CH_4 and $\delta^{13}C$. In general, $\delta^{13}C$ values for unoxidized CH_4 in the anaerobic zone range from about –57 to –60; with oxidation, these values can undergo a positive shift to –35 or more (Bogner et al, 1996, and references cited therein). An additional possibility is to use combined $\delta^{13}C$ and δD methods since the hydrogen isotope has a larger relative fractionation factor than does the carbon. Recent studies have addressed these issues as well as expanded methods, models and uncertainties for isotopic approaches to CH_4 oxidation in landfill cover soils (De Visscher et al, 2004; Mahieu et al, 2006; Chanton et al, 2008).

It is expected that future methodological and modelling improvements will permit more robust quantification of incremental CH_4 oxidation through landfill cover soils. In situations where static chamber measurements yield 'negative' fluxes, indicating that methanotrophs are capable of oxidizing all of the CH_4 transported from the landfill below and also oxidizing additional CH_4 out of the atmosphere, the stable carbon isotopic methods are not applicable. In these cases, the static chamber is quantifying the rate of atmospheric CH_4 oxidation/uptake, and data should be reported accordingly. As discussed above, where fluxes are positive, isotopic measurements comparing the $\delta^{13}C$ for emitted CH_4 compared to CH_4 in the anaerobic zone can be used to quantify the fractional CH_4 oxidation – typically, fractional CH_4 oxidation is significantly higher than the 10 per cent allowed by IPCC national inventory methodologies for annual reporting (IPCC, 2006; Chanton et al, 2008), based on Czepial et al (1996b), as discussed in more detail below. A recent review reported that the average percentage of CH_4 oxidized in cover soil was 35 ± 6 per cent across a variety of landfill sites around the globe (Chanton et al, 2009).

Some researchers have relied on $CH_4:CO_2$ ratios as an indicator of CH_4 oxidation in landfill soil gas profiles. This approach is not recommended because, in addition to the production of CO_2 via methanogenesis and CH_4 oxidation, CO_2 is also produced and consumed by many other subsurface and

Table 11.1 Field comparison of three techniques for landfill CH_4 emissions: Deviation between measured CH_4 release rate and controlled CH_4 release during blind field test (Wisconsin, October 2008)

Total CH_4 release rates (range for trials 1–4; each trial with 1–3 release points)	Deviation of total measured CH_4 flux versus total rate of controlled CH_4 release (range for trials 1–4)		
	FLUXSENSE (mobile plume/FTIR)	WMX/Arcadis (VRPM)	NPL (DIAL)
1.01–3.28g CH_4 sec^{-1} released from 1–3 points on 40 × 40m area in 300 × 500m evaluation area	+ 4 to +19%	−3 to −21% for closest vertical plane −42 to 48% for farthest vertical plane	−21 to +19%
(approx. 59–177g CH_4 m^{-2} d^{-1} for 40 × 40m release area only, or 0.06–0.19g CH_4 m^{-2} d^{-1} for 300 × 500m evaluation area)	SD for technique 4–9%	SD for technique 18–33%	SD for technique 24–31%

Note: Wind speed ranged from 3.8 to 4.4m s^{-1}. SD = standard deviation.
Source: Babillotte et al (2009)

near-surface processes (soil respiration, organic matter oxidation, activity of soil invertebrates, photosynthesis) (Kuzyakov, 2006). In fact, from laboratory incubations, no significant correlation was observed between CH_4 oxidation and the corresponding CO_2 production in peat soils (Moore and Dalva, 1997). A further complication is that CO_2 is more soluble than CH_4 and thereby more readily partitioned to soil moisture (Stumm and Morgan, 1987; Zabowski and Sletten, 1991). In non-landfill soils, elevated CO_2 soil gas concentrations have also been observed to suppress microbial respiration (Koizumi et al, 1991).

During the last few years, the development of robust engineering designs for landfill 'biocovers' has indicated that the limits of natural CH_4 oxidation in landfill settings could be expanded with engineered solutions (Barlaz et al, 2004; Huber-Humer, 2004; Abichou et al, 2006b; Stern et al, 2007; Bogner et al, 2010) In general, the common elements of landfill biocovers have included a gas dispersion layer of coarse materials above the waste (to reduce the variability in CH_4 fluxes to the base of the cover) overlain by compost or other organic substrates which have been shown to have high CH_4-oxidizing capacity. Issues requiring further investigation include the possibility of in situ methanogenesis under saturated conditions in organic cover materials, as well as N_2O emissions from N-enriched organic materials such as sewage sludge compost. Incidentally, N_2O emissions from landfills are considered an insignificant source globally (Bogner et al, 1999; Rinne et al, 2005). However,

N_2O emissions may be locally important where there is abundant N, high moisture, and restricted aeration; this includes cover soils amended with sewage sludge (Börjesson and Svensson, 1997) or where aerobic or semi-aerobic landfilling practices are implemented (Tsujimoto et al, 1994).

At larger scales, there have been few research studies that have attempted to quantity the contribution of landfill CH_4 to regional air quality. In large part this is due to the dispersed nature of landfill sources and the complexity of multiple CH_4 sources in most areas, including wide ranges in emission rates and overlapping isotopic signatures with other sources (such as wetlands). A 'top-down' study for sources of atmospheric CH_4 in London used stable carbon isotopes to model the relative contribution of several potential sources, including landfill CH_4 (Lowry et al, 2001). Zhao et al (2009) estimated the landfill contribution to atmospheric CH_4 for a region in central California. However, directly quantifying the contribution of landfill CH_4 to regional atmospheric CH_4 remains a significant challenge for the future.

Existing and evolving tools and models for landfill methane generation, oxidation and emissions

Historically, at the time of the first full-scale landfill gas recovery and utilization projects (1975–1980), predictive tools were developed to estimate the theoretical quantity of recoverable CH_4 for commercial gas utilization projects (for example see EMCON Associates, 1980). These tools generally had two forms: (1) empirical 'rule of thumb' calculations for the annual recoverable CH_4 based on the mass of waste in place, or (2) theoretical first-order kinetic models of various forms that were used to predict the annual expected CH_4 generation and recovery from annual landfilling rates. The original first-order models were site-specific. These models were fine-tuned and field validated with values for CH_4 yields and kinetic coefficients that could replicate annual CH_4 recovery at particular sites (for example the Scholl Canyon model for a southern California site). It is important to emphasize that, at that time, the focus was on tools for predicting commercially recoverable CH_4 with the tacit assumption that both fugitive emissions and lateral migration would be minimized when active gas extraction was implemented.

From the mid- to late 1980s and into the early 1990s, there were still very few field measurements for landfill CH_4 emissions in the published literature, although chamber techniques, as well as calculations based on gas concentration profiles and pressure gradients, were beginning to be applied. However, there were two developments that led to the extensive use of the first-order kinetic models (with default values) to be applied specifically to landfill CH_4 emissions, as opposed to their previous application to CH_4 generation and recovery. These two developments included the mandated implementation of new comprehensive air-quality regulatory programmes in many developed countries and the 1988 formation of the IPCC, which, by 1991, included the development of the IPCC National Greenhouse Gas Inventories Programme (IPCC-NGGIP). The thinking at the time was that tools based on first-order kinetic models for landfill CH_4 generation provided a conservative value (i.e. overestimate) for landfill CH_4 emissions for both purposes. In most cases, the 'validation' of the modelling approaches involved a comparison of modelled results to landfill gas recovery data (Peer et al, 1993). Comprehensive Dutch studies in the early 1990s examined a variety of single and multi-component zero-order, first-order and second-order models, concluding that a multi-component first-order model (validated using data from nine full-scale Dutch landfills) was able to model gas recovery within 30–40 per cent of the actual measured recovery (Van Zanten and Scheepers, 1994).

In general, the IPCC Programme requires the development of methods and reporting of national emissions for UNFCCC member countries, with annual reporting requirements for the highly developed countries and less frequent reporting requirements for other countries. As landfills were recognized as one of the many sources of atmospheric CH_4 (with early emissions estimates as high as 70Tg CH_4 yr^{-1}: Bingemer and Crutzen, 1987), it was therefore necessary to develop, in the absence of comprehensive field measurement programmes, standardized methods for estimating national landfill CH_4 emissions. The thinking at the time was that both simple mass balance tools (where CH_4 generation was assumed to occur in the same year as waste disposal) and first-order kinetic models with conservative default values (which were more suitable for developed countries with available waste data) could be used to provide conservative (high) estimates for CH_4 emissions to the atmosphere.

The comprehensive 1996 IPCC National Inventory Revised Guidelines (IPCC, 1996) allowed both the use of a simple mass balance method as the Tier I default method as well as a Tier II first-order kinetic model (called FOD, or first-order decay model in this context). There were country-to-country differences in the national FOD models deployed under Tier II. Moreover, by 1996, the first study had been published which estimated annual fractional CH_4 oxidation – Czepiel et al (1996b), working at the 17ha Nashua New Hampshire landfill without engineered gas recovery, calculated a value of 11 per cent for the Nashua site, based on field measurements of emissions, supporting laboratory incubations for CH_4 oxidation, and the application of an annual climatic model. Thus, for developed countries, the 1996 guidelines allowed two subtractions from the estimated annual landfill CH_4 generation: a 10 per cent subtraction for CH_4 oxidation and the reported landfill CH_4 recovery for either flaring or landfill gas utilization projects.

The latest 2006 National Inventory Revised Guidelines include a standardized Tier I multi-component FOD model for landfill CH_4 emissions (EXCEL) with default input values for various waste fractions for landfills over a wide range of climatic conditions. Tier II encourages country-specific values. Tier III permits scaling-up of site-specific studies to regional estimates. Tier IV allows more complex site-specific modelling tools. The fraction of landfilled organic carbon that is biodegradable but not expected to degrade in the landfill (typically a conservative estimate of 50 per cent: Bogner, 1992; Barlaz, 1998) is reported as an information item within the waste sector, but credited to the harvested wood products sector (IPCC, 2006).

Some caveats are in order. In general, although the regulatory first-order kinetic models and the current Tier I/Tier II IPCC methodologies function as standardized tools, they cannot, without modification, be expected to replicate CH_4 generation or recovery at specific field sites. Historically, first-order kinetic models have continued to be applied for initial estimates for commercial landfill gas recovery projects, then fine-tuned with respect to actual gas recovery when data are available from full-scale systems. Within Kyoto compliance mechanisms, the Tier I FOD models are widely applied for 'baseline' estimates within the approved consolidated landfill gas methodology (ACM0001) for the Kyoto Protocol Clean Development Mechanism (CDM). This 'flexible mechanism' permits entities in developed countries with Kyoto obligations to financially support greenhouse gas reduction projects in developing countries. However, it is important to emphasize that for landfill gas CDM projects, the actual CERs (certified emission reductions) are credited not on the model results but on rigorous site-specific quantification of the CH_4 recovered and destroyed. However, there are also several 'avoided methane to landfill' CDM methodologies where CERs from alternative waste management projects (composting, incineration, mechanical-biological treatment) are credited solely on the basis of the waste composition and the model results. Given the many issues with the site-specific application of the FOD models for these 'avoided methane' applications, these methodologies need to be reconsidered with regard to their application after the end of the first Kyoto commitment period (2012).

Numerous field and laboratory studies during the last decade have improved our understanding of the actual range of landfill CH_4 emission rates but added to the complexity of interpreting combined CH_4 emissions and oxidation processes at specific sites within the context of Equation 15. With respect to oxidation, specifically, the available modelling tools for landfill applications were recently summarized by De Visscher and Spokas, as contributors to Scheutz et al (2009). In general, CH_4 oxidation in landfill cover soils is a complex process involving simultaneous transport and microbial oxidation processes. Several types of models currently exist. These include simulation models for CH_4 oxidation (for example Czepiel et al, 1996b) where the model simulates properties which cannot be measured directly. Also, existing models can be used to better understand specific aspects of oxidation processes in landfill cover soils; for example, Hilger et al (1999) and Wilshusen et al (2004) used models to understand how the formation of EPSs (extra-polymeric substances) affects transport and oxidation processes. Mahieu et al (2005) used models to understand stable isotope fractionation effects. Rannaud et al (2007) used a model to estimate the depth of CH_4 oxidation. Finally, models can be used for prediction or design (for example De Visscher and Van Cleemput, 2003; Park et al, 2004; Rannaud et al, 2007). Typically, existing models have been calibrated with laboratory data and then used to predict or better understand field conditions.

Existing CH_4 oxidation models for landfill settings generally fall into three types: empirical models, process-based models and the collision model. Empirical models (for example Czepiel et al, 1996b; Park et al, 2004) are assemblies of empirical equations based on measured data. These require little information on the fundamental microbial and transport processes but should not be extrapolated to other sites where field measurements are lacking. Process-based models (for example Hilger et al, 1999; Stein et al, 2001) combine mass transport equations with CH_4 oxidation kinetics in a numerical scheme. Process-based models are potentially the most realistic models, but they generally require a large number of parameters that are difficult to obtain in field settings, which limits their usability. The collision model is an attempt to balance theoretical and empirical modelling by representing the processes occurring in a landfill cover as collisions between gas molecules and the soil (Bogner et al, 1997a).

The latest generation of landfill emission models is departing from the modelling of CH_4 generation to simulating the actual soil and microbial processes controlling emissions. Currently, a model is being developed to provide an improved field-validated methodology for landfill CH_4 emissions in California in the context of the state greenhouse gas inventory (Bogner et al, 2009; Spokas et al, 2009). This model is a Tier IV model under the current IPCC National Inventory Guidelines (2006) because it uses more complex site-specific methods, after which the results are summed to provide regional totals. This model also accounts for seasonal differences in climatic variables (air temperature, precipitation, solar radiation) and soil microclimate (soil temperature and moisture for layered soils) with respect to their quantitative influence on the seasonality of CH_4 oxidation rates in cover soils. Using this model, Figure 11.4 illustrates the comparative effect of seasonal oxidation on CH_4 emissions for a 60cm clay soil cover at a hypothetical Midwestern US landfill.

Conclusions, trends and broader perspectives

Landfill CH_4 is a small contributor to global anthropogenic greenhouse gas emissions. The broader implementation of cost-effective landfill gas recovery and utilization systems can achieve additional reductions in landfill CH_4 emissions. Moreover, landfill gas utilization can assist local communities by providing renewable energy benefits to offset fossil fuel use. Methanotrophic CH_4 oxidation in cover soils, in association with the transport limitations provided by the cover materials, provides a secondary biological control on emissions that is, of course, limited by the capacity of soils to develop and continuously maintain methanotrophic consortia under field conditions. Enhanced ('biocover') methanotrophic oxidation in cover soils can be assisted

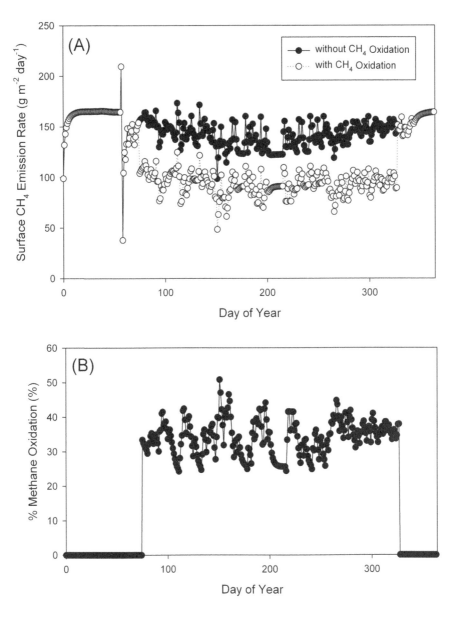

Figure 11.4 Model simulations for landfill CH_4 emissions with and without CH_4 oxidation using the model described in Spokas et al (2009): (A) variability in the seasonal methane emissions for a hypothetical Midwestern US landfill (Chicago, IL) with and without methane oxidation; and (B) the resulting percentage of CH_4 oxidized in the cover material

Note: The cover was assumed to consist of 60cm clay loam with an active gas recovery system. Methane concentration at the base of the cover was assumed to be 45 per cent. The spike in the emissions at day 57 was due to soil thawing.
Source: Unpublished author's data

by site-specific designs to promote oxidation in cover soils. Other important considerations to limit emissions include landfill design, operations and maintenance practices including, where feasible, the implementation of economical horizontal gas collection systems concurrent with filling. In both the EU and the US, annual national landfill CH_4 emissions have decreased since 1990 (Deuber et al, 2005; US EPA, 2008). This is due to increased landfill CH_4 recovery and utilization as well as policy measures to reduce the mass of biodegradable waste landfilled (such as the EU landfill directive 1999/31/EC). For developing countries, however, rates of landfill CH_4 emissions are expected to increase concurrently with increased landfilling. This increase could be mitigated through increased rates of CH_4 recovery via incentives such as the current CDM and anticipated post-Kyoto policies and measures expected to be implemented during the next two to three years.

In addition to the direct reduction of landfill CH_4 emissions via engineered gas recovery systems, the retardation of emissions by engineered cover systems, and the optimization of in situ rates of methanotrophic CH_4 oxidation, one must also be aware of complementary waste management practices to reduce emissions. These include: (1) alternative waste management practices such as composting or incineration that reduce landfill CH_4 generation; and (2) recycling, reuse and waste minimization practices that reduce waste generation. In the context of integrated waste management, it is important to preserve choices for local officials to make informed decisions regarding local waste management practices. Those decisions can benefit from consideration of multiple technical and non-technical issues, including waste quantities and characteristics, local costs and financing issues, regulatory constraints, and infrastructure requirements including available land area, collection and transport considerations. Most waste sector technologies are mature, cost-effective and bestow significant co-benefits for public health and environmental protection (Bogner et al, 2009).

In addition to CH_4 transport and oxidation, there are numerous other C and N cycle processes in landfill cover soils which influence observed soil gas profiles and measured fluxes. These include aerobic respiration (resulting in CO_2 production and flux), plant photosynthesis (CO_2 uptake) and, especially where N is abundant, nitrification and denitrification processes that can produce gaseous intermediates of soil N cycling (N_2O, NO, N_2). Although these processes have been infrequently studied in landfill settings, nevertheless, they further complicate field measurement and the understanding of gaseous fluxes and CH_4 oxidation in landfill settings. At vegetated sites with an engineered gas recovery system and low CH_4 fluxes, the observed CO_2 flux may be dominated by root zone respiration rather than CH_4 oxidation or the direct transport of landfill gas (Bogner et al, 1997a, 1999). For field studies where it is important to quantity the various CO_2 fluxes, one can choose among a variety of applicable techniques (for example, see Panikov and Gorbenko, 1992).

Field studies to date have quantified landfill CH_4 emissions at a limited number of humid, temperate, semi-arid and subtropical sites, but extensive emission measurements in tropical environments are generally lacking. In addition, there have been relatively few comprehensive field campaigns over complete annual cycles, none of which have used multiple methods for emissions. Existing data have shown that emission and oxidation rates vary spatially and temporally over several orders of magnitude. Therefore, additional field data and improved modelling tools, including models currently under development that incorporate seasonal meteorological and soil microclimate variability (Spokas et al, 2009), are needed for better understanding and predictability of emissions over various spatial scales.

References

Abichou, T., Chanton, J., Powelson, D., Fleiger, J., Escoriaza, S., Lei, Y. and Stern, J. (2006a) 'Methane flux and oxidation at two types of intermediate landfill covers', *Waste Management*, vol 26, no 11, pp1305–1312

Abichou, T., Mahieu, K., Yuan, L., Chanton, J. and Hater, G. (2006b) 'Effects of compost biocovers on gas flow and methane oxidation in a landfill cover', *Waste Management*, vol 29, pp1595–1601

Babillotte, A., Lagier, T., Taramni, V. and Fianni, E. (2008) 'Landfill methane fugitive emissions metrology field comparison of methods', in *Proceedings from the 5th Intercontinental Landfill Research Symposium*, www.ce.ncsu.edu/iclrs/Presentations.html.

Babillotte, A., Green, R., Hater, G. and Watermolen, T. (2009) 'Field intercomparison of methods to measure fugitive methane emissions on landfills', in *Proceedings from the First International Greenhouse Gas Measurement Symposium*, 22–25 March 2009, San Francisco, CA, Air & Waste Management Association, Pittsburgh, PA

Barlaz, M. (1998) 'Carbon storage during biodegradation of municipal solid waste components in laboratory-scale landfills', *Global Biogeochemical Cycles*, vol 12, pp373–380

Barlaz, M. A., Schaefer, D. M. and Ham, R. K. (1989a) 'Inhibition of methane formation from municipal refuse in laboratory scale lysimeters', *Applied Biochemistry and Biotechnology*, vol 20/21, pp181–205

Barlaz, M. A., Schaefer, D. M. and Ham, R. K. (1989b) 'Bacterial population development and chemical characteristics of refuse decomposition in a simulated sanitary landfill', *Applied Environmental Microbiology*, vol 55, pp55–65

Barlaz, M. A., Ham, R. K. and Schaefer, D.M. (1990) 'Methane production from municipal refuse: A review of enhancement techniques and microbial dynamics', *CRC Critical Reviews in Environmental Control*, vol 19, no 6, pp557–584

Barlaz, M. A., Green, R. B., Chanton, J. P., Goldsmith, C. D. and Hater, G. R. (2004) 'Evaluation of a biologically active cover for mitigation of landfill gas emissions', *Environmental Science and Technology*, vol 38, pp4891–4899

Bingemer, H. G. and Crutzen, P. J. (1987) 'The production of CH_4 from solid wastes', *Journal of Geophysical Research*, vol 92, no D2, pp2182–2187

Bogner, J. (1992) 'Anaerobic burial of refuse in landfills: Increased atmospheric methane and implications for increased carbon storage', *Ecological Bulletin*, vol 42, pp98–108

Bogner, J. and Lagerkvist, A. (1997) 'Organic carbon cycling in landfills: models for a continuum approach', in *Proceedings of Sardinia '97 International Landfill Symposium*, published by CISA, University of Cagliari, Sardinia

Bogner, J. and Spokas, K. (1993) 'Landfill CH_4: Rates, fates, and role in global carbon cycle', *Chemosphere*, vol 26, pp366–386

Bogner, J., Vogt, M., Moore, C. and Gartman, D. (1987) 'Gas pressure and concentration gradients at the top of a landfill', in *Proceedings of GRCDA 10th International Landfill Gas Symposium* (West Palm Beach, Florida), Governmental Refuse Collection and Disposal Association, Silver Spring, MD

Bogner, J., Sweeney, R., Coleman, D., Huitric, R. and Ririe, G. T. (1996) 'Using isotopic and molecular data to model landfill gas processes', *Waste Management and Research*, vol 14, pp367–376

Bogner, J., Spokas, K. and Burton, E. (1997a) 'Kinetics of methane oxidation in landfill cover materials: Major controls, a whole-landfill oxidation experiment, and modeling of net methane emissions', *Environmental Science and Technology*, vol 31, pp2504–2614

Bogner, J., Meadows, M. and Czepiel, P. (1997b) 'Fluxes of methane between landfills and the atmosphere: Natural and engineered controls', *Soil Use and Management*, vol 13, pp268–277

Bogner, J., Spokas, K. and Burton, E. (1999) 'Temporal variations in greenhouse gas emissions at a midlatitude landfill', *Journal of Environmental Quality*, vol 28, pp278–288

Bogner, J., Abdelrafie-Ahmed, M., Diaz, C., Faaij, A., Gao, Q. Hashimoto, S., Mareckova, K., Pipatti, R. and Zhang, T. (2007) 'Waste Management', in B. Metz, O. R. Davidson, P. R. Bosch, R. Dave and L. A. Meyer (eds) *Climate Change 2007: Mitigation. Contribution of Working Group III to the Fourth Assessment Report of the Intergovernmental Panel on Climate Change*, Cambridge University Press, Cambridge, UK and New York, NY

Bogner, J., Spokas, K., Chanton, J., Franco, G. and Young, S. (2009) 'A new field-validated inventory methodology for landfill methane emissions' in *Proceedings of SWANA Landfill Gas Symposium* (Atlanta, Georgia), Solid Waste Association of North America, Silver Spring, MD

Bogner, J., Chanton, J., Blake, D., Abichou, T. and Powelson, D. (2010) 'Effectiveness of a Florida landfill biocover for reduction of CH_4 and NMHC Emissions', *Environmental Science and Technology* (in press)

Börjesson, G. (1997) 'Methane oxidation in landfill cover soils', PhD thesis, Swedish University of Agricultural Sciences, Uppsala, Sweden

Börjesson, G. and Svensson, B. (1997) 'Nitrous oxide release from covering soil layers of landfills in Sweden', *Tellus B*, vol 49, pp357–363

Chanton, J. P. (2005) 'The effect of gas transport on the isotope signature of methane in wetlands', *Organic Chemistry*, vol 36, pp753–768

Chanton, J. P. and Liptay, K. (2000) 'Seasonal variation in methane oxidation in landfill cover soils as determined by an in situ stable isotope technique', *Global Biogeochemical Cycles*, vol 14, pp51–60

Chanton, J. P., Rutkowski, C. M. and Mosher, B. M. (1999) 'Quantifying methane oxidation from landfills using stable isotope analysis of downwind plumes', *Environmental Science and Technology*, vol 33, pp3755–3760

Chanton, J. P., Powelson, D. K., Abichou, T. and Hater, G. (2008) 'Improved field methods to quantify methane oxidation in landfill cover materials using stable carbon isotopes', *Environmental Science and Technology*, vol 42, pp665–670

Chanton, J. P., Powelson, D. K. and Green, R. B. (2009) 'Methane oxidation in landfill cover soils, is a 10 per cent default value reasonable?', *Journal of Environmental Quality*, vol 38, pp654–663

Christensen, T. H., Kjeldsen, P. and Lindhardt, B. (1996) 'Gas-generating processes in landfills', in T. H. Christensen, R. Cossu, and R. Stegmann (eds) *Landfilling of Waste: Biogas*, E & FN Spoon, London, pp27–50

Christophersen, M. and Kjeldsen, P. (2001) 'Lateral gas transport in soil adjacent to an old landfill: Factors governing gas migration', *Waste Management and Research*, vol 19, no 2, pp144–159

Christophersen, M., Holst, H., Chanton, J. and Kjeldsen, P. (2001) 'Lateral gas transport in soil adjacent to an old landfill: Factors governing emission and methane oxidation', *Waste Management and Research*, vol 19, pp595–612

Czepiel, P. M., Mosher, B., Harriss, R., Shorter, J. H., McManus, J. B., Kolb, C. E., Allwine, E. and Lamb, B. (1996a) 'Landfill CH_4 emissions measured by enclosure and atmospheric tracer methods', *Journal of Geophysical Research*, vol 101, pp16711–16719

Czepiel, P. M., Mosher, B., Crill, P. M. and Harriss, R. C. (1996b) 'Quantifying the effect of oxidation on landfill methane emissions', *Journal of Geophysical Research*, vol 101, pp16721–16729

Czepiel, P. M., Shorter, J. H., Mosher, B., Allwine, E., McManus, J. B., Harriss, R. C., Kolb, C. E. and Lamb, B. K. (2003) 'The influence of atmospheric pressure on landfill methane emissions', *Waste Management*, vol 23, pp593–598

D'Addario, E., Pappa, R., Pietrangel, B. and Valdiserri, M. (1993) 'The acidogenic digestion of the organic fraction of municipal solid waste for the production of liquid fuels', *Water Science and Technology*, vol 27, no 2, pp92–183

Deuber, O., Cames, M., Poetzsch, S. and Repenning, J. (2005) 'Analysis of greenhouse gas emissions of European countries with regard to the impact of policies and measures', report by Öko-Institut to the German Umweltbundesamt, Berlin

De Visscher, A. and Van Cleemput, O. (2003) 'Simulation model for gas diffusion and methane oxidation in landfill cover soils', *Waste Management*, vol 23, pp581–591

De Visscher, A., De Pourcq, I. and Chanton, J. (2004) 'Isotope fractionation effects by diffusion and methane oxidation in landfill cover soils', *Journal of Geophysical Research-Atmospheres*, vol 109, doi:10.1029/2004JD004857

EMCON Associates (1980) *Methane Generation and Recovery from Landfills*, Ann Arbor Science Publishers, Ann Arbor, Michigan, US

Fairweather, R. and Barlaz, M. (1998) 'Hydrogen sulfide production during decomposition of landfill inputs', *Journal of Environmental Engineering*, vol 124, pp353–361

Forster, P., Ramaswamy, V., Artaxo, P., Berntsen, T., Betts, R., Fahey, D. W., Haywood, J., Lean, J., Lowe, D. C., Myhre, G., Nganga, J., Prinn, R., Raga, G. M. S. and Van Dorland, R. (2007) 'Changes in atmospheric constituents and in radioactive forcing', in S. Solomon, D. Qin, M. Manning, Z. Chen, M. Marquis, K. B. Averyt, M. Tignor and H. L. Miller (eds) *Climate Change 2007: The Physical*

Science Basis. Contribution of Working Group I to the Fourth Assessment Report of the Intergovernmental Panel on Climate Change, Cambridge University Press, Cambridge, UK and New York, NY

Galle, B., Samuelsson, J., Svensson, B. and Börjesson, G. (2001) 'Measurements of CH_4 emissions from landfills using a time correlation tracer method based on FTIR absorption spectroscopy', *Environmental Science and Technology*, vol 35, no 1, pp21–25

Gebert, J. and Gröngröft, A. (2006) 'Passive landfill gas emission – influence of atmospheric pressure and implications for the operation of methane-oxidising biofilters', *Waste Management*, vol 26, pp245–251

Graff, C., Spokas, K. and Mercet, M. (2002) 'The use of geostatistical models in the determination of whole landfill emission rates', in *Proceedings of the 2nd Intercontinental Landfill Symposium*, Asheville, NC, 13–16 October 2002

Halvadakis, C. P., Robertson, A. P. and Leckie, J. O. (1983) 'Landfill methanogenesis: Literature review and critique', Technical Report No. 271, Dept. of Civil Engineering, Stanford University, Stanford, California

Harris, S., Smith, R. and Suflita, J. (2007) 'In situ hydrogen consumption kinetics as an indicator of subsurface microbial activity', *FEMS Microbiology Ecology*, vol 60, pp220–228

Hilger, H. A., Liehr, S. K. and Barlaz, M. A. (1999) 'Exopolysaccharide control of methane oxidation in landfill cover soil', *Journal of Environmental Engineering*, vol 125, pp1113–1123

Hovde, D. C., Stanton, A. C., Meyers, T. P. and Matt, D. R. (1995) 'CH_4 emissions from a landfill measured by eddy correlation using a fast-response diode laser sensor', *Journal of Atmospheric Chemistry*, vol 20, pp141–162

Huber-Humer, M. (2004) 'Abatement of landfill methane emissions by microbial oxidation in biocovers made of compost', PhD thesis, University of Natural Resources and Applied Life Sciences Vienna, Austria

IPCC (Intergovernmental Panel on Climate Change) (1996) *1996 IPCC Guidelines for National Greenhouse Gas Inventories*, IPCC/IGES, Hayama, Japan

IPCC (2006) *2006 IPCC Guidelines for National Greenhouse Gas Inventories*, IPCC/IGES, Hayama, Japan

Ishigaki, T., Yamada, M., Nagamori, M., Ono, Y. and Inoue, Y. (2005) 'Estimation of methane emission from whole waste landfill site using correlation between flux and ground temperature', *Environmental Geology*, vol 48, pp845–853

Kjeldsen, P. and Fischer, E. V. (1995) 'Landfill gas migration – Field investigations at Skellingsted landfill, Denmark', *Waste Management and Research*, vol 13, pp467–484

Koizumi, H., Nakadai, T., Usami, Y., Satoh, M., Shiyomi, M. and Oikawa, T. (1991) 'Effect of carbon dioxide concentration on microbial respiration in soil', *Ecology Research*, vol 6, pp227–232

Kuzyakov, Y. (2006) 'Sources of CO_2 efflux from soil and review of partitioning methods', *Soil Biology and Biochemistry*, vol 38, no 3, pp 425–448

Latham, B. and Young, A. (1993) 'Modellisation of the effects of barometric pressure on landfill gas migration', in T. H. Christensen, R. Cossu and R. Stegmann (eds) *Proceedings of Sardinia '93: Fourth International Landfill Symposium*, CISA, Environmental Sanitary Engineering Centre, Cagliari, Italy

Lee, S., Xu, Q., Booth, M., Townsend, T., Chadik, P. and Bitton, G. (2006) 'Reduced sulfur compounds in gas from construction and demolition debris landfills', *Waste Management*, vol 26, pp526–533

Lewis, A. W., Yuen, S. T. S. and Smith, A. J. R. (2003) 'Detection of gas leakage from landfills using infrared thermography – applicability and limitations', *Waste Management and Research*, vol 21, pp436–447

Liptay, K., Chanton, J., Czepiel, P. and Mosher, B. (1998) 'Use of stable isotopes to determine methane oxidation in landfill cover soils', *Journal of Geophysical Research*, vol 103, pp8243–8250

Lowry, D., Holmes, C. W., Rata, N. D., O'Brien, P. and Nisbet, E. G. (2001) 'London methane emissions: Use of diurnal changes in concentration and δ13C to identify urban sources and verify inventories', *Journal of Geophysical Research*, vol 106, no D7, pp7427–7248

Mahieu, K., De Visscher, A., Vanrolleghem, P. A. and Van Cleemput, O. (2005) 'Improved quantification of methane oxidation in landfill cover soils by numerical modelling of stable isotope fractionation', in T. H. Christensen, R. Cossu and R. Stegmann (eds) *Proceedings of Sardinia '05: Tenth International Waste Management and Landfill Symposium (3–7 October 2005)*, CISA, Environmental Sanitary Engineering Centre, Cagliari, Italy

Mahieu, K., De Visscher, A., Vanrolleghem, P. A. and Van Cleemput, O. (2006) 'Carbon and hydrogen isotope fractionation by microbial methane oxidation: Improved determination', *Waste Management*, vol 26, pp389–398

McBain, M. C., McBride, R. A. and Wagner-Riddle, C. (2005) 'Micrometeorological measurements of N_2O and CH_4 emissions from a municipal solid waste landfill', *Waste Management and Research*, vol 23, pp409–419

Monni, S., Pipatti R., Lehtilä A., Savolainen, I. and Syri S. (2006) *Global Climate Change Mitigation Scenarios for Solid Waste Management*, VTT Publications, Technical Research Centre of Finland, Espoo

Moore, T. R. and Dalva, M. (1997) 'Methane and carbon dioxide exchange potentials of peat soils in aerobic and anaerobic laboratory incubations', *Soil Biology and Biochemistry*, vol 29m, no 8, pp1157–1164

Morris, J. (2001) 'Effects of waste composition on landfill processes under semi-arid conditions', PhD thesis, University of the Witwatersrand at Johannesburg, South Africa

Mosher, B. W., Czepiel, P., Harriss, R., Shorter, J., Kolb, C., McManus, J. B., Allwine, E. and Lamb, B. (1999) 'CH_4 emissions at nine landfill sites in the northeastern United States', *Environmental Science and Technology*, vol 33, no 12, pp2088–2094

Nastev, M., Therrien, R., Lefebvre, R. and Gélinas, P. (2001) 'Gas production and migration in landfills and geological materials', *Journal of Contaminant Hydrology*, vol 52, pp187–211

Nozhevnikova, A. N., Lifshitz, A. B., Lebedev, V. S. and Zavarin, G. A. (1993) 'Emissions of CH_4 into the atmosphere from landfills in the former USSR', *Chemosphere*, vol 26, pp401–417

Panikov, N. S. and Gorbenko, A. J. (1992) 'The dynamics of gas exchange between soil and atmosphere in relation to plant–microbe interactions: Fluxes measuring and modelling', *Ecological Bulletin*, vol 42, pp53–61

Park, S. Y., Brown, K. W. and Thomas, J. C. (2004) 'The use of biofilters to reduce atmospheric methane emissions from landfills: Part I. Biofilter design', *Water, Air, and Soil Pollution*, vol 155, pp63–85

Peer, R. L., Thorneloe, S. A. and Epperson, D. L. (1993) 'A comparison of methods for estimating global methane emissions from landfills', *Chemosphere*, vol 26, pp387–400

Pohland, F. G. and Al-Yousfi, B. (1994) 'Design and operation of landfills for optimum stabilization and biogas production', *Water Science Technology*, vol 30, pp117–124

Poulsen, T. G. (2005) 'Impact of wind turbulence on landfill gas emissions' in T. H. Christensen, R. Cossu and R. Stegmann (eds) *Proceedings of Sardinia '05: Tenth International Waste Management and Landfill Symposium*, 3–7 October 2005, CISA, Environmental Sanitary Engineering Centre, Cagliari, Italy

Rannaud, D., Cabral, A. R., Allaire, S., Lefebvre, R. and Nastev, M. (2007) 'Migration d'oxygène et oxidation du methane dans les barriers d'oxydation passive installées dans les sites d'enfouissement', in *Comptes rendus de la 60e Conférence géotechnique canadienne*, 21–24 October 2007, Ottawa, Canada

Rinne, J., Pihlatie, M., Lohila, A., Thum, T., Aurela, M., Tuovinen, J-P., Laurila, T. and Vesala, R. (2005) 'N_2O emissions from a municipal landfill', *Environmental Science and Technology*, vol 39, pp7790–7793

Rogner, H-H., Zhou, D., Bradley, R., Crabbé, P., Edenhofer, O., Hare, B., Kuijpers, L. and Yamaguchi, M. (2007) 'Introduction', in B. Metz, O. R. Davidson, P. R. Bosch, R. Dave and L. A. Meyer (eds) *Climate Change 2007: Mitigation. Contribution of Working Group III to the Fourth Assessment Report of the Intergovernmental Panel on Climate Change*, Cambridge University Press, Cambridge, UK and New York, NY, pp95–116

Scheutz, C., Bogner, J., Chanton, J., Blake, D., Morcet, M. and Kjeldsen, P. (2003) 'Comparative oxidation and net emissions of methane and selected non-methane organic compounds in landfill cover soils', *Environmental Science and Technology*, vol 37, pp5150–5158

Scheutz, C., Bogner, J., Chanton, J. P., Blake, D., Morcet. M., Aran, C. and Kjeldsen, P. (2008) 'Atmospheric emissions and attenuations of non-methane organic compounds in cover soils at a French landfill', *Waste Management*, vol 28, pp1892–1908

Scheutz, C., Kjeldsen, P., Bogner, J., De Visscher, A., Gebert, J., Hilger, H., Huber-Humer, M. and Spokas, K. (2009) 'Microbial methane oxidation processes and technologies for mitigation of landfill gas emissions', *Waste Management and Research*, vol 27, pp409–455

Segers, R. (1998) 'Methane production and methane consumption: A review of processes underlying wetland methane fluxes', *Biogeochemistry*, vol 41, pp23–51

Spokas, K. and Bogner, J. (2010) 'Limits and dynamics of methane oxidation in landfill cover soils', *Waste Management* (in press).

Spokas, K., Graff, C., Morcet, M. and Aran, C. (2003) 'Implications of the spatial variability of landfill emission rates on geospatial analyses', *Waste Management*, vol 23, pp599–607

Spokas, K., Bogner, J., Chanton, J., Morcet, M., Aran, C., Graff, C., Moreau-le-Golvan, Y., Bureau, N. and Hebe, I. (2006) 'Methane mass balance at three landfill sites: What is the efficiency of capture by gas collection systems?', *Waste Management*, vol 26, pp516–525

Spokas, K., Bogner, J., Chanton, J. and France, G. (2009) 'Developing a new field-validated methodology for landfill methane emissions in California', in T. H. Christensen, R. Cossu and R. Stegmann (eds) *Proceedings of Sardinia '09: Twelfth International Waste Management and Landfill Symposium*, 5–9 October 2009, CISA, Environmental Sanitary Engineering Centre, Cagliari, Italy

Stein, V. B., Hettiaratchi, J. P. A. and Achari, G. (2001) 'Numerical model for biological oxidation and migration of methane in soils', *Practice Periodical of Hazardous Toxic and Radioactive Waste Management*, vol 5, pp225–234

Stern, J., Chanton, J., Abichou, T., Powelson, D., Yuan, L., Escoriza, S. and Bogner, J. (2007) 'Use of a biologically active cover to reduce landfill methane emissions and enhance methane oxidation', *Waste Management*, vol 27, pp1248–1258

Stumm, W. and Morgan, J. J. (1987) *Aquatic Chemistry*, John Wiley & Sons, New York, NY

Trégourès, A., Beneito, A., Berne, P., Gonze, M. A., Sabroux, J. C., Pokryszka, Z., Savanne, D., Tauziede, C., Cellier, P., Laville, P., Milward, R., Arnaud, A., Levy, F. and Burkhalter, R. (1999) 'Comparison of seven methods for measuring methane flux at a municipal solid waste landfill site', *Waste Management and Research*, vol 17, pp453–458

Themelis, N. J. and Ulloa, P. A. (2007) 'Methane generation in landfills', *Renewable Energy*, vol 32, no 7, pp1243–1257

Tsujimoto, Y., Masuda, J., Fukuyama, J. and Ito, H. (1994) 'N_2O emissions at solid waste disposal sites in Osaka City', *Air Waste*, vol 44, pp1313–1314

US EPA (Environmental Protection Agency) (2006) *Global Anthropogenic Non-CO_2 Greenhouse Gas Emissions: 1990–2020*, Office of Atmospheric Programs, Climate Change Division, Washington, DC, www.epa.gov/nonco2/econ-inv/pdfs/global_emissions.pdf

US EPA (2008) *Inventory of US Greenhouse Gas Emissions and Sinks 1990–2006*, (USEPA #430-R-08-005) Office of Atmospheric Programs, Climate Change Division, Washington, DC, www.epa.gov/climatechange/emissions/usinventoryreport.html

Van Zanten, B. and Scheepers, M. (1994) *Modelling of Landfill Gas Potentials*, Report prepared for International Energy Agency (IEA) Expert Working Group on Landfill Gas, published by Technical University of Lulea, Lulea, Sweden

Vieitez, E. R. and Ghosh, S. (1999) 'Biogasification of solid wastes by two-phase anaerobic fermentation', *Biomass and Bioenergy*, vol 16, no 5, pp299–309

Webster, R. and Oliver, M. A. (1992) 'Sample adequately to estimate variograms of soil properties', *Journal of Soil Science*, vol 43, pp177–192

Wilshusen, J. H., Hettiaratchi, J. P. A., De Visscher, A. and Saint-Fort, R. (2004) 'Methane oxidation and formation of EPS in compost: Effect of oxygen concentration', *Environmental Pollution*, vol 129, pp305–314

Zabowski, D. and Sletten, R. S. (1991) 'Carbon dioxide degassing effects on the pH of spodosol soil solutions', *Soil Science Society of America Journal*, vol 55, pp1456–1461

Zhao, C., Andrews, A. E., Bianco, L., Eluszkiewicz, J., Hirsch, A., MacDonald, C., Nehrkorn, T. and Fischer, M. L. (2009) 'Atmospheric inverse estimates of methane emissions from Central California', *Journal of Geophysical Research*, vol 114, no D16302, pp1–13

Zilioli, E., Gomarasca, M. A. and Tomasoni, R. (1992) 'Application of terrestrial thermography to the detection of waste disposal sites', *Remote Sensing of the Environment*, vol 40, pp153–160

12
Fossil Energy and Ventilation Air Methane

Richard Mattus and Åke Källstrand

Introduction

Global CH_4 emissions from energy-related sources are currently estimated to be between 74 and 77Tg each year, with 30–46Tg arising from coal mining (Denman et al, 2007) and so on a par with that released from landfill sites (Chapter 11). This is excluding energy-related emissions attributable to biomass burning (Chapter 7). In the US, energy-related CH_4 emissions constituted some 35 per cent of national CH_4 emissions in 2007 (US EPA, 2009). The bulk of these emissions arise from CH_4 release during fossil fuel extraction, transport, handling and in the use phase. Extraction-related losses are dominated by those from coal mining.

This chapter will briefly review CH_4 emissions from gas- and oil-related sources of CH_4 (see Chapter 13 for a more detailed discussion of relevant mitigation strategies and their cost effectiveness) before focusing on coal mines and a rapidly developing technology – that of intercepting ventilation air methane (VAM) – that has the potential to substantially reduce emissions from this source.

Natural gas losses

With natural gas being composed of >90 per cent CH_4, its loss to the atmosphere during extraction, processing and supply can represent a significant component of local and national CH_4 emissions budgets. Globally, such natural gas-related CH_4 emissions are estimated to result in the emission of 25–50Tg CH_4 yr^{-1} (Wuebbles and Hayhoe, 2002) and on a par, if not greater than, losses due to coal mining. In 2007 in the US, CH_4 emissions from natural gas systems totalled 104Tg CO_2-eq, making this source of anthropogenic CH_4 the third largest after enteric fermentation and landfills (US EPA, 2009). Globally, emissions are projected to increase by 54 per cent between 2005 and 2020, with the largest increases occurring in Brazil (>700 per cent) and China (>600 per cent).

As well as incidental emissions occurring at the point of extraction, significant amounts of CH_4 may be lost during the above-ground transfer of CH_4, often via leaks in pipelines, during venting and at compressor stations. Such losses are especially prevalent where the extraction, storage and supply infrastructure is aged or poorly maintained.

It is estimated, that in the 1990s, around 6 per cent of the natural gas piped across Russia was lost due to leaks, with regional leakage rates reportedly ranging from between 1 and 15 per cent depending on the quality of the extraction, handling and distribution systems. Losses in developed-world nations are commonly much lower, at between 1 and 2 per cent of production (Wuebbles and Hayhoe, 2002). Due to the high pressures involved, losses at compressor stations can be substantial, with the main sources being leaks at valves, the deliberate venting of instruments and incomplete combustion of CH_4 in the compressor engines. Such fugitive emissions can be difficult to detect and intercept before they escape from the compressor station, but where airflows can be well controlled much of this CH_4 can be effectively captured.

More efficient extraction techniques, upgraded infrastructure and instigation of direct inspection and maintenance of equipment can all therefore help to reduce such natural gas-related CH_4 emissions (up to 80 per cent of fugitive emissions for the latter strategy) (US EPA, 2006). The substitution of natural gas with compressed air in the pneumatic control devices used in gas distribution systems may also provide substantial emissions reductions. Some flaring may also be employed at the point of extraction or processing to convert unwanted CH_4 (that at relatively low concentrations or in excess of the processing capacity) to carbon dioxide and water vapour, but in many areas an expansion of capacity and improvement in infrastructure can allow such CH_4 to be utilized.

Oil-related emissions

The geological formation of oil can result in large CH_4 deposits (as natural gas) being closely associated with the oil reserves. During drilling and subsequent extraction, the trapped CH_4 may therefore be released to the atmosphere in large quantities. Oil itself only contains trace amounts of CH_4 and so the bulk (97 per cent) of related CH_4 emissions is likely to occur at the oil fields, rather than during refinement (3 per cent) and transportation (1 per cent) (US EPA, 2006). Global estimates of CH_4 emissions from oil and its derivatives range from 6 to 60Tg per year, with the US EPA estimating that oil production was responsible for some 57 million tonnes CO_2-eq in 2000, making it the 11th largest anthropogenic source of CH_4 globally (US EPA, 2006). In 2007 in the US, petroleum systems were responsible for emissions of 28.8Tg CO_2-eq as CH_4, with an additional 2.3 and 1Tg CO_2-eq being emitted as CH_4 from mobile combustion and petrochemical production, respectively (US EPA, 2009). Globally, CH_4 emissions from this source are projected to increase by ~100 per cent between 2005 and 2020.

Well-targeted collection of the CH_4 associated with oil extraction activities can vastly reduce emissions from this source. Mitigation of CH_4 emissions associated with oil production commonly centres on injection of captured CH_4 back into the oil field (which can enhance oil recovery), flaring off of CH_4 as CO_2 and water vapour (though limited due to costs, especially at offshore sites), or utilization of captured CH_4 as an additional fuel source where concentrations are sufficiently high and the infrastructure exists (US EPA, 2006).

Coal bed methane

During the geological process of coal formation, called 'coalification', CH_4 is formed and much of this may then remain trapped within the coal seam and the surrounding strata until released by mining operations. For each tonne of coal formed many thousands of cubic feet of CH_4 may also be formed (Thakur, 1996). Generally, the deeper the coal seam and the higher the carbon content of the coal, then the greater the amounts of CH_4 that are produced and trapped, with more than 90 per cent of fugitive CH_4 emissions from the coal sector arising from underground mining (US EPA, 2006). In 2000, emission factors for 56 so-called 'gassy' mines (those with high CH_4 concentrations) in the US ranged from 57 to 6000 cubic feet of CH_4 per mine per year. As technological improvements allow the exploitation of ever-deeper and more 'gassy' coal deposits, there is a risk that coal mining-related CH_4 emissions will increase in the future without improved mitigation (US EPA, 2006).

Methane emissions from underground mines are in the range of 10–25m^3 tonne^{-1} coal compared to values of 0.3–2.0m^3 tonne^{-1} for surface mines. Globally, this source is estimated to be between 30 and 46Tg CH_4 each year (Denman et al, 2007). In 2007 in the US, coal mining CH_4 emissions stood at 57.6Tg CO_2-eq, with a further 5.7Tg CO_2-eq arising from CH_4 emissions from abandoned coal mines (US EPA, 2009).

A proportion of the CH_4 associated with coal is also found in the pores of the coal itself and so is released when the coal is pulverized during processing, by diffusion or during combustion after it is extracted. Methane emissions from stationary combustion in the US, of which coal combustion is the primary source, were 6.6CO_2-eq in 2007 (US EPA, 2009). In shallow and open cast mines, the trapped CH_4 is often released directly to the atmosphere during mining. In deeper mines, CH_4 at relatively low concentrations (~1 per cent v/v) is more often released via ventilation shafts to prevent potentially dangerous CH_4 concentrations building up and it is on this dominant emissions pathway and its potential mitigation to which this chapter now focuses.

Potential role in climate change mitigation

As stated in previous chapters, the relatively short atmospheric lifetime of CH_4 provides an opportunity to achieve rapid mitigation of climate forcing in the short term, and may therefore prove crucial during the decade or two required to develop alternative sources of energy and more cost-effective mitigation technologies aimed at reducing CO_2 emissions.

On a global basis, emissions from livestock and manure constitute around one third of all anthropogenic CH_4 emissions (Chapters 9 and 10). However, these are distributed over a very large number of individual emission sources, making effective mitigation potentially very difficult. While a single cow may emit 50–100kg of CH_4 per year, a single coal mine shaft can emit 50,000 tonnes of CH_4 annually. As such, coal mines represent a very strong point source at which mitigation efforts may be very efficiently targeted.

Ventilation air methane

Methane in air is explosive at concentrations between 5 per cent and 15 per cent. Safety is achieved by balancing concentrations sufficiently below or sufficiently above this range. To address this coal mine CH_4 safety issue, large volumes of ventilation air are pushed through mines in order to dilute the CH_4 released down to safe levels i.e. below the lower explosion limit (LEL). This leads to concentrations of VAM of 1 per cent or lower. However, the high volumes of ventilation air being emitted may still lead to very high absolute emissions of CH_4 to the atmosphere.

In order to reduce the amount of CH_4 released into the ventilation air in the first place, advanced techniques for efficient drilling and pre-drainage of CH_4 have been developed. Two key mitigation strategies currently employed for this. The first, 'degasification' or gas drainage, involves a network of vertical wells being drilled before (up to ten years prior to mining) and/or after mining operations and high-quality CH_4 being extracted (commonly at concentrations of 30–90 per cent) for energy generation. Significant cost savings may be made using this strategy, with an estimated 57 per cent of the gas thus captured being suitable for injection into the natural gas pipeline network (US EPA, 2006).

The second, that of 'enhanced degasification', employs similar strategies to that of basic degasification, but with more advanced drilling and processing technologies, such as dehydrators and nitrogen removal systems. Such enhanced systems may improve recovery efficiency by 20 per cent compared to basic systems, with an estimated 77 per cent of the CH_4 captured being suitable for injection into the natural gas pipeline network (US EPA, 2006).

The amount of CH_4 that can be pre-drained depends on many factors, including the permeability of the coal and that of the surrounding strata. The process of excavation relieves local pressure, causing CH_4 in the surrounding rock to migrate and enter into the ventilation air. With extensive pre-drainage of CH_4, some mines have shown that it is possible to reduce the total amount of CH_4 released into the ventilation air by around 50 per cent. In many cases however, coal mine CH_4 concentrations are too low to make degasification strategies commercially viable (i.e. too low to feed into the natural gas pipeline network) and emissions to the atmosphere in the form of VAM become the prime target for mitigation.

The fact that CH_4 destruction by oxidation – the primary process employed in mitigating VAM emissions – can generate significant energy makes it commercially interesting to use CH_4 emissions from coal mines as a source of energy, either for district heating or electricity generation. This has already been the case for some landfills (Chapter 11) and at some sites with coal mine degasification (see below). The key problem in addressing the area of VAM to date has been the amount of air that must be processed in order to oxidize sufficient quantities of CH_4.

Mitigation of VAM emissions

In order to find ways to reduce the large emissions of VAM, many different technologies have been explored (Methane to Markets, 2009). One successful strategy is to burn some of the ventilation air CH_4 for energy generation, providing a significant energy bonus in addition to avoiding losses of VAM to the atmosphere. In rare cases where a large power plant can be located close to a ventilation shaft, this is likely to be the best option, but the size of ductwork required to transport the approximately 1 million m^3 h^{-1} of air common to these systems disqualifies locations at any distance away from the mine ventilation shaft. Other VAM technologies include the use of ventilation air as combustion air for gas engines – where the main fuel is that of drainage gas, use in recuperative and catalytic lean-burn gas turbines, and destruction using tailored thermal and catalytic oxidation processes (Somers and Schultz, 2008).

Successful demonstration of VAM processing

The only VAM technology that has so far proven to be commercially feasible on a large-scale basis is a system at the West Cliff Colliery of BHP Billiton, Australia (Somers and Schultz, 2008). The installation is called WestVAMP (West Cliff Colliery Ventilation Air Methane Plant) and is based on the VOCSIDIZER technology developed, patented and supplied by MEGTEC Systems.

Using VAM at a concentration of 0.9 per cent, high-grade steam is generated and used to drive a conventional power plant steam turbine. The plant is thereby using a fuel consisting of ventilation air with more than 99 per cent air. It is processing 20 per cent of the ventilation air from the mine shaft, while driving a conventional 6MW(e) steam turbine and reducing emissions by an estimated 250,000 tonnes CO_2-eq per year.

Operated during a few months in 1994, at the Thoresby Mine in the UK, MEGTEC first demonstrated that their VOCSIDIZER technology could efficiently abate VAM emissions. In principle, the VOCSIDIZER is a very well-insulated steel box with a bed consisting of a heat-exchange-efficient ceramic bed. Initially, the centre portion is heated to 1000°C. Then, the ventilation air containing CH_4 in dilute form is passed through the bed. At the natural

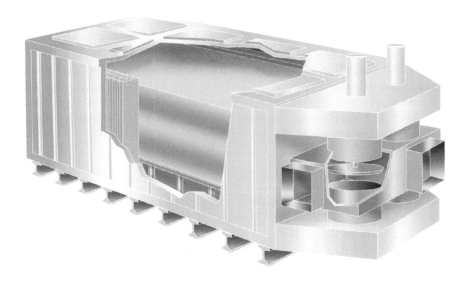

Figure 12.1 Cut-away illustration of the VOCSIDIZER

Source: MEGTEC

oxidizing temperature of CH_4 the gas is oxidized, releasing energy that, by efficient heat exchange, is conserved and used to heat the next incoming portion of ventilation air up to VAM oxidizing temperature. In order to keep the hot oxidizing zone in the centre of the equipment, the direction of flow is alternated. A cut-away illustration is shown in Figure 12.1.

In 2001 to 2002, a second demonstration was made at the Appin Colliery of BHP in Australia (partly funded by ACARP, the Australian Coal Association Research Program). This trial demonstrated two important aspects, namely: (1) stability of operation – that the process could handle the natural swings in VAM concentration in a coal mine ventilation shaft (in order to demonstrate this, the installation was successfully operated for 12 months); and (2) energy recovery – the energy released in the VOCSIDIZER oxidation process was successfully utilized to generate steam.

The ultimate demonstration of the VOCSIDIZER technology was made in the WestVAMP project at the West Cliff Colliery of BHP Billiton in Australia. This VAM power plant was officially inaugurated on 14 September 2007 after several months of successful operation. The project was partly funded by the Australian Greenhouse Office and has two revenue streams – the value of the electricity generated and 'carbon credits' sold within the New South Wales trading scheme. The WestVAMP installation had, by 2009, generated over 80GWh of electricity and 500,000 carbon credits (tonnes CO_2-eq) traded on the local New South Wales trading scheme. In the first fiscal year of operation, BHP Billiton reported plant availability to be at 96 per cent, including two

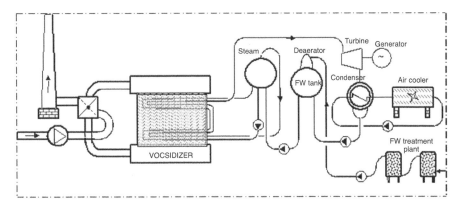

Figure 12.2 Principle process diagram of a VAM power plant based on the VOCSIDIZER

Source: MEGTEC

planned maintenance shutdowns (see Booth, 2008). The plant can be described as a conventional steam-based power plant, using the VOCSIDIZER as an unconventional furnace capable of operating on an extremely lean fuel. A principle process diagram of a VAM power plant based on the VOCSIDIZER technology is shown in Figure 12.2.

The VOCSIDIZER technology for VAM processing has also been demonstrated by CONSOL Energy in West Virginia – with support from the US EPA (United States Environmental Protection Agency) and the US Department of Energy – and by the ZHENGZHOU Mining Group in the Henan Province of the People's Republic of China. The Chinese installation is the first VAM processing plant of any kind in the world to be formally approved by the UNFCCC for the generation of carbon credits.

Technically, the VOCSIDIZER technology is essentially based on the very efficient heat exchange between the air being processed and the internal configuration of the system. The efficiency of the system means that a VAM concentration of only 0.2 per cent is required to maintain the oxidizing process. There is no combustion chamber – instead the oxidation takes place inside a special ceramic bed. There are also no 'hot spots' or periods of open flame combustion, meaning that significant production of nitrogen oxides (NO_x) is avoided.

The potential for VAM processing

As the air volumes that must be processed for abating VAM emissions are very large, the modularity of the processing system is important. VOCSIDIZERs are arranged in groups of four units on two levels in a so-called 'VAM cube', each capable of processing 250,000m^3 h^{-1} of ventilation air. Large installations consist of multiple VAM cubes, each with a footprint of approximately 500m^2 (Figure 12.3).

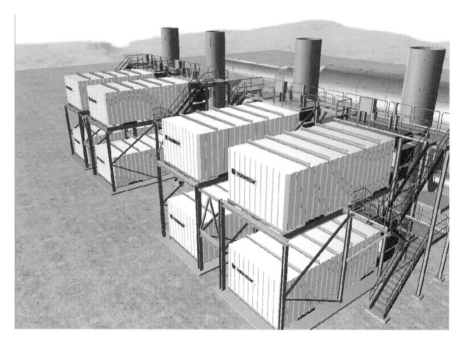

Figure 12.3 An installation of two VOCSIDIZER 'VAM cubes' for processing 500,000m³ h⁻¹ of coal mine ventilation air

Source: MEGTEC

As the system is an air-processing installation, the CH_4 reduction capacity relates directly to the VAM concentration, where a 0.3 per cent CH_4 concentration 'generates' annual emission reductions per 'VAM cube' of around 80,000 tonnes of CO_2-eq, 0.6 per cent CH_4 around 160,000 tonnes CO_2-eq and 0.9 per cent CH_4 around 240,000 tonnes CO_2-eq. An installation processing 1 million m³ h⁻¹ of coal mine ventilation air (equivalent to the capacity of four VAM cubes) can therefore achieve annual emission reductions of the order of 1 million tonnes of CO_2-eq.

Despite being in an extremely dilute form, the amount of energy contained in the mine ventilation air can be very high. From a VAM cube processing approximately 250,000m³ h⁻¹ of ventilation air, the thermal energy recoverable from 0.6 per cent VAM is around 10MW of heat energy. Increasing or decreasing the VAM concentration by 0.3 per cent means increasing or decreasing the recoverable energy content by around 7MW.

As previously mentioned, the main limitation to energy generation using VAM VOCSIDIZER technology is not the extremely dilute nature of the 'fuel', but rather the requirement for the system to be located in close vicinity to the mine ventilation shaft.

The carrier of the thermal energy produced from the VAM can be hot water, hot oil or steam, depending on the intended use of the recovered energy.

This can include space heating, cooling (by driving absorption chillers) or for the generation of electricity. Normally, the thermal energy requirement (for example space heating in buildings) near the mine ventilation shaft is very limited. Therefore, in most cases, it is sufficient to apply secondary heat exchange to the exhaust ducts that leave the VOCSIDIZER equipment, to provide for the relatively small heat energy demands on-site (for example hot water for miners' showers and heating of buildings) – a single VAM cube processing ventilation air with an average VAM concentration of 0.6 per cent can generate 8MW of 70°C water.

VAM opportunities for climate change mitigation

It is estimated that if VAM oxidizer technology was employed at all sites where VAM concentrations were >0.15 per cent around 97 per cent of all CH_4 emissions arising from coal mine ventilation air could be avoided (US EPA, 2006). In 2003, the global VAM market was estimated at 160 million tonnes CO_2-eq at a net present value cost of only \$3 per tonne of CO_2-eq (US EPA, 2003), with a projection that catalytic oxidation of VAM has the potential to mitigate 0.94 million tonnes CO_2-eq yr^{-1} in the US alone by 2020 (a 24 per cent reduction compared to baseline emissions) (US EPA, 2006). Other mitigation strategies for addressing coal mine-related CH_4 emissions, such as degasification and enhanced degasification, appear to have more limited potential for achieving large emissions reductions in the next decade or two (US EPA, 2006), but all have the potential to play a significant role in the suite of measures available to reduce emissions globally.

Conclusions

Methane has a GWP of 21–25 on a 100-year time horizon, but with its relatively short atmospheric lifetime (about ten years) this GWP increases substantially when shorter time horizons are employed (see Chapter 1, Box 1.1). So, while focusing on decarbonizing the global energy structure in the medium to long term, large reductions in CH_4 emissions from point sources such as coal mines provide a potentially very effective way of mitigating climate change in the short to medium term (10–20 years). Coal mine VAM emissions provide an excellent opportunity for just such a response, allowing significant emission reductions to be achieved through addressing a small number of powerful point sources.

References

Booth, P. (2008) *West VAMP BHP Illawara Coal*, US Coal Mine Methane Conference, Pittsburgh, www.epa.gov/cmop/docs/cmm_conference_oct08/08_booth.pdf

Denman, K. L., Chidthaisong, A., Ciais, P., Cox, P. M., Dickinson, R. E., Hauglustaine, D., Heinze, C., Holland, E., Jacob, D., Lohmann, U., Ramachandran, S., da Silvas Dias, P. L., Wofsy, S. C. and Zhang, X. (2007) 'Couplings between changes in the climate system and biochemistry', in S. Solomon, D. Qin, M. Manning, Z. Chen, M. Marquis, K. B. Averyt, M. Tignor and H. L. Miller (eds) *Climate Change 2007: The Physical Science Basis*, Cambridge University Press, Cambridge, pp499–587

Methane to Markets (2009) *Methane Technologies for Mitigation and Utilization*, Methane to Markets Programme, www.methanetomarkets.org/m2m2009/documents/events_coal_20060522_technology_table.pdf

Somers, J. M. and Schultz, H. L. (2008) 'Thermal oxidation of coal mine ventilation air methane', in K. G. Wallace Jr. (ed) *12th US/North American Mine Ventilation Symposium 2008*, Reno, Nevada, USA, pp301–306, available at www.epa.gov/cmop/docs/2008_mine_vent_symp.pdf

Thakur, P. C. (1996) 'Global coal bed methane recovery and use', *Fuel and Energy Abstracts*, vol 37, p180

US EPA (United States Environmental Protection Agency) (2003) *Assessment of the Worldwide Market Potential for Oxidizing Coal Mine Ventilation Air Methane*, EPA 430-R-03-002, United States Environmental Protection Agency, Air and Radiation, Washington, DC

US EPA (2006) *Global Mitigation of Non-CO_2 Greenhouse Gases*, EPA 430-R-06-005, United States Environmental Protection Agency, Office of Atmospheric Programs, Washington, DC

US EPA (2009) *Inventory of Greenhouse Gas Emissions and Sinks: 1990–2007*, United States Environmental Protection Agency, Washington, DC, www.epa.gov/methane/sources.html

Wuebbles, D. J. and Hayhoe, K. (2002) 'Atmospheric methane and global change', *Earth-Science Reviews*, vol 57, pp177–210

13
Options for Methane Control

André van Amstel

Introduction

Emissions of greenhouse gases other than carbon dioxide are responsible for about 50 per cent of radiative forcing, the determinant of global warming (IPCC, 2007). This means that emission reductions of such gases, like CH_4 and N_2O, are important and should be taken into account when developing policy strategies for reducing the risk of dangerous climate change.

Emission reductions of CH_4 can be very effective in reducing climate forcing compared to CO_2, because each unit of CH_4 is 21–25 times more powerful than CO_2 over a 100-year time horizon. Given the fact that CH_4 can often be recovered and used for energy generation, many options can be cost effective because of the benefits of the energy sold. In this chapter cost estimates of technical reduction measures will be given that will be used in the integrated assessment to estimate the overall reduction costs for different CH_4 reduction strategies.

Which options are available to reduce methane emissions and what are the costs of the options?

Various authors have made estimates of the costs of measures per tonne of CH_4 emission reduction. With some technical measures, CH_4 emissions can be significantly reduced and experience of these measures has been developed in actual CH_4 reduction projects. A body of literature is available on measures to reduce CH_4 (for example AEAT, 1998; IEA, 1999; Hendriks and de Jager, 2000; De la Chesnaye and Kruger, 2002; Graus et al, 2003; Gallaher et al, 2005; Delhotal et al, 2005; Harmelink et al, 2005). Measures are considered cheap if they are below $20 per tonne. Some measures even generate money and can be implemented for –$200 to $0 per tonne of reduced CH_4. These are called 'no-regret' measures. Most measures, however, are more expensive and can only be implemented at $20–500 per tonne of CH_4 reduced. Very expensive options are available at $500 per tonne or more. Van Amstel (2005) made a first global estimate of costs.

The IPCC (2007) showed that climate is already changing and that these changes are likely to accelerate in the next 50 years as a result of human activities. To limit these changes, greenhouse gas emissions should be reduced rapidly worldwide. The UNFCCC has a stated objective to stabilize atmospheric concentrations of greenhouse gases at a level that will avoid dangerous climate change. The science, technology and expertise are potentially available to meet this goal. Proposals to limit atmospheric CO_2 concentrations to prevent the most damaging effects of climate change have often focused on stabilization targets of 550ppm – almost doubling the pre-industrial average concentration of 280ppm. The current CO_2 concentration is 386ppm, with a growth rate of ~2ppm yr^{-1}.

Greenhouse gas emissions are reported to the Climate Convention and the Kyoto Protocol. IPCC (2006) developed guidelines for national inventories of emissions. In general, greenhouse gas emissions from fossil fuels can be controlled by energy efficiency measures, energy conservation and decarbonization of the economy, which includes fuel switching to renewables. Short-term options are carbon capture and storage, fuel switching from coal to gas (i.e. from a high to a lower carbon-density fuel) or from fossil to biomass fuels (biomass is a carbon-dense fuel but emitted CO_2 is taken up by the vegetation in a closed short-term cycle) (Leemans et al, 1998; Pacala and Socolow, 2004). Some even advocate a large increase in nuclear power generation around the world, but the problems of nuclear waste storage and security remain unresolved in many areas.

Pacala and Socolow (2004) have suggested that the policy gap between a business as usual (BAU) scenario and a stabilization scenario could be tackled stepwise through stabilization wedges, using current technologies. In a BAU scenario, carbon emissions are increasing by 1.5 to 2 per cent per year. In their view, every wedge contains a package of measures that reduces the emissions substantially. Wedges can be achieved from, for example, energy efficiency, the decarbonization of the supply of electricity and fuels by means of fuel shifting, carbon capture and storage, nuclear energy and renewable energy. With energy conservation and efficiency improvements, large reductions in carbon dioxide emissions have already been achieved in some sectors and regions. In industrialized countries, the autonomous energy efficiency improvement (AEEI) of the total economy was 2 per cent per year over the last four decades, with a levelling off during the last 20 years (Schipper, 1998).

Van Amstel (2009) developed scenarios for 2100, and estimated climate change with methane reduction measures. Every possible measure to reduce the other greenhouse gases, including CH_4, should be added to the greenhouse gas technology portfolio. Methane emitted to the atmosphere is 21–25 times more powerful than CO_2 and, as we have seen in previous chapters, it can often be captured and used for energy generation. For CH_4, this means tracking down and removing all leaks in the exploration, mining and supply sectors of the fossil fuel industry. It means abandoning wasteful venting and flaring in the oil and gas industry and capturing coal mine ventilation air CH_4.

Pre-mining degasification in 'gassy' coal mines must be promoted to increase safety and to capture and use the CH_4 for energy generation. A market for ventilation air CH_4 use and oxidation is emerging in many of the world's coal-producing countries with an equipment sales market of more than $8.4 billion (EPA, 2003; Gunning, 2005; Mattus, 2005). Harvesting biogenic CH_4 from landfills, manure fermentation and wastewater treatment plants, for example (IEA, 2003; Maione et al, 2005; Ugalde et al, 2005), should also be promoted wherever possible.

Below, I discuss different options that are already being deployed at an industrial scale and that could be scaled-up further. Methane can also be seen as the perfect carrier for hydrogen in the hydrogen economy (it is much cheaper than metal hydrates). This chapter then continues with a conceptual section focused on the way to determine the costs of emission reductions and a technical section in which the reduction measures are described.

On the determination of characteristics and costs of emission reduction options

In this section, a number of issues important for the assessment of greenhouse gas emission reduction options will be considered. I start with some definitions and then discuss the possibilities to estimate the costs of reduction technologies. The options can be evaluated using supply curves.

Two types of measures can be distinguished that reduce (with respect to a given baseline development) the emissions of greenhouse gases: efficiency improvements and volume measures. Efficiency improvement is defined as decreasing the specific emissions. Volume measures are a reduction of the human activity itself.

Specific emissions are defined as the emission per unit of human activity, such as emissions per amount of coal produced, or emissions per unit of meat or milk produced. Emissions are calculated with the general formula: volume of activity × emission per unit of activity × abatement per unit of emission. Total costs in the target year are calculated from the reduction of emissions compared to a baseline development without reduction measures. In general, the activity indicator is a physical measure relevant for the economic value of a human activity. Examples are the amount of fossil fuels produced, the size of the livestock herd or the amount of meat or milk produced, or the amount of municipal waste produced. The choice of the activity indicator is not unambiguous: sometimes a simple indicator must be chosen because of a lack of statistics. Here, only efficiency improvements are described. These are often described as technological options; however, this does not always mean that technologies are applied. Also good housekeeping can be considered an efficiency improvement (Harmelink et al, 2005).

Several types of emission reduction potentials can be distinguished:

- The technological potential is the amount by which it is possible to reduce greenhouse gas emissions or improve energy efficiency by implementing a technology or practice that has already been demonstrated. The emission reduction is calculated with respect to a baseline scenario development.
- The economic potential is the portion of technological potential for greenhouse gas emission reductions or energy efficiency improvements that could be achieved cost effectively through the creation of markets, reduction of market failures, increased financial and technological transfers. The achievement of economic potential requires additional policies and measures to break down market barriers. Measures are cost effective when the benefits of the measures are larger than the costs (including interest and depreciation).
- The market potential is the portion of the economic potential for greenhouse gas emission reductions or energy efficiency improvements that could be achieved under forecast market conditions, assuming no new policies and measures.

An assessment of emission reduction technologies within a certain sector in general consists of the following steps:

- breakdown of the sectoral emissions into categories and/or processes;
- inventory of possible emission reduction options per category and process;
- determination of characteristics of the various emission reduction options;
- evaluation of the set of emission reduction options (for instance by construction of a supply curve or evaluation against baseline scenarios with the help of an integrated model).

Balanced cost estimates are based on real projects where CH_4 emissions have been reduced by technical measures. These estimates take into account all costs of a project, including investment, operation and maintenance, labour and the depreciation of capital. It is important that cost estimates are comparable between countries and regions, so methods between research groups should be comparable. There are different ways to estimate the costs of abatement measures. Marginal abatement cost curves are estimated by Gallaher et al (2005), in which the reduction of costs over time has been taken into account. Here a static approach has been chosen, as first adopted by Blok and de Jager (1994), and applied in The Netherlands by Harmelink et al (2005). The 'Energy Modelling Forum 21' (EMF21) also applied static marginal abatement cost curves for non-CO_2 greenhouse gases for worldwide abatement (Delhotal et al, 2005). The data used in the EMF21 study are country- or region-specific marginal abatement cost curves based on country- or region-specific labour rates, energy system infrastructure and the latest emission data. A static analysis has limitations. First, the static approach to the abatement cost assessment does not account for technological change over time, which would reduce the cost of abatement and increase the efficiency of the abatement options. Second, the static EMF21 approach used only limited regional data. Difficulties arise with the cost estimates when multipurpose projects are developed. Then it is difficult to attribute the investment costs to CH_4 reduction because the investment was done also for other purposes. For example, wastewater and sewage treatment plants are developed to improve the health and living conditions of the people. Anaerobic digesters can be flared or the CH_4 used for cogeneration to reduce CH_4 emissions from biomass or liquid effluents with high organic content. Because most centralized systems automatically either flare or capture and use CH_4 for safety reasons, add-on abatement technology for existing wastewater treatment plants does not exist. As a result, potential emission reductions depend on large-scale structural changes in wastewater management. For this reason the cost of CH_4 reduction in sewage treatment is difficult to estimate. Overriding economic and social factors influence wastewater treatment practices throughout the world. The benefits of installing wastewater systems in developing countries for the purpose of disease reduction greatly outweigh potential benefits associated with CH_4 reduction. It would be misleading to imply that carbon taxes would be the driving force behind investment decisions that influence CH_4 emissions from wastewater. So cost estimates must be treated with caution.

Here the method as described by Blok and de Jager (1994) for specific costs is used. In this method the costs per tonne of reduction of CH_4 (or other greenhouse gas) per sub-sector are calculated with the following formula:

$$C_{spec} = (a.I + OM - B)/ER \qquad (16)$$

in which: C_{spec} = the specific emission reduction costs (in US$_{1990}$ per tonne of avoided CH_4); a = an annuity factor depending on interest (or discount) rate r and depreciation period n; $a = r/(1 - (1 + r)^{-n})$; I = the (additional) initial investment required for the measure in US$; a.I = the annualized capital costs in US$ per year; OM = the (additional) annual operation and maintenance costs associated with the measure in US$ per year; B = benefits associated with the measure in US$ per year (for example avoided energy costs); ER = emission reduction associated with the measure in tonne of avoided CH_4 per year. Also:

$$\text{Payback time} = I/(B - OM) \qquad (17)$$

Costs of measures are expressed in constant US$_{1990}$ per tonne of emission reduction for CH_4. The specific costs can be negative if the benefits associated with the measure are sufficiently large. Thus money can be earned by implementing beneficial measures. The costs calculated this way are meant to be the life-cycle costs, i.e. the total costs taking into account the technical lifespan of the equipment required for the measure. Thus the depreciation period is taken equal to the technical lifespan of equipment. The prices are market prices. The real interest rate is taken, i.e. the market interest rate corrected for inflation. This interest is 3 to 6 per cent generally. The sum of the emission reductions with negative net costs is called the economic potential. The specific costs can be used as a selection criterion for the set of measures chosen in a certain scenario. Total costs calculated for a selected set of measures give a first approximation of the national economic costs of the measures on the gross national product (assuming perfect competition). However the discount rate r that should be chosen is still a matter of discussion, and might be higher than the real interest rate. In public policy decision-making the social discount rate is sometimes chosen higher (at 10 per cent) because it should reflect the rate of return that could be realized through private spending on consumption and investment, assuming the same level of risk (Callan and Thomas, 2000). Also the technical lifetime as the period of depreciation might be too long. In reality measures are taken if payback times are five years or even shorter. However, the method by Blok and de Jager (1994) is adopted here because it is widely accepted.

Below, the available technical measures and reduction potential and costs for CH_4 will be described in the next section for the various important CH_4 sources.

Technical reduction potential for methane

Sources of methane emissions from natural gas and oil infrastructure

Natural gas and oil infrastructure accounts for over 20 per cent of global anthropogenic CH_4 emissions. Methane gas emissions occur in all sectors of the natural gas and oil industries, from drilling and production, through to processing and transmission, to distribution and even end use as a fuel. The natural gas infrastructure is composed of five major segments: production, processing, transmission, storage and distribution. The oil industry CH_4 emissions occur primarily from field production operations, such as venting gas from oil wells, oil storage tanks and production-related equipment. Table 13.1 provides a summary of country-specific CH_4 emissions from oil and gas for 1990 and 2000 and expected emissions for 2010 for some of the largest producers.

Methane emission reductions from oil
During the production of oil, associated gas is pumped up to the surface. In many producing countries this gas is either flared or vented. In these cases it is a wasted resource and it is often more cost effective to capture this gas for marketing or for energy generation at the production site. This valuable resource is vented and flared in large quantities in Nigeria and several production sites in the Middle East. Investments are needed to produce liquefied petroleum gas (LPG) for the local market or more distant markets.

Methane emission reductions from natural gas
Methane emissions from fossil fuel exploration could potentially increase in the near future. While looking for unconventional fossil fuel resources, CH_4 hydrates have recently attracted increasing attention. As each cubic metre of

Table 13.1 Country reported CH_4 emissions from oil and gas for 1990 and 2000, and a scenario for 2010

Country	Gg CO_2-equivalent		
	1990	2000	2010
Russia	335.3	252.9	273.5
US	121.2	116.4	138.7
Ukraine	71.6	60.2	39.4
Venezuela	40.2	52.2	68.0
Uzbekistan	27.2	33.7	42.9
India	12.9	24.4	54.9
Canada	17.1	23.3	23.8
Mexico	11.1	15.4	22.1
Argentina	8.0	13.7	30.5
Thailand	2.9	8.6	15.9
China	0.9	1.5	4.9

Source: Van Amstel, 2009

CH_4 hydrate contains up to 170m³ of CH_4 gas, the CH_4 hydrate resource potential in some regions is huge. At the moment, characterization of the regional resource potentials is taking place (Hunter, 2004). Technology to bring this resource to the surface is, however, still lacking. Therefore it is important to develop methods that are absolutely 'leak-free' to prevent hydrate CH_4 from substantially adding to global CH_4 emissions should these deposits be exploited.

During the production of natural gas the main CH_4 emission sources are:

- natural gas treatment operations, like drying of the gas with glycol and removal of H_2S or CO_2 with amine solutions;
- separation of natural gas condensate and measurement of its volume;
- safety systems with CH_4 purge streams in order to prevent air infiltration into the system.

The mainstream gas is dehydrated with glycol. During decompression of the glycol, gas is released and either vented or flared (especially on offshore platforms). During maintenance works, or through safety valves, gas is also released and again vented or flared. Flaring was only recently introduced at 'flare tips' located well away from the platforms because of security reasons. It is presumed that, on newly built platforms, measures will be taken to reduce flaring and venting to a minimum. On platforms that are in use, measures can be taken at little cost to increase the on-site use of this 'waste' gas. Achieving CH_4 emissions of below 10g GJ^{-1} are feasible, as proven by some offshore operations in Norway (OLF, 1994). To reduce CH_4 losses from such fossil fuel systems the following options are available (see also Table 13.4).

Option 1 Increased inspection and maintenance
Before repair or maintenance, emissions occur when parts of the system are depressurized. These gases may be recompressed and rerouted to the system. This might be done using a transportable compressor unit. Recompression is economically attractive when the amounts recompressed are over 65,000m³. Onshore recompression is often economically viable with specific costs of –$$_{1990}$200 tonne^{-1} of avoided CH_4 per year. Offshore, this is harder to achieve since all equipment has to be flown in by helicopter (De Jager et al, 1996).

Option 2 Increased on-site use of otherwise vented CH_4 at oil and gas production sites
Reduction of CH_4 emissions can be achieved by the increased on-site use or marketing of 'waste' gas during production of oil and gas. Increased on-site use in production is a relatively cheap measure with net costs of $10 per tonne CH_4 emission reduction. Increased marketing by liquefaction and transport to distant markets is even less expensive with –$100 per tonne of reduced CH_4 emission. In theory, a ~50 per cent reduction in emissions can be achieved via this strategy between 1990 and 2025, with 60 per cent being in the former Soviet states and the Middle East (De Jager et al, 1996).

Option 3 Increased flaring instead of venting
During production of oil and gas, venting and flaring takes place from the safety system at the production site. During 'flaring', gas is burned and CO_2 emission takes place, while during 'venting' gas is emitted unburned. Because CH_4 has 21–25 times the global warming potential of CO_2 on a mass basis, venting can result in greater climate forcing than flaring. Thus, increased flaring instead of venting reduces this problem. Costs are low if flares are installed in the development phase or during the start of operations (De Jager et al, 1996). If flares have to be installed, the costs are $400 per tonne of avoided CH_4. Increased flaring, instead of venting, can in theory reduce net emissions in CO_2-eq by 10 to 30 per cent between 1990 and 2025 in OECD (Organisation for Economic Co-operation and Development) countries, and by 50 per cent in oil- and gas-producing regions such as the former Soviet states, the Middle East and Eastern Europe.

Option 4 Accelerated pipeline modernization
Leaking of CH_4 is higher in old pipeline systems or in permafrost regions with low levels of maintenance. Accelerated pipeline modernization is expensive if pipes are underground. In the former Soviet states and the Middle East, by 2025, 90 per cent of leakages can be controlled by modernization of pipes. In other regions by 2025, a reduction of 20–40 per cent in leakages can be attained. Estimated costs are $1000 per tonne of CH_4 reduction if underground, and $500 if above-ground pipes are replaced (De Jager et al, 1996).

Option 5 Improved leak control and repair
In OECD countries, about three to five leaks are found per km of mains gas supply. In older systems this number could be much higher. Improved control and repair can radically reduce leakages by 2025, with reductions of up to 90 per cent in Eastern Europe, Russia, Middle East, India and China, and of 40 per cent in the other regions. This measure has medium costs, with around $200 per tonne of CH_4 reduction (De Jager et al, 1996).

Methane emission reduction from coal mining

In deep underground mines CH_4 is trapped in the coal beds and the surrounding rock strata. High pressure of the overlying strata prevents this CH_4 from being released. Once mining starts, this CH_4 can escape and fill the excavated space. Coal mine CH_4 is a safety hazard and therefore mine shafts are ventilated (Chapter 12). It is emitted into the atmosphere through ventilation air containing on average about 1 per cent of CH_4. After mining, a small but not negligible part of the CH_4 is released from the coal by post-mining activities such as breaking, crushing and drying. Global emission factors for deep mining are between 10 and $25m^3$ per tonne coal, not including post-mine activities. Global emission factors for surface mines are between 0.3 and $2.0m^3$ per tonne, not including post-mine activities. Emissions from post-mining activities are $4–20m^3$ per tonne (Kruger, 1993). Emissions can be mitigated before, during and after mining.

Before mining, CH_4 can be captured by drilling from the surface into the coal strata and capturing the CH_4 before mining starts. This can be done from half a year to several years before mining starts. During mining, ventilation air can be used as combustion air. After mining, CH_4 emissions can be reduced by recovering CH_4 from the areas of fractured coal and rock that is left behind, for example by using 'gob-wells'.

There are two main methods of underground mining: 'room and pillar' and 'longwall mining'. In the 'room and pillar' approach, coal deposits are mined by cutting a network of rooms into the coal seam and leaving behind pillars of coal to support the roof of the mine. 'Longwall mining' means cutting successive strips of coal from a face, typically 100–250m in length. Roadways are driven into each side of the panel of coal to access the face and the roof strata are allowed to collapse behind the face. Usually the face is established as the remote end of the panel and is retreated back towards the main trunk roadways (AEAT, 1998).

Current European CH_4 mine-draining practice is to drill inclined holes into the strata above the seam being mined. These are drilled from the panel roadways in advance of the face. Gas emission starts as the face approaches and passes beneath the holes. In the case of retreat mining, the holes then pass into the caved area behind the face, where access is no longer possible. In this way, a considerable amount of air is drawn in and mixed with the CH_4. This means that the gas, when it reaches the surface contains 50 per cent or less of

CH_4. This has cost implications, in that the system must be designed to handle a higher total flow. It may also cause problems at the utilization stage.

Another method, used in the US, is to drill vertical holes from the surface into the target seam and drain off gas prior to mining. This is a way of obtaining gas independently of any mining operation. Coal bed methane (CBM), as it is called, is extracted in large quantities from the Black Warrior and San Juan basins in the US. Because there is minimal opportunity for dilution, the gas is near-pure CH_4 and needs only minimal treatment. This system is only practical in 'gassy' mines with high coal permeability. The potential for degassing European coals is very limited, as European coals have a high rank and very low permeability. Another approach developed in Germany is to drive in a 'super adjacent heading'. This is a special roadway some distance above the face and parallel to its run, which then acts as a collection point. Overall, the following mitigation strategies options can be summarized.

Option 6 Pre-mining degasification
This measure can be taken for safety reasons. It recovers CH_4 before the coal is mined. It is an emission reduction option if CH_4 is then used for energy generation. Wells can be drilled in advance of mining, within the mine or from the surface. In typical mining regions such as China, Eastern Europe and Russia, a 90 per cent reduction in venting losses can be achieved by 2025, and in the US a 70 per cent reduction is possible, at an estimated cost of $40 per tonne of CH_4 emission reduction. In other regions, a reduction of 30–50 per cent can be achieved by 2025 using this strategy (IEA, 1999).

Option 7 Enhanced gob-well recovery
This measure can be taken to reduce CH_4 emissions from the 'gob' of the coal mine. This is the place where the roof has collapsed after mining and it may require additional drilling. Methane can again be used for energy generation. By the year 2025 about 50 per cent of CH_4 emissions from used coal mines can be recovered this way, at around $10 per tonne of CH_4 reduction (IEA, 1999).

Option 8 Ventilation air utilization (see also Chapter 12)
Ventilation is necessary in underground mines for safety reasons and large fans are used to blow air through the mines. The CH_4 content of the air must be below 5 per cent, and is frequently below 0.5 per cent to comply with relevant regulations. Ventilation air can be used as combustion air in turbines or boilers. Technical and economic feasibility has been demonstrated in Germany, Australia and in the UK. If introduced in other countries CH_4 emissions can be reduced by up to 80 per cent in typical mining regions, and by 10 to 20 per cent in the other regions. Estimated costs are $10 per tonne CH_4 reduction (IEA, 1999). All the major coal producers have some recovery and utilization of mine CH_4. For Germany the overall utilization rate in 1995 was 36 per cent. For the UK the recovery rate is estimated to be 11 per cent. In Spain, recovery rates are below 5 per cent (AEAT, 1998).

Methane emission reduction from enteric fermentation in livestock

Cattle, sheep, goats (ruminants) and horses and pigs (pseudo-ruminants) produce CH_4 as part of their normal digestive process. In cattle it is estimated that about 4 to 7 per cent of gross energy intake is lost as CH_4 (Van Amstel et al, 1993). Animals in modern intensive livestock feedlots with high protein supplements emit 4 per cent of gross energy intake, animals in range systems without supplements emit up to 7 per cent of gross energy intake. The worldwide spectacular growth of herds since the 1950s is responsible for the increase of this CH_4 source. Reductions per unit of meat or milk can be expected with increasing productivity improvements, especially in developing regions. Absolute reductions in the future are expected to be small. The following reduction options can be described.

Option 9 Improved production efficiency
Methane is produced as part of the normal digestion process in ruminant animals. Methane emissions per unit of milk or beef can be reduced if breeding programmes result in animals with high milk or meat production and good reproduction. Production efficiency is already very high in the US, Canada, OECD Europe and Japan, no reduction is expected here. In the other regions this can be improved. In those regions a CH_4 reduction of 20 per cent is expected without extra costs (Blok and de Jager, 1994).

Option 10 Improved feeding
In many regions, feeding of livestock can be improved. An increase in protein-rich feed and increasing digestibility of roughage reduces CH_4 per unit of product. No reductions are expected in the OECD countries. A reduction of about 10 per cent is expected in the other regions. Costs are low, with $5 per tonne of CH_4 reduction, but availability of high-quality feed is a problem in those regions (EPA, 1998).

Option 11 Production-enhancing agents
These agents have been proved to improve production in the US, but are not generally accepted by the public in Europe. Examples are hormones such as bovine somatotropin bST and anabolic steroids. In the European Union these hormones are banned. If used in other regions a ~5 per cent reduction of emissions may take place. This option is expensive, with an estimated $400 per tonne of avoided CH_4 (AEAT, 1998).

Option 12 Reducing animal numbers
In some regions in the world animal density is so high that environmental problems are a consequence. Examples are manure surpluses, ammonia emissions and acid deposition on natural areas, forests and surface water, also leading to groundwater pollution. Reducing animal numbers could be the solution. This is not a technical measure but a volume measure and may result

autonomously if environmental regulations are too costly for farmers. In the European Union, dairy animal numbers were reduced after milk quotas were introduced. Reductions in CH_4 emissions of 20 to 30 per cent may be achieved in OECD countries as a result of economic circumstances and as a result of these quota systems. Costs of reducing animal numbers are taken as zero.

Option 13 Increased rumen efficiency
Rumen fermentation is controlled by rumen acidity and turnover rate. These are controlled by diet and other nutritional characteristics, such as level of intake, feeding strategies, forage length and quality, and the ratio between forage and concentrates. Although significant advances in the knowledge of effects of various combinations of factors on microbial growth have been made in recent years, there is still insufficient information available to identify and control the interactions in the rumen that will result in optimal rumen fermentation.

Feeding ruminants on diets containing high levels of readily fermented non-structural carbohydrate has been shown to minimize CH_4 production by reducing the protozoal population and lowering rumen acidity. This, however, may lead to depressed rumen fermentation, which may then lower the feed energy conversion into animal product and may affect the animal's health. Using diets without much forage is therefore not considered a sustainable method to control CH_4 emissions from ruminants. A number of possible but costly options have been identified for rumen manipulation with additives: hexose partitioning, propionate precursors, direct-fed microbials that oxidize CH_4, genetic engineering and the immunogenic approach. In Table 13.2 an overview is given of technical reduction potentials in the European Union and costs of the rumen manipulation (AEAT, 1998). Costs are high with $3000–6000 per tonne of avoided CH_4 (AEAT, 1998).

Table 13.2 Reduction and costs in the European Union of increased rumen efficiency

Measure	Reduction per head (%)	Ktonne reduction in 2020 in EU			Costs $/t CH_4		Available from year
		Dairy	Non-dairy	Total	Dairy	Non-dairy	
Propionate precursors	25	69	270	339	2729	5686	2005
Hexose partitioning	15	41	162	203	n/a	n/a	2010
Methane oxidizers	8	22	86	108	n/a	n/a	2010
Probiotics	8	21	81	102	5440	11,332	Now
High genetic merit	20	364	n/a	364	0	0	Now

Source: AEAT (1998)

Methane from manure

If manure is kept in liquid or slurry storage, at temperatures higher than about 15°C and storage periods longer than 100 days, methanogenic bacteria will produce CH_4 in significant quantities (Zeeman, 1994). While, if manure is kept in liquid or slurry storage, at temperatures of 10°C or less and storage periods less than 100 days, CH_4 will not be produced in significant quantities (Zeeman, 1994). The amount of CH_4 produced is dependent on the manure composition and manure management systems. Methane emission reduction measures can be directed at preventing conditions where fermentation starts, or at stimulation of controlled fermentation with CH_4 recovery for energy generation. Small-scale manure digesters are widespread in China and India. Large-scale systems can be found in Denmark. The following reduction options can be described.

Option 14 Dry storage
Methane is produced during anaerobic decay of the organic material in livestock manure. It is emitted from manure if fermentation takes place over longer periods in anaerobic liquid or slurry storage. The optimum temperature is 15–35°C. Manure can also be stored dry and cool, to prevent fermentation. Manure management varies in different regions of the world. Therefore a reduction is only expected in regions with high animal densities where liquid/slurry storage is practised. A reduction of about 10 per cent can be achieved in those regions, with a cost of US$200 per tonne of avoided CH_4 (AEAT, 1998).

Option 15 Daily spreading
With daily spreading of manure on the land, anaerobic degradation can be minimized by avoiding anaerobic soil conditions. Ammonia and N_2O loss should be taken into account with this option. Up to 90 per cent of ammonia is lost to the atmosphere within 12 hours of spreading, therefore it is important to incorporate the manure into the soil to minimize this loss. Switching to daily spreading of manure can reduce related CH_4 emissions by 10–35 per cent in Europe. However it is a labour intensive and therefore a costly measure, costing ~$2000 per tonne of avoided CH_4 emission (AEAT, 1998).

Option 16 Large-scale digestion and biogas recovery
Advanced farm-scale and village-scale digesters are in use in some European countries with high intensity livestock farming, for example Denmark and Germany. Manure is transported from the farms to the farm-scale or village-scale digester. The technology is proven and the economic performance is acceptable. If introduced in all OECD countries and China, a 10 per cent reduction would be expected in CH_4 emissions from manure in these regions. Costs are ~$1000 per tonne CH_4 for large-scale digesters (AEAT, 1998).

Option 17 Small-scale digestion and biogas recovery

Small-scale digesters are found in temperate and tropical regions to enhance the anaerobic decomposition of the manure and to maximize CH_4 production and recovery. In OECD countries they are not used. Fermented manure is a high-quality fertilizer and an important product in resource-constrained regions. A reduction of 10 per cent of manure-related CH_4 emissions is expected in these regions, if introduced widely in India, China, East Asia and Latin America. Costs are ~$500 per tonne of avoided CH_4 emission for small-scale digesters. Benefits are the waste reduction and small-scale energy production for households and farms. A more detailed cost estimate of these options is given in Table 13.3.

Table 13.3 Costs of measures to reduce CH_4 emissions from animal manure in Europe in $/t CH_4

Measure	Cool climate			Temperate climate		
	Pigs	Dairy	Beef	Pigs	Dairy	Beef
Daily spreading	2264	4124	5505	645	1183	1579
Large-scale digestion, UK	450	828	1106	75	138	184
Large-scale digestion, Denmark	2287	4211	7878	500	921	1723
Small scale, Germany (heat and power)	286	526	703	48	88	117
Small scale, Germany (heat only)	95	175	234	16	29	39
Italy (CHP)	181	334	446	30	56	74

Source: AEAT (1998)

Methane emission reductions from wastewater and sewage treatment

Emissions from wastewater systems are still highly uncertain. The OECD countries have invested in wastewater treatment plants with CH_4 recovery. The problem is mainly concentrated in regions without this technology. At present for the wastewater sector, industrial wastewater in developing regions is considered to be the main CH_4 source (Thorneloe, 1993; Doorn and Liles, 2000). Two possible routes for reduction of CH_4 emissions can be followed. Aerobic treatment technologies can be selected, which prevents the formation of CH_4. Alternatively, anaerobic treatment technologies can be used, with CH_4 recovery for energy generation (Lexmond and Zeeman, 1995). Anaerobic treatment is commercially available. Fuel can be produced from the CH_4 for about $1 per GJ (Lettinga and van Haandel, 1993). This emissions reduction option is summarized below.

Option 18 Increased on-site use of biogas
Sewage treatment can result in CH_4 emissions if stored and treated under anaerobic conditions and if the CH_4 is released instead of being recovered for energy generation. In the OECD countries, CH_4 is often already recovered in closed systems and only a 10 per cent reduction in emissions is expected through increased on-site biogas use. In other regions, larger reductions – of up to 80 per cent – are expected if these closed systems are introduced. The costs are variable, being between $50 and $500 per tonne of CH_4 reduction (Byfield et al, 1997).

Methane from landfills

Methane is formed in landfills by methanogenic microorganisms (methanogens). The CH_4 generation is higher with a high organic matter content. The CH_4 is formed over a shorter time period where there is a higher content of easily degradable material, such as fruit and vegetable waste. The CH_4 emission shows a time lag after the point at which waste is landfilled. The CH_4 emission in time can be estimated by a simple first-order decay function. Research in The Netherlands, in which measured amounts are compared with estimates, shows that the uncertainty of the first-order decay function is 22 per cent. It also shows that the accuracy of the estimate is not improved if more detailed estimation functions are applied. About $1.87m^3$ landfill gas (with 57 per cent CH_4) is formed per kg of organic carbon (Oonk et al, 1994). The most important mitigation strategies are to reduce, reuse and recycle waste. Those are volume-related measures and not covered in this context. Three technical options can be formulated for CH_4 emission reductions from landfills. The first is to optimize waste gas formation and to recover as much as possible for energy use. This is one of the most promising and profitable options worldwide for CH_4 reductions. The waste gas is recovered from boreholes or gas trenches. To optimize waste gas recovery an impermeable clay layer cover is needed. The second option is to separately collect organic waste to be fermented in closed systems with CH_4 recovery for energy generation. The third option is controlled gasification of organic waste and wood. Small-scale CHP systems can be fuelled with organic waste and wood from gardens and village parks. The following reduction options can be described.

Option 19 Landfill gas recovery and use for heat
Methane is formed in landfills during anaerobic decomposition of the organic material in the waste. Gas can be recovered by drilling holes in the landfill and collecting the gas. It can be used directly at the site or distributed to off-site buildings and industries. This option is beneficial, with an estimated cost of –$50 per tonne of avoided CH_4 (Meadows et al, 1996).

Option 20 Landfill gas recovery and upgrading
Landfill gas can also be upgraded to natural gas quality for use in a gas grid. This option is profitable with a 'cost' of –$200 per tonne of avoided CH_4. If

not enough gas is available for use it can be flared. These options are a success in the OECD countries. A 50 per cent reduction in CH_4 emissions from landfill is possible in these regions using this strategy. Power generation costs ~$23 per tonne CH_4. Flaring costs ~$44 per tonne reduced CH_4. Improved capping of landfills at $600 per tonne reduced CH_4, though expensive, is important for an early start of landfill gas recovery.

Option 21 Controlled gasification of waste
Gasification of waste is a proven technology. It seems to be the option with the highest profit, at about $350 per tonne of CH_4 reduction. Introduced in the OECD regions, expected reductions are some 80 per cent of CH_4 from landfills (De Jager et al, 1996).

Option 22 Reduction of biodegradable waste to landfill
Waste to landfill can be reduced by recycling, open composting, closed composting or incineration. Paper recycling is the most profitable option with a 'cost' of –$2200 per tonne reduced CH_4.

Option 23 Reduction of biodegradable waste to landfill by composting or incineration
Open composting is expensive with a cost of ~$1000 per tonne reduced CH_4. Incineration of waste costs ~$1423 per tonne reduced CH_4 and 'closed composting' costs as much as $1800 per tonne reduced CH_4.

Option 24 Fermentation of organic waste
Instead of landfilling organic waste, it can be separately collected and fermented in closed systems. This option is in study in Europe and the US. At the moment it is more expensive than landfill gas recovery. If introduced, a 50 per cent reduction is expected in CH_4 emissions from landfills in OECD Europe and North America, at a cost of $500 per tonne of CH_4 reduction. AEAT (1998) estimated the cost of this option in Europe at ~$1860 per tonne reduced CH_4.

Methane from wetland rice

Methane emissions from wetland rice have increased since the 1950s, with increasing cultivation area. Reduction strategies have to be found that do not interfere with yields. Denier van der Gon (2000) has suggested some strategies. These options have to be further explored. Intermittent draining is suggested, but this can only be applied in areas with abundant water.

Option 25 Intermittent draining and other cultivation practices
Methane is formed during the growing season in the flooded soils of rice fields. Present knowledge indicates that various cultivation practices could potentially achieve significant reductions. These include cultivar selection, water management and nutrient application and soil management. Potentially, a

reduction of 30 per cent in CH_4 emissions from typical rice regions can be achieved with low costs. In other regions, a 5 per cent reduction may be possible by 2025. The costs are low if water is abundant with a cost of ~$5 per tonne of avoided CH_4 (Byfield et al, 1997).

Trace gases from biomass burning

Biomass burning is a source for various trace gases such as CO, CH_4, NO_x and VOCs (Chapter 7). Especially high concentrations are measured during the smouldering of biomass (Delmas, 1994). One option to reduce CH_4 emissions is therefore to increase the burning efficiency of biomass.

Option 26 Improved burning of traditional fuelwood

Emissions of trace gases like CH_4, non-methane volatile organic carbons (NMVOCs), CO and NO_x in traditional biomass burning are relatively high because of the burning conditions (for example the slow smouldering of moist biomass). Modern technology for biomass burning leads to lower emissions. One such promising technology is biomass integrated gasifier/intercooled steam-injected gas turbine (BIG/ISTIG) technology. If introduced, trace gas emissions from traditional fuelwood could be reduced by 90 per cent in all regions at low cost. A profit of $100 per tonne of avoided CH_4 was estimated in Byfield et al (1997).

Option 27 Reduced deforestation

Reduced biomass burning often also leads to lower rates of deforestation. It is estimated that this measure costs ~$200 per tonne of avoided CH_4 emission (Byfield et al, 1997).

Option 28 Reduced agricultural waste and savanna burning

In most European countries, agricultural waste burning has been abandoned or is prohibited. However, in tropical countries it still occurs and can lead to both air pollution and CH_4 emission problems. Alternative handling of agricultural waste for example storage and application can reduce burning rates. This costs ~$150 per tonne of avoided CH_4 according to Byfield et al (1997).

Most cost estimates originated from studies in industrialized countries. Costs of measures taken in other world regions represent a real gap in current knowledge. In the integrated analysis in the next section two scenarios are formulated. Measures are taken to reduce CH_4 emissions in all world regions. Three CH_4 emission reduction packages are formulated for each scenario. Six resulting CH_4 reduction strategies are analysed. Here the costs of reductions are calculated off-line.

Methane emission control costs

I have chosen a cost-effectiveness analysis because it is difficult to estimate the benefits of reduced climate change. It is a sector-level analysis in combination with the detailed bottom-up cost-effectiveness estimation methodology. A baseline without climate policies is needed in such an analysis. The baseline by definition gives the emissions of greenhouse gases in the absence of climate change interventions. The baseline is critical to the assessment of the costs of climate change mitigation because it determines the potential for future greenhouse gas reduction as well as the costs of implementing these reduction policies. The baseline also has a number of important implicit assumptions about future economic policies at the macroeconomic and sectoral levels, including sectoral structure, resource intensity, prices and technology choice. In my analysis, the 'P1 and Q1' scenarios, as described in detail below, are considered the baselines against which the costs will be analysed.

Scenarios

Two contrasting scenarios and three methane emission reduction packages are developed. Each scenario is for the long run (2000–2100). The scenarios depict different possible futures and the reduction packages show different ways of combining methane abatement options. An existing integrated assessment model (IMAGE) is used to analyse the consequences of six emission reduction strategies for future air temperature. A reduction strategy is taken here as a scenario plus a methane abatement package.

The scenarios P and Q are in general agreement with scenarios as described in the IPCC *Special Report on Emission Scenarios* A1B-IMAGE and B1-IMAGE (Table 13.4) (IPCC, 2000). For the respective scenarios (P and Q), P1 and Q1 here contain no CH_4 abatement (i.e. they are the baselines to which reduction strategies are compared), P2 and Q2 contain moderate methane abatement, and P3 and Q3 contain maximum methane abatement. The storylines of the P1 and Q1 baseline scenarios, and of the CH_4 reduction strategies P2 and Q2 and P3 and Q3, are given below.

Scenario P describes a prosperous world with an economic growth of 3 per cent per year, with relatively low population growth, resulting in 8.7 billion people in 2050 and 7.1 billion in 2100. Present trends in globalization and liberalization are assumed to continue, in combination with a large technological change through innovations. This leads to relatively high economic growth in both the industrialized and the non-industrialized regions of the world. Affluence, in terms of per capita gross regional product, converges among world regions, although the absolute differences in affluence are growing. Increasing affluence results in a rapid decline in fertility. The global economy expands at an average annual rate of 3 per cent to 2100, reaching around $525 trillion. This is about the same as average global growth since 1850. Global average income per capita reaches about $21,000 by 2050 contributing to a great improvement in

Table 13.4 Basic assumptions of scenarios from the *Special Report on Emission Scenarios*

Scenario	A1Biofuels	B1	A2	B2
Population 2020	7.5 billion	7.6 billion	8.2 billion	7.6 billion
Population 2050	8.7 billion	8.7 billion	11.3 billion	9.3 billion
Population 2100	7.1 billion	7.1 billion	15.1 billion	10.4 billion
World GDP in 2020 (10^{12} US$_{1990}$)	56	53	41	51
World GDP in 2050 (10^{12} US$_{1990}$)	181	136	82	110
World GDP in 2100 (10^{12} US$_{1990}$)	525	330	243	235
Resource base	Includes unconventional (oil, hydrates etc.)	Identified resources	Includes unconventional (oil, hydrates etc.)	Identified resources

Note: GDP = gross domestic product.
Source: IPCC (2000)

health and social security for the majority of people. High income translates into high car ownership, sprawling suburbia and dense transport networks. Increasing service and information orientation leads to a significant decline in intensity of energy and materials. Energy and mineral resources are abundant in this scenario because of the rapid technical progress. This reduces resource use per unit of output, but also the economically recoverable reserves. Methane emissions from fossil sources are growing. Final energy intensity (end-use energy per GDP) decreases at a rate of 1.3 per cent per year. With the rapid increase in income, dietary patterns are assumed to shift initially to a high meat and milk diet with related increasing animal numbers and increasing methane emissions from animals and manure, but this may decrease subsequently with increasing emphasis on the health of the ageing society. Growth could produce increased pressure on the global resources. Conservation of natural areas is changing in management of natural resources.

Scenario Q is a contrasting scenario with a much lower demand in all sectors because of the lower GDP development (<3 per cent per year growth). The Q scenario describes a world where the modernization of the OECD countries has spread to the other regions over the period 2000 to 2100. A shift takes place from an economy predominantly relying on heavy industry towards a society with an economy predominantly based on services with an increasing dematerialization of production, recycling of materials and increasing energy efficiency. As in scenario P, the world population starts to drop after 2050 from 8.7 billion downward to 7.1 billion people. Combined with a population that peaks in 2050 and returns to numbers below the expected 2020 world population, this leads to a moderate increase of CO_2 and non-CO_2 greenhouse gas emissions and resulting concentrations in the atmosphere for Q. Increasing affluence leads to better living conditions, birth control and health care. Reduced fertility leads to a stabilizing global population in the middle of this century and a gradual decrease of global population between 2050 and 2100. Urbanization is halted or even reversed to more decentralized living supported by the information revolution. Affluence measured as the per capita gross regional product converges among world regions at a faster rate than in the BAU IPCC92 scenarios. The economy becomes more oriented to services and information exchange. As a consequence there is an increasing decline in the energy intensity and materials intensity of production. Renewable resources increasingly replace fossil fuels. The methane emissions from fossil sources will decrease compared to scenario P. The growing energy demand in the tropical regions and the high degree of energy efficiency measures makes electricity the most important energy carrier. Technology transfer from OECD countries to less developed regions is very successful in Q, and the industry in the more populous regions such as India and China is converted to comply with the highest pollution prevention standards in the world. Fuel desulphurization is becoming standard. The energy efficiency of power production and production in iron and steel and the chemical industry in India and China is increasing at a high rate of at least 1–1.5 per cent per year. In power production the conversion efficiency increases by at least 10 per cent between 2000 and 2100 to 48 per cent for coal, 53 per cent for oil and 58 per cent for gas. Transmission losses in power lines decrease to 8 per cent in all regions in 2100. Lower energy and material intensity in manufacturing results in falling energy demands in industry. Technology transfer from the industrialized countries to the less industrialized regions is accelerated to combat pollution. Materials recycling becomes a global business. Increasing recycling of waste reduces landfill methane emissions. To solve congestion, public transport is boosted by large investments. Examples are subways in larger cities, bicycling lanes and clean electric buses. Fast trains connect the larger cities. Air traffic is largely intercontinental. Private cars remain important, but saturation occurs in use at lower than present-day levels in the US. Hybrid and electric cars are increasing because of low petrol use, low noise and reduced pollution. Bicycling also increases. The rapid expansion of telecommunication and information technology gives the less developed regions large opportunities. Cell phones and satellite systems become the means of communication in Africa and Latin America. The growth of mega-cities is slowing down. Governments understand that in metropolitan areas large investments are needed in public transport to reduce urban pollution. The quality of systems to collect garbage and of landfill management needs improving. Landfill gas recovery and use is improved, but especially in the methane reduction strategies. Traditional burning of biomass is abandoned and biomass is increasingly used to produce liquid fuels or in BIG/ISTIG technology. Especially for the land-rich regions such as Latin America and Africa the production costs of biomass-derived fuels drop to $2–3/GJ. The land surface that is used for biomass fuels is at least 800 million ha or the size of Brazil to meet demands. Plantations for biomass or biofuels show a strong increase in all regions. Trade in biomass-derived liquid fuels between regions is increasing. Cultivation occurs on surplus agricultural land and does not lead to additional deforestation. In earlier scenario analysis with a low energy supply system (Leemans et al, 1998), plantation forests were assumed to encroach on virgin tropical forests with a risk of declining biodiversity. Globally, the pressure due to the major increase in the demand for food and fodder in the period 1990–2030 is almost completely compensated by an increase in productivity. In agriculture, average yields in cereals for example in non-OECD countries increase by a factor of four. In OECD countries, these increase by a factor of two. Forest area is declining until 2030 but expanding after 2030. More efficient agriculture and improved production relieve the pressure on pristine forests. The improved production in agriculture leads to an expansion of forest area of 30 per cent in the period between 2030 and 2100. Large forest reserves are implemented and developed for eco-tourism. The number of dairy cattle shows a decline in this century because the improvement in production efficiency is faster than the growth of milk consumption. The number of animals slaughtered for beef increases in the period 1995 to 2060 as a result of increasing consumption of meat notwithstanding the increasing animal productivity. The consumer trend is away from the Western-style diet with high meat consumption, as people become aware of its implications on land use and health. This results in lower livestock numbers and related reductions in methane emissions. Farmers shift to sustainable practices. As a result, fertilizer use starts declining. This reduces nitrous oxide emissions from high-input agriculture. Subsistence agriculture and fuelwood use rapidly decline. Self-sufficiency in food production is increasing but food trade remains large in a safe world. Logging becomes sustainable and most wood is produced from plantations. In some regions, production of commercial biofuels is increasing. Large pristine forest areas are converted to conservation areas to safeguard biodiversity. Promoting compact cities and major transport and communication corridors controls human settlements. Current infrastructure is improved rather then extended.

The total costs of the efforts needed to limit CH_4 emissions in a growing economy under the P2, P3 and Q2, Q3 scenarios are then calculated by comparing with the baselines of P1 and Q1.

Assumptions about costs of measures

Often a cost curve of measures is made to illustrate that profitable and cheap options will be chosen first and the more expensive options will be taken later. In general, an assumption is made that costs of the same measures will be reduced when taken later in time. I have taken a different approach. Reduction options have been introduced simultaneously in different sectors. I assumed that the cheapest options (i.e. <$50 per tonne CH_4) for each sector would be chosen before 2025. I assumed that the more expensive options would be introduced in P3 and Q3 (maximum CH_4 reduction) only after 2025.

Biomass burning

For a reduction of CH_4 emissions from biomass burning, a cost estimate is used of $200 per tonne of reduced CH_4 for reduced deforestation for all regions and for all years from Byfield et al (1997).

Waste and savanna burning

For agricultural waste and savanna burning a cost estimate of $150 per tonne of reduced CH_4 is used, based on reduced burning of agricultural waste and savanna in all regions and all years from Byfield et al (1997).

Landfills

For landfills it is assumed that the most profitable measures could only be taken after development of the gasification technique. Therefore a 'cost' is assumed of –$50 in 1990 and 2000 for the introduction of landfill gas recovery for heat generation (where a negative 'cost' denotes a net profit), based on information from AEAT (1998) and Blok and de Jager (1994). A 'cost' is assumed of –$200 per tonne of reduced CH_4 in 2025 and 2050 for landfill gas recovery and upgrading, based on Meadows et al (1996) and AEAT (1998). A 'cost' is assumed of –$350 in 2075 and 2100 for controlled gasification of waste, based on Blok and de Jager (1994). Because of a lack of information no distinction between regions could be made, therefore it is assumed that the costs are the same in all regions.

Sewage treatment

For CH_4 reductions in sewage treatment a steady cost increase is assumed from $50 in 1990 to $500 in 2100 for all regions, based on information from Byfield et al (1997) for the increased on-site use of biogas. A cost of $50 is assumed in 1990–2010 in OECD regions where sewage treatment plants have already been build, and $100 in 2000, $200 in 2025, $300 in 2050, $400 in 2075 and $500 in 2100 in non-OECD regions. The reason for the increase in costs is that the investment costs for sewage treatment plants is rather high and sewage treatment plants will have to be built predominantly in the non-OECD regions. As noted earlier, wastewater and sewage treatment plants are mainly developed to improve the health and living conditions of the people. Anaerobic digesters can be flared or the CH_4 used for cogeneration to reduce CH_4 emissions from biomass or liquid effluents with high organic content. Because most centralized systems automatically either flare or capture and use CH_4 for safety reasons, add-on abatement technology for existing wastewater treatment plants does not exist. As a result, potential emission reductions depend on large-scale structural changes in wastewater management. For this reason, the cost of CH_4 reduction in sewage treatment is difficult to estimate. Overriding economic and social factors influence wastewater treatment practices throughout the world. The benefits of installing wastewater systems in developing countries for the purpose of disease reduction greatly outweigh potential benefits associated with CH_4 reduction. It would be misleading to imply that costs of CH_4 measures would be the only driving force behind investment decisions that influence CH_4 emissions from wastewater.

Rice paddies and wetlands

A reduction in CH_4 emissions from rice fields is relatively easy to achieve. Costs for CH_4 reduction in wetland rice are only $5 per tonne CH_4 for intermittent draining and other cultivation practices, based on information from Byfield et al (1997). Methane reduction from natural wetlands is not practised, although reclaiming wetlands is a feasible but expensive measure, only to be used to increase agricultural lands. Drainage of natural wetlands also risks enhancement of CO_2 losses from these systems. No significant net emissions reductions from this latter strategy are expected therefore.

Enteric fermentation

The reduction in CH_4 emissions from ruminant animals is a combined effect of increased genetic merit resulting in more efficiency in meat and milk production and reduced animal numbers. The costs of genetic improvement are very low, with an estimated $5 per tonne reduced CH_4 for improved feeding (EPA, 1998). Production-enhancing agents are more expensive (about $400, according to AEAT, 1998) and also unacceptable to many consumers in Europe. Products to increase rumen efficiency are in the experimental stage. They are even more expensive (about $3000 to $6000 according to AEAT, 1998), are also unacceptable to many consumers and have not yet been introduced.

Animal waste

Methane emission reduction by anaerobic digestion in animal waste management systems with biogas recovery and use is expensive. Costs are variable depending on climate and an overall cost of $500 is assumed in all OECD regions. Costs are higher – with $1000 – in non-OECD regions, where manure digestion can only be introduced at relatively high investment costs.

Fossil fuel exploitation

For CH_4 emissions from fossil fuel exploitation it is assumed that the most profitable measures would be taken first. Therefore, increased maintenance is assumed at a 'cost' of –$200 per tonne CH_4 in 1990 and increased on-site use of otherwise vented gas at a cost of –$100 in 2000 and 2025. Other measures are taken later in time at a cost of $100 in 2050, $200 in 2075 and $300 in 2100. In 1990, the introduction of improved inspection and maintenance is assumed. In 2000 and 2025 extra measures are taken to increase on-site gas use from vents and flares. The more expensive measures are taken between 2050 and 2100. Cost estimates are based on AEAT (1998) and De Jager et al (1996). The cost development is based on my own assumptions.

The content of the six CH_4 reduction packages is summarized in Table 13.5 along with the assumptions made on the costs of measures. It is assumed that

Table 13.5 Overview of options and costs that were used in the six methane reduction strategies in this study

Source of CH$_4$	Description of option	US$$_{1990}$/ tonne of avoided CH$_4$ per year	Reference
Oil and gas production	Increased inspection and maintenance	−200	De Jager et al (1996)
	Increased on-site use of otherwise vented methane at oil and gas production sites offshore	−100–10	De Jager et al (1996)
	Increased flaring instead of venting	200–400	De Jager et al (1996)
Oil and gas transmission	Accelerated pipeline modernization	500–1000	De Jager et al (1996)
Gas distribution	Improved leak control and repair	200	De Jager et al (1996)
Coal mining	Pre-mining degasification	40	IEA (1999)
	Enhanced gob-well recovery	10	IEA (1999)
	Ventilation air use	10	IEA (1999)
Cattle enteric fermentation	Improved production efficiency	0	Blok and de Jager (1994)
	Improved feeding	5	EPA (1998)
	Production enhancing agents	400	AEAT (1998)
	Reducing animal numbers	0	Blok and de Jager (1994)
	Increase rumen efficiency	3000–6000	AEAT (1998)
Manure	Dry storage	200	AEAT (1998)
	Daily spreading	2000	AEAT (1998)
	Large-scale digestion and biogas recovery	1000	AEAT (1998)
	Small-scale digestion and biogas recovery	500	AEAT (1998)
Sewage treatment	Increased on-site use of biogas	50–500	Byfield et al (1997)
Landfills	Reduction of biodegradable waste to landfill by paper recycling	−2200	Meadows et al (1996)
	Controlled gasification of waste	−350	De Jager et al (1996)
	Landfill gas recovery and upgrading	−200	Meadows et al (1996)
	Landfill gas recovery and use for heat	−50	Meadows et al (1996)
	Reduction of biodegradable waste to landfill by composting or incineration	1000–1800	Meadows et al (1996)
Rice	Intermittent draining and other cultivation practices	5	Byfield et al (1997)
Biomass burning	Improved burning of traditional fuelwood	−100	Byfield et al (1997)
	Reduced deforestation	200	Byfield et al (1997)
	Reduced agricultural waste and savanna burning	150	Byfield et al (1997)

the more expensive options are taken after 2025 and that measures costing more than $500/tonne avoided CH_4 were too expensive for adoption. Table 13.6 gives an overview of the costs of mitigation by source sector and its assumed variation over time (1990–2100).

Table 13.6 Costs of reduction measures in US$$_{1990}$ tonne^{-1} of CH_4 yr^{-1} in the period 1990–2100 as input to IMAGE

Source	1990	2000	2025	2050	2075	2100
Biomass burning	200	200	200	200	200	200
Agricultural waste burning	150	150	150	150	150	150
Savanna burning	150	150	150	150	150	150
Landfills	−50	−50	−200	−200	−350	−350
Sewage	50	100	200	300	400	500
Wetland rice	5	5	5	5	5	5
Animals	5	5	5	5	5	5
Animal waste	500	500	500	500	500	500
Fossil fuel exploitation	−200	−100	−100	100	200	300

Source: Based on own assumptions Van Amstel (2009) and Blok and de Jager (1994); De Jager et al (1996); Meadows et al (1996); Byfield et al (1997); EPA (1998); IEA (1999); and AEAT (1998)

Results

Twenty-seven CH_4 mitigation options have been identified in this analysis. At least nine profitable zero cost options are apparently available, mainly in coal, oil and gas production. The most expensive options involve increasing rumen efficiency and manure management. The cost estimates vary, but at least a common methodology is used to arrive at broadly comparable results. The overall costs per sector for six CH_4 reduction strategies in the scenarios are now estimated. Costs of reduction strategies are estimated against baseline scenarios.

Cost estimates of six reduction strategies

The total cost estimates for the different reduction packages are estimated relative to the baseline reduction strategies P1 and Q1. The cost estimates are given in terms of US$$_{1990}$ for six time steps between 1990 and 2100. The estimates beyond 2025 must be interpreted with caution because the reduction strategies P1 and Q1 are based on a number of important implicit assumptions about future economic policies at the macroeconomic and sectoral levels,

Table 13.7 Worldwide costs in US$$_{1990}$ thousands of CH$_4$ abatement in reduction strategies P2 and Q2

P2	1990	2000	2025	2050	2075	2100
Biomass burning	0	−2000	44,200	−15,000	17,200	−10,200
Agricultural waste	0	80,550	625,950	1,355,100	1,499,250	1,406,700
Savanna burning	0	900	360,300	700,350	1,069,200	1,552,650
Landfills	0	0	−3,558,400	−10,788,400	−26,329,800	−28,511,000
Sewage	0	0	0	0	0	0
Wetland rice	0	−25	−5	−45	440	625
Animals	0	−10	−250	165	50	−245
Animal waste	0	−100	900	−2200	−2500	300
Leakage	0	0	0	0	0	0
Total sectors	0	79,315	−2,527,305	−8,750,030	−23,746,160	−25,561,170
Q2	**1990**	**2000**	**2025**	**2050**	**2075**	**2100**
Biomass burning	0	−12,800	−4200	51,800	44,200	4800
Agricultural waste	0	79,500	493,650	1,329,450	1,423,200	1,242,750
Savanna burning	0	0	300,450	822,450	1,260,750	1,897,200
Landfills	0	0	−2,476,400	−10,36,5800	−24,123,750	−27,186,600
Sewage	0	0	0	0	0	0
Wetland rice	0	75	−40	770	−500	185
Animals	0	70	−20	−410	−700	−220
Animal waste	0	200	−1400	−2100	−2400	−800
Leakage	0	0	0	0	0	0
Total sectors	0	67,045	−1,687,960	−8,163,840	−21,399,200	−24,042,685

Note: Negative costs are profits.

including sectoral structure, resource intensity, prices and technology choice. Costs of moderate abatement in P2/Q2 are given in Table 13.7 and costs of maximum abatement in P3/Q3 are given in Table 13.8.

To put these costs in perspective, 1 per cent global GDP increase equals $250 billion. The profits in P2 in 2100 are about 0.1 per cent GDP. The costs are lower than 0.1 per cent GDP in all years according to these tables. Total costs in 2000 are $67–93 million. In 2025 total costs are negative in all strategies, so profits can be made by reducing CH$_4$ of between $1.7 billion and $6.7 billion. After 2050, in the moderate CH$_4$ strategies P2 and Q2, profits can also be made. In 2050, in the maximum abatements strategies P3 and Q3, costs have to be incurred because some expensive options are included. After 2050,

Table 13.8 Worldwide costs in US$₁₉₉₀ thousands of CH_4 abatement in reduction strategies P3 and Q3

P3	1990	2000	2025	2050	2075	2100
Biomass burning	0	−11,400	39,400	53,000	21,000	0
Agricultural waste	0	0	300	−3150	−7500	47,550
Savanna burning	0	150	455,400	877,050	1,496,550	2,331,450
Landfills	0	0	−4,448,400	−13,486,200	−36,861,650	−42,767,550
Sewage	0	0	1,873,600	6,191,700	11,033,600	15,031,500
Wetland rice	0	0	8125	40,640	52,785	57,430
Animals	0	35	60	690	2230	−975
Animal waste	0	−400	3000	1300	1300	−600
Leakage	0	80,900	−3,816,400	7,902,700	21,690,400	30,019,500
Total sectors	0	69,285	−5,884,915	1,577,730	−2,571,285	4,718,305
Q3	1990	2000	2025	2050	2075	2100
Biomass burning	0	12,600	1000	3800	79,400	4800
Agricultural waste	0	79,800	493,350	1,329,450	1,423,050	1,242,900
Savanna burning	0	−900	373,950	1,031,400	1,724,250	2,847,900
Landfills	0	0	−3,095,800	−12,956,800	−33,170,900	−40,779,200
Sewage	0	0	1,444,400	6,191,700	10,616,800	15,031,500
Wetland rice	0	75	45	46,140	57,300	66,625
Animals	0	175	−620	−1385	455	1360
Animal waste	0	600	−4500	−5100	1900	1200
Leakage	0	0	−5,924,800	9,210,200	14,190,600	11,583,300
Total sectors	0	92,350	−6,712,975	4,849,405	−5,077,145	−9,999,615

Note: Negative costs are profits.

costs are variable in the maximum CH_4 abatement strategies P3 and Q3 because both the profits in landfill CH_4 mitigation and the costs in the sewage sector and fossil fuel industry are high. However, costs are negative in all measures where CH_4 is captured and used for energy generation. Costs become higher after 2050 in the sewage sector because more regions are including sewage treatment, and in addressing leakage in the fossil fuel industry because increasingly expensive options are deployed.

Costs to reduce CH_4 from landfills and the fossil fuel industry depend to a large extent on the value of the CH_4 for the energy companies involved in utilizing any captured CH_4. Measures will be less profitable in a market with abundant alternatives. Recently, over-capacity of electricity production in different regions in the world resulted in lower prices for CHP and in the price for alternative fuels. In such a situation, investment in technical measures for CH_4 reduction from landfills and the fossil fuel industry will be more difficult to obtain. It is very difficult to look far into the future to predict the profits from CH_4 capture. Here, only a first attempt has been made and the results beyond 2025 must be considered as very preliminary. From Table 13.7 it seems that, overall, profitable reductions in CH_4 are possible in the moderate CH_4 reduction packages. The profitable reductions in the maximum CH_4 abatement packages in Table 13.8 seem to be outweighed by the more expensive options, resulting in high costs in 2050, reduced costs in 2075, and high costs in 2100 for P3, but profits for Q3.

Methane capture from landfills can be very profitable because of the potential for energy generation and related revenues. Sewage treatment with increased on-site use of CH_4 is very expensive because the energy that is generated is not sold to another party. In sewage treatment, the investment costs in terms of CH_4 mitigation alone are very high and sewage treatment has yet to be introduced in many regions in the world. However, the benefits of investment will probably also include improved human health and reduced water pollution. Methane capture in the coal, oil and gas sectors is very expensive in the longer term because the cheap/very profitable options have already been taken before 2025.

Conclusions

We calculated the technological potential for CH_4 reduction – the amount by which it is possible to reduce greenhouse gas emissions or improve energy efficiency by implementing a technology or practice that has already been demonstrated. The emission reduction is calculated with respect to a baseline scenario of development.

The economic potential is the proportion of technological potential for CH_4 emission reductions or energy efficiency improvements that could be achieved cost effectively through the creation of markets, reduction of market failures, increased financial and technological transfers. The achievement of economic potential requires additional policies and measures to break down market barriers. Measures are cost effective when the benefits of the measures are larger than the costs (including interest and depreciation). We could not calculate the economic potential of CH_4 reductions because it is difficult to assess the direction of future policies aimed at stimulating CH_4 reductions through the reduction of market failures. It is also difficult to assess the future price of oil, which helps to dictate the willingness to search for alternatives.

Our research question was: which options are available to reduce CH_4 emissions? We concluded that, in total, 27 options can be selected based on demonstrated technology that can be deployed immediately. Another research question was: what are the overall costs of the options? From calculations of the total reduction costs of the maximum CH_4 reduction strategies P3 and Q3 we can conclude that options to reduce CH_4 from landfills are very promising indeed. Options to reduce CH_4 from leakage in the fossil fuel industry are more expensive after 2025. Options to reduce CH_4 from sewage treatment are expensive because investment costs are high and CH_4 can rarely be sold to a third party.

Overall, it can be concluded that CH_4 emission reductions are relatively cheap in 2050 and 2100 with less than 0.1 per cent of GDP. The benefits include both global climate change mitigation and an improved public health through improvements in local air quality. Methane emissions reductions are all profitable until 2025 and again in 2075 and 2100 for Q3. Options to reduce leakage in oil and gas will become more expensive after 2025. The emission reductions of CH_4 after 2025 are very profitable for landfill gas. This option is so profitable that it reduced the overall costs of all options. Sewage treatment can become more profitable if CH_4 can be sold to third parties.

References

AEAT (AEA Technology) (1998) *Options to Reduce Methane Emissions*, AEAT-3773, issue 3, European Commission, Brussels

Blok, K. and de Jager, D. (1994) 'Effectiveness of non-CO_2 greenhouse gas reduction technologies', *Environmental Monitoring and Assessment*, vol 31, pp17–40

Byfield, S., Marlowe, I. T., Barker, N., Lamb, A., Howes, P. and Wenborn, M. J. (1997) *Methane from Other Anthropogenic Sources*, AEA Technology, Culham, Oxford, UK

Callan, S. J. and Thomas, J. M. (2000) *Benefit-cost Analysis in Environmental Decision Making*, The Dryden Press, Orlando, CA

De Jager, D., Oonk, J., van Brummelen, M. and Blok, K. (1996) *Emissions of Methane by the Oil and Gas System: Emission Inventory and Options for Control*, Ecofys, Utrecht, The Netherlands

De la Chesnaye, F. C. and Kruger, D. (2002) 'Stabilizing global methane emissions. A feasibility assessment', in J. van Ham, A. P. M. Baede, R. Guicherit and J. G. F. M. Williams-Jacobse (eds) *Non-CO_2 Greenhouse Gases: Scientific Understanding, Control Options and Policy Aspects*, Proceedings of the Third International Symposium, Maastricht, The Netherlands, Millpress, Rotterdam, Netherlands, pp583–588

Delhotal, K. C., de la Chesnaye, F. C., Gardiner, A., Bates, J. and Sankovski, A. (2005) 'Mitigation of methane and nitrous oxide from waste, energy and industry', *Multigas Mitigation and Climate Change*, Special Issue no 3, *The Energy Journal*, pp45–62

Delmas, R. (1994) 'An overview of present knowledge on methane emissions from biomass burning', *Fertilizer Research*, vol 37, pp181–190

Denier van der Gon, H. A. C. (2000) 'Changes in methane emissions from rice fields from 1960 to 1990: Impacts of modern rice technology', *Global Biogeochemical Cycles*, vol 14, pp61–72

Doorn, M. and Liles, D. (2000) 'Quantification of methane emissions from latrines, septic tanks and stagnant open sewers in the world', in J. van Ham (ed) *Non-CO_2 Greenhouse Gases*, Kluwer Academic Publishers, Dordrecht, pp83–89

EPA (US Environmental Protection Agency) (1998) *Inventory of US Greenhouse Gas Emissions and Sinks 1990–1996*, EPA 236-R-98-006, EPA, Washington, DC

EPA (2003) *Assessment of Worldwide Market Potential for Oxidizing Coal Mine Ventilation Air Methane*, EPA, Washington, DC

Gallaher, M. P., Petrusa, J. E. and Delhotal, C. (2005) 'International marginal abatement costs of non-CO_2 greenhouse gases', *Environmental Sciences*, vol 2, pp327–339

Graus, W., Harmelink, M. and Hendriks, C. A. (2003) *Marginal GHG-abatement Curves for Agriculture*, Ecofys, Utrecht, The Netherlands

Gunning, P. M. (2005) 'The methane to markets partnership: An international framework to advance recovery and use of methane as a clean energy source', *Environmental Sciences*, vol 2, pp361–367

Harmelink, M. G. M., Blok, K. and ter Avest, G. H. (2005) 'Evaluation of non-CO_2 greenhouse gas emission reductions in the Netherlands in the period 1990–2003', *Environmental Sciences*, vol 2, pp339–351

Hendriks, C. A. and de Jager, D. (2000) 'Global methane and nitrous oxide emissions: Options and potential for reduction', in J. van Ham et al (eds) *Non-CO_2 Greenhouse Gases*, Kluwer Academic Publishers, Dordrecht, pp433–445

Hunter, R. (2004) 'Characterization of the Alaska north slope gas hydrate resource potential: Fire in the ice', *The National Energy Technology Laboratory Methane Hydrate Newsletter*, Spring

IEA (International Energy Agency) (1999) *Technologies for the Abatement of Methane Emissions*, Report SR7, International Energy Agency Greenhouse Gas Research & Development Programme, IEA, Cheltenham, UK

IEA (2003) *Non-CO_2 Greenhouse Gas Network: Greenhouse Gas Reduction in the Agricultural Sector*, Report PH4/20, International Energy Agency Greenhouse Gas Research & Development Programme, IEA, Cheltenham, UK

IPCC (Intergovernmental Panel on Climate Change) (2000) *Special Report on Emissions Scenarios*, Nakicenovic, N., Alcamo, J., Davis, G., de Vries, B., Fenhann, J., Gaffin, S., Gregory, K., Grübler, A. et al, Working Group III, Intergovernmental Panel on Climate Change (IPCC), Cambridge University Press, Cambridge, 595pp, available at www.grida.no/climate/ipcc/emission/index.htm

IPCC (2006) *IPCC Guidelines for National Greenhouse Gas Inventories*, Prepared by the National Greenhouse Gas Inventories Programme, H. S. Eggleston, L. Buendia, K. Miwa, T. Ngara and K. Tanabe (eds), IGES, Japan

IPCC (2007) *Climate Change 2007: The Physical Science Basis. Contribution of Working Group I to the Fourth Assessment Report of the Intergovernmental Panel on Climate Change*, S. Solomon, D. Qin, M. Manning, Z. Chen, M. Marquis, K. B. Averyt, M. Tignor and H. L. Miller (eds), Cambridge University Press, Cambridge and New York

Kruger, D. (1993) 'Working group report: Methane emissions from coal mining', *IPCC Workshop on Methane and Nitrous Oxide*, National Institute for Public Health and the Environment, RIVM Bilthoven, The Netherlands, pp205–219

Leemans, R., van Amstel, A. R., Battjes, C., Kreileman, E. and Toet, S. (1998) 'The land cover and carbon cycle consequences of large scale utilization of biomass as an energy source', *Global Environmental Change*, vol 6, no 4, pp335–357

Lettinga, G. and van Haandel, A. C. (1993) 'Anaerobic digestion for energy production and environmental protection', in T. B. Johansson, H. Kelly, A. K. N. Reddy and R. H. Williams (eds), *Renewable Energy*, Island Press, Washington, DC, pp817–839

Lexmond, M. J. and Zeeman, G. (1995) *Potential of Controlled Anaerobic Waste Water Treatment in Order to Reduce the Global Emissions of the Greenhouse Gases Methane and Carbon Dioxide*, Wageningen University Technology Report 95-1, The Netherlands

Maione, M., Arduini, I., Rinaldi, M., Mangani, F. and Capaccioni, B. (2005) 'Emission of non-CO_2 greenhouse gases from landfills of different age located in central Italy', *Environmental Sciences*, vol 2, pp167–177

Mattus, R. (2005) 'Major coal mine greenhouse gas emission converted to electricity – first large scale installation', *Environmental Sciences*, vol 2, pp377–382

Meadows, M. P., Franklin, C., Campbell, D. J. V., Wenborn, M. J. and Berry, J. (1996) *Methane Emissions from Land Disposal of Solid Waste*, AEA Technology, Culham, UK

OLF (The Norwegian Oil Industry Association) (1994) 'Environmental Programme Phase II Summary Report', Stavanger, Norway

Oonk, H., Weenk, A., Coops, O. and Luning, L. (1994) *Validation of Landfill Gas Formation Models*, TNO-MEP, Apeldoorn, The Netherlands

Pacala, S. and Socolow, R. (2004) 'Stabilization wedges: Solving the climate problem for the next 50 years with current technologies', *Science*, vol 305, pp968–972

Schipper, L. (1998) *The IEA Energy Indicators Effort: Extension to Carbon Missions as a Measure of Sustainability*, IPCC Expert Group Meeting on Managing Uncertainty in National Greenhouse Gas Inventories, 13–15 October 1998, Maison de la Chimie, Paris

Thorneloe, S. A. (1993) 'Methane from waste water treatment', in A. R. van Amstel (ed) *International IPCC Workshop on Methane and Nitrous Oxide, Methods in National Emission Inventories and Options for Control*, RIVM, Bilthoven, The Netherlands, pp115–130

Ugalde, T. M., Kaebernick, M. Slattery, A. M. W. J. and Russell, K. (2005) 'Dwelling at the interface of science and policy: Harnessing the drivers of change to reduce greenhouse gas emissions from agriculture', *Environmental Sciences*, vol 2, pp305–315

Van Amstel, A. R. (2005) 'Integrated assessment of climate change with reductions of methane emissions', *Environmental Sciences*, vol 2, pp315–327

Van Amstel, A. R. (2009) 'Methane: Its role in climate change and options for control', Thesis, Wageningen University

Van Amstel, A. R., Swart, R. J., Krol, M. S., Beck, J. P., Bouwman, A. F. and van der Hoek, K. W. (1993) *Methane, the Other Greenhouse Gas*, Research and policy in the Netherlands, RIVM, Bilthoven, The Netherlands

Zeeman, G. (1994) 'Methane production and emissions in storages for animal manure', *Fertilizer Research*, vol 37, pp207–211

14
Summary

André van Amstel, Dave Reay and Pete Smith

Methane and climate change

The natural greenhouse effect is one of the reasons that we can thrive on this earth. Greenhouse gases such as CO_2, CH_4 and N_2O are transparent for the short-wave radiation from the sun, but they absorb part of the long-wave radiation (heat) emitted from the earth back into space. Without the natural blanket of greenhouse gases in the atmosphere (i.e. if the atmosphere contained only oxygen and nitrogen) the average temperature on earth would be −18°C instead of the more comfortable +15°C of today. The enhanced greenhouse effect is caused by increased emissions of greenhouse gases due to human activities, and this causes an increase in global mean surface and tropospheric temperature.

Methane is the most abundant organic trace gas in the atmosphere and, after carbon dioxide, the second most important greenhouse gas emitted by human activities. Average global concentrations of CH_4 have more than doubled since pre-industrial times, from 700ppb by volume to 1750ppbv. The concentration over the Northern Hemisphere is on average higher with 1800ppbv. Over dominant source areas like Western Europe concentrations occasionally increase to 2500ppbv.

Climate control

Methane emissions are already being addressed globally, along with those of other greenhouse gases, as a result of the UNFCCC, signed during the Earth Summit in Rio de Janeiro in 1992, and the subsequent Kyoto Protocol in 1997. Although the contribution of CH_4 to enhanced global warming is less than that of CO_2, CH_4 is very interesting in terms of mitigation policy because of its short lifetime in the atmosphere (~10 years) and high GWP (~25). Significant reductions in anthropogenic CH_4 emissions can therefore have a considerable effect in terms of reduced climate forcing within a few decades.

United Nations Framework Convention on Climate Change

The UNFCCC calls for the stabilization of greenhouse gas concentrations in the atmosphere at a level that would prevent dangerous anthropogenic interference with the climate system. Such a level is to be achieved within a timeframe sufficient to allow ecosystems to adapt naturally to climate change, to ensure that food production is not threatened and to enable economic development to proceed in a sustainable manner. As a first step towards achieving this objective, industrialized countries were required, but failed, to bring their greenhouse gas emissions back to 1990 levels by 2000. Most OECD countries, however, had adopted national emissions reductions targets that are in line with this requirement. Implicitly, a comprehensive approach was adopted taking into account all sources and sinks of all greenhouse gases. All industrialized countries that are a party to the Convention also have to report their national greenhouse gas emissions and their adopted response policies. The emissions and climate policies are reported in the National Communications of the Parties to the Convention. Inventories of emissions and sinks are required annually and independent experts nominated by Parties to the Convention review these National Communications.

Kyoto Protocol

Further reductions of greenhouse gases after 2000 were negotiated in Japan in 1997 in the Kyoto Protocol. An average reduction of greenhouse gas emissions of 5 per cent between 1990 and the commitment period of 2008 to 2012 was agreed between the industrialized nations. Europe agreed to a reduction of 8 per cent, Japan 7 per cent and the US 6 per cent. The US later withdrew from the Kyoto Protocol, but the Kyoto Protocol came into force in February 2005 following ratification by Russia.

A major sticking point in international negotiations over how to best reduce global greenhouse gas emissions has been money, with some economic models predicting very large implementation costs for the measures outlined as part of the Kyoto Protocol. However, where reductions are not just confined to reducing carbon dioxide emissions, but instead include the 'non-CO_2' greenhouse gases, such as CH_4, the predicted price of reductions falls considerably (Reilly et al, 1999).

In the Kyoto Protocol, a 'net flux' approach is adopted for a basket of greenhouse gases, including CO_2, CH_4, N_2O, hydro fluorocarbons, perfluorocarbons and sulphur hexafluoride (SF_6). In this net flux approach, CO_2 emitted from deforestation is counted as an emission, but carbon dioxide sequestered in forests that are planted after 1990 can be subtracted from the emissions. Further sink categories such as soils are still under negotiation.

National greenhouse gas emission inventories

To facilitate the reporting and review within the framework of the Climate Convention, credible and comparable data from countries are needed. Therefore, the IPCC in collaboration with the UNEP, WMO, the International Energy Agency (IEA) and the OECD have developed draft Guidelines for National Inventories of Greenhouse Gas Emissions and Sinks. These Guidelines were officially adopted by the Parties to the Convention as the common methodology for national inventories. The draft guidelines have been widely discussed and tested for some years by experts of many countries in order to achieve consensus about the methods. Based on this, IPCC revised these Guidelines in 1996. The Subsidiary Body for Scientific and Technical Advice (SBSTA) recommended this update to be used for inventories from industrialized countries. Good Practice Guidelines have also been prepared by the IPCC for the reporting of national greenhouse gas emissions and sinks under the Kyoto Protocol. As inventories are inherently uncertain, quality assessment and control of the annual inventories plays an important role. New IPCC Guidelines were released in 2006 (IPCC, 2006). In these guidelines, uncertainty management and quality control is an integral part of the inventory methodology. Emissions of CO_2 from fossil fuels can be quantified relatively easily using IEA statistics on energy and default emission factors, based on the carbon content of fuels.

Methane inventories

For CH_4, comprehensive inventory methods are currently at a relatively early stage of development, and results still have wide uncertainty ranges. Part of this problem is associated with the difficulty in translating local flux measurement results into emissions estimates for larger areas, such as countries or continents. Another part of the problem is related to the complexity of processes involved in biogenic production of CH_4, for example, by microorganisms in anaerobic soils (Chapter 2). Emissions are related to soil type and environmental conditions. Human interference with the soil system is influencing emissions. For example, flooding is known to have varying effects on the emissions of CH_4 depending on duration of flooding, temperature and soil carbon content (Chapter 8). Thus, because of the dependency of emissions on local climate, soil and management conditions, extrapolation of local emission results is difficult.

Uncertainty ranges in national inventories are about 5 to 10 per cent for carbon dioxide from fossil fuels, 50 to 100 per cent for CO_2 from land use-related sources and sinks, and 100 per cent for N_2O from soils. For CH_4 these uncertainties are also high, being at about 30 to 35 per cent for most sources. Emission inventories rely on statistical information and emission factors. Emission factors can be derived from field-scale measurements and appropriate methods for upscaling to the national levels. IPCC has made a great and commendable effort to develop IPCC Guidelines for National Emission Inventories over the last few decades. Many countries have started measurement campaigns for non-CO_2 greenhouse gases to reduce uncertainties. The uncertainties are likely to be reduced over the coming years in national inventories. Improvements can be made in the national inventories by measurements, improved statistics and better upscaling. Improved reporting and documentation may increase the confidence in the country estimates.

Van Amstel (2009) has compared the official CH_4 inventory estimates and the authoritative data source that is the EDGAR database, and found that the main reasons for differences therein were a result of the different emission factors and activity data used. Eventually, this kind of comparison will, we hope, contribute to the validation and verification of both national inventories and EDGAR, and so contribute to the improvement of methodologies to estimate CH_4 budgets.

Satellites
The European Space Agency launched Envisat on 1 March of 2002. The SCIAMACHY instrument on board Envisat shows real-time CH_4 concentration fields and profiles for the troposphere for the first time in history. The results can help provide a clearer picture in time and space of global CH_4 emissions and concentrations, and thus improve the 'a priori' estimates of modellers. A reduction in uncertainties has since been achieved because local measurements at ground stations can now be verified with measurements from space.

The future of methane and climate change

Throughout this book, the various chapter authors have attempted to provide a view of how CH_4 emissions may vary in the future in response to both human activities and to climate change. Of the responses to climatic change, elevated temperatures at high latitudes leading to enhanced CH_4 emissions from wetlands, and to the large amounts of CH_4 stored in clathrates potentially becoming unstable, would appear to represent the climate feedback of most concern and greatest uncertainty.

For human activities, it is apparent that there exist myriad opportunities for improved mitigation of CH_4 emissions in the coming years and decades. Van Amstel (2005, 2009) has conducted an integrated analysis of the impacts on 21st-century climate change that result from a scenario of unabated versus abated emissions of CH_4. The analysis was based on model runs by the IMAGE integrated assessment model. In the IMAGE model, a set of scenarios was developed in close cooperation with the IPCC to assist the climate negotiations for the Kyoto Protocol. The IMAGE model was used because it included information on major processes that determine uncertainties that are not included in other models. The analysis showed that CH_4 emissions could be reduced in the future at relatively low cost, while still playing a significant role in reducing climate change and sea level rise. By 2100, the analysis of CH_4 abatement alone (i.e. without mitigation of other greenhouse gas emissions) indicates that the projected temperature increase would be half a degree lower than that without CH_4 abatement, and that sea level rise would be reduced by four centimetres.

Conclusions

Significant reductions in global CH_4 emissions are both technologically feasible and, in many cases, very cost-effective strategies for climate change mitigation. Their wider implementation in coming years and decades will largely depend on the policy and market signals delivered by the UNFCCC Conference of the Parties in Mexico in 2010 and South Africa in 2011, but failing to make full use of the potential for CH_4 mitigation globally will inevitably make effective mitigation of climate change through reduction of CO_2 emissions alone all the more difficult. The scientific community can provide improved CH_4 flux estimates, reduce uncertainties and enhance our understanding of key climate change feedback mechanisms – such as CH_4 emissions from high latitude wetlands and clathrate deposits. The technology to deliver deep cuts in CH_4 emissions from a host of important sectors is already available. To put CH_4 mitigation at the heart of a robust and well-integrated framework for tackling global climate change, improved national and international policy is required to facilitate rapid technology transfer and provide the financial incentives that will ensure that the myriad potential opportunities for the effective mitigation of CH_4 emissions around the world are made real.

References

Amstel, A. R. van (2005) 'Integrated assessment of climate change with reductions of methane emissions', *Environmental Sciences*, vol 2, pp315–326

Amstel, A. R. van (2009) 'Methane: Its role in climate change and options for control', Masters dissertation, Wageningen University, Wageningen, The Netherlands

IPCC (Intergovernmental Panel on Climate Change) (2006) *2006 IPCC Guidelines for National Greenhouse Gas Inventories*, S. Eggleston, L. Buendia, K. Miwa, T. Ngara and K. Tanabe (eds), Institute for Global Environmental Strategies, Kanagawa, Japan, www.ipcc-nggip.iges.or.jp and www.ipcc.ch

Reilly, J., Prinn, R., Harisch, J., Fitzmaurice, J., Jacoby, H., Kicklighter, D., Melillo, J., Stone, P., Sokolov, A. and Wang, C. (1999) 'Multi-gas assessment of the Kyoto Protocol', *Nature*, vol 401, pp549–555

Contributors

Editors

Dave Reay, Senior Lecturer in Carbon Management, School of GeoSciences, Crew Building, The King's Buildings, West Mains Road, Edinburgh EH9 3JN, Scotland, UK. Tel: +44(0)131 6507723, Fax: +44(0)131 6620478, Email: David.Reay@ed.ac.uk

Pete Smith, Royal Society-Wolfson Professor of Soils & Global Change, Institute of Biological and Environmental Sciences, School of Biological Sciences, University of Aberdeen, Cruickshank Building, St Machar Drive, Aberdeen AB24 3UU, Scotland, UK. Tel: +44 (0)1224 272702, Fax: +44 (0)1224 272703, Email: pete.smith@abdn.ac.uk

André van Amstel, Assistant Professor, Department of Environmental Science, Environmental Systems Analysis Group, Wageningen University, Droevendaalsesteeg 4, 6708 PB Wageningen, The Netherlands. Tel: +31 317 484815, Fax: +31 317 419000, Email: andre.vanamstel@wur.nl

Chapter contributors

David E. Bignell, Professor of Zoology, School of Biological and Chemical Sciences, Queen Mary University of London, Mile End Road, London E1 4NS, UK. Tel: +44 (0)20 7882 3008, Email: d.bignell@qmul.ac.uk

Jean E. Bogner, President Landfills+ Inc., Landfills +, Inc., 1144 N. President, Wheaton, IL 60187, USA. Tel: 01-630-665-0872, Fax: 01-630-665-0826, Email: jbogner@landfillsplus.com

Torben R. Christensen, Professor, GeoBiosphere Science Centre, Lund University, Sölvegatan 12, 22362 Lund, Sweden. Email: torben.christensen@nateko.lu.se

Harry Clark, Climate Land & Environment Section Manager, AgResearch, Private Bag 11008, Palmerston North 4442, New Zealand. Tel: +64 6 351 8111, Email: harry.clark@agresearch.co.nz

Franz Conen, Research Fellow, Institute of Environmental Geosciences, Department of Geosciences, University of Basel, Bernoullistrasse 30, CH – 4056 Basel, Switzerland. Tel: +41 61 267 04 81, Email: franz.conen@unibas.ch

Hendrik Jan van Dooren, Wageningen UR Livestock Research, Edelhertweg 15, 8219 PH Lelystad, The Netherlands. Email: hendrikjan.vandooren@wur.nl

Miriam H. A. van Eekert, Senior Researcher LeAF (Lettinga Associates Foundation), Bomenweg 2, 6700 AM Wageningen, The Netherlands. Email: miriam.vaneekert@wur.nl

Giuseppe Etiope, Senior Researcher, INVG – Istituto Nazionale di Geofisica e Vulcanologia, Sezione Roma 2, Via Vigna Murata 605, 00143 Roma, Italy. Tel: +39 0651860394, Fax: +39 0651860338, Email: giuseppe.etiope@ingv.it

Åke Källstrand, Development Manager, MEGTEC Systems AB, Box 8063, SE-402 78 Gothenburg, Sweden

Francis M. Kelliher, Professorial Research Fellow, AgResearch, Lincoln Research Centre, Private Bag 4749, Christchurch 8140, New Zealand and Lincoln University, Soil and Physical Sciences Department, PO Box 84, Lincoln 7647, New Zealand. Email: Frank.Kelliher@agresearch.co.nz

Frank Keppler, Research Associate, Department of Atmospheric Chemistry, Max Planck Institute for Chemistry, Joh.-Joachim-Becher-Weg 27, Mainz 55128, Germany. Tel: +49 6131 305–316, Fax: +49 6131 305–511, Email: frank.keppler@mpic.de

Joel S. Levine, Senior Research Scientist, Science Directorate, NASA Langley Research Center, Hampton, Virginia 23681-2199, USA. Tel: 757-864-5692, Fax: 757-864-6326, Email: joel.s.levine@nasa.gov

Marjo Lexmond, Director, LeAF (Lettinga Associates Foundation), Bomenweg 2, 6700 AM Wageningen, The Netherlands and Wageningen University, sub-department of Environmental Technology, Bomenweg 2, 6700 AM Wageningen, The Netherlands. Email: marjo.lexmond@wur.nl

Richard Mattus, Managing Director, MEGTEC Systems AB, Box 8063, SE-402 78 Gothenburg, Sweden. Email: RMattus@megtec.se

Andy McLeod, Senior Lecturer, School of GeoSciences, Crew Building, The King's Buildings, West Mains Road, Edinburgh EH9 3JN. Tel: +44 (0) 131 650 5434, Fax: +44 (0) 131 662 0478, Email: Andy.McLeod@ed.ac.uk

Caroline M. Plugge, Assistant Professor, Laboratory of Microbiology, Wageningen University, Dreijenplein 10, 6703 HB Wageningen, The Netherlands. Tel: +31-317-483752, Fax: +31-317-483829, Email: Caroline.Plugge@wur.nl

Keith A. Smith, Senior Honorary Professorial Fellow, School of GeoSciences, Crew Building, The King's Buildings, West Mains Road, Edinburgh EH9 3JN, UK. Fax: +44 (0) 131 662 0478, Email: Keith.Smith@ed.ac.uk

Kurt Spokas, Soil Scientist, Soil and Water Management Research, University of Minnesota, 1991 Upper Buford Circle, Saint Paul, MN 55108-6024, USA. Tel: (612) 626-2834, Fax: (651) 649-5175, Email: Kurt.Spokas@ars.usda.gov

Alfons J. M. Stams, Professor of Microbiology, Laboratory of Microbiology, Wageningen University, Dreijenplein 10, 6703 HB Wageningen, The Netherlands. Tel: +31-317-483101, Fax: +31-317-483829, Email: fons.stams@wur.nl

Kazuyuki Yagi, National Institute of Agro-Environmental Sciences, 3-1-1 Kannondai, Tsukuba, Ibaraki 305, Japan. Email: kyagi@affrc.go.jp

Grietje Zeeman, Senior Researcher, LeAF (Lettinga Associates Foundation), Bomenweg 2, 6700 AM Wageningen, The Netherlands and Associated Professor, Wageningen University, sub-department of Environmental Technology, Bomenweg 2, 6700 AM Wageningen, The Netherlands. Email: Grietje.Zeeman@wur.nl

Acronyms and Abbreviations

AD	anaerobic digestion
AEEI	autonomous energy efficiency improvement
ALGAS	Asia Least Cost Greenhouse Gas Abatement Strategy
ALMA	airborne laser methane assessment
AR4	Fourth Assessment Report
BAU	business as usual
BIG/ISTIG	biomass integrated gasifier/intercooled steam-injected gas turbine
BMP	biochemical methane potential
BOD	biological oxygen demand
C	carbon
CBM	coal bed methane
CCN	cloud condensation nuclei
CDM	Clean Development Mechanism
CER	certified emission reduction
CH_4	methane
$CHCl_3$	chloroform
CHP	combined heat and power
CO	carbon monoxide
CO_2	carbon dioxide
CO_2-eq	carbon dioxide equivalents
COD	chemical oxygen demand
COS	carbonyl sulphide
CSTR	continuously stirred tank reactor
CV	coefficient of variation
DIAL	differential absorption lidar
DM	dry matter
DMI	dry matter intake
DS	dry solids
EF	emission factor
EGSB	expanded granular sludge bed
EMF21	Energy Modelling Forum 21
EPS	extra-polymeric substance
FAO	United Nations Food and Agriculture Organization

Fd	ferredoxin
FOD	first-order decay
FTIR	Fourier transform infrared
GCM	Global Circulation Model
GDP	gross domestic product
GE	gross energy
GWP	global warming potential
H_2	hydrogen
H_2S	hydrogen sulphide
H_4MPT	tetrahydromethanopterin
HRPM	horizontal radial plume mapping
HRT	high-rate tank
HS-CoM	coenzyme M
HS-CoB	coenzyme B
IBP	International Biological Program
IDW	inverse distance weighting
IEA	International Energy Agency
IPCC	Intergovernmental Panel on Climate Change
kJ	kilojoule
LEL	lower explosion limit
LPG	liquefied petroleum gas
LW	live weight
MCF	methane correction factor
ME	metabolizable energy
MEGAN	Model of Emissions of Gases and Aerosols from Nature
MFR	methanofuran
Mha	million hectare
N	nitrogen
N_2O	nitrous oxide
ng	nanogram
NGGIP	National Greenhouse Gas Inventories Programme
NH_3	ammonia
nm	nanometre
NMHC	non-methane hydrocarbon
NPP	net primary productivity
OECD	Organisation for Economic Co-operation and Development
OH	hydroxyl
P	phosphorus
Pa	pascal
ppb	parts per billion
ppm	parts per million
ROS	reactive oxygen species
SBSTA	Subsidiary Body for Scientific and Technical Advice
SCIAMACHY	scanning imaging absorption spectrometer for atmospheric chartography

SO_2	sulphur dioxide
SO_4^-	non-volatile sulphate
SRT	slow-rate tank
T	temperature
TFI	Task Force on National Greenhouse Gas Inventories
Tg	teragram
TPM	total particulate matter
TPS	total petroleum system
UASB	upflow anaerobic sludge blanket
UNEP	United Nations Environment Programme
UNFCCC	United Nations Framework Convention on Climate Change
US EPA	United States Environmental Protection Agency
UV	ultraviolet
V	volume
VAM	ventilation air methane
VES	Veolia Environmental Services
VOC	volatile organic compound
VRPM	vertical radial plume mapping
VS	volatile solids
WestVAMP	West Cliff Colliery Ventilation Air Methane Plant
WHO	World Health Organization
WMO	World Meteorological Organization
WMX	Waste Management, Inc.

Index

Abies lasiocarpa (subalpine fir) 85
abiogenic seepage 49
acetate 17–18, 118
Achillea millefolium (yarrow) 78
AEEI (autonomous energy efficiency improvement) 212
aerobic methane formation 74, 78, 83, 87, 91–92
aerobic waste treatment 152, 153, 160
agricultural waste burning 227, 232
agriculture 127, 151
    see also biomass burning; manure; rice cultivation; ruminants; waste
air quality 187
air treatment, manure 160
Akiyama, H. 125
alga (*Chlamydomonas reinhardtii*) 79
Alnus glutinosa (black alder) 87–88
America see US
amino acids, mineralization 20
anaerobic digestion 10, 152, 166
    see also methanogenesis
anaerobic food chain 23–24
anaerobic mineralization 14, 20, 28
    see also methanogenesis
anaerobic waste treatment 151–170
    anaerobic digestion 10, 152, 166
    biogas production 152–153, 158
    emission mitigation 158
    human waste 162
    manure 10, 156–160, 168–169, 223–224
    sludge 165–167
    technology 153–156
    wastewater 10, 160–165, 169–170
Angelidaki, I. 159
animals see ruminants
anthropogenic sources, methane 6, 8–11, 151
antibiotics, ruminants 146

ants 62
    see also termites
Apicotermitinae 64
Apium graveolens (celery) 88
Arabidopsis thaliana (thale cress) 79
archaea 14–18
arctic tundra 32
Argentina 137
Arrhenius, S. 27
Artemesia absinthum (wormwood) 78
Asia 106–110, 116, 170
atmospheric concentrations, methane 1–2, 242
Australia 137, 140, 205, 206–207
autonomous energy efficiency improvement (AEEI) 212

Barlaz, M. 177
barley (*Hordeum vulgare*) 80
Bartlett, K. B. 32
basil (*Occimum basilicum*) 88
Bastviken, D. 32
batch reactors 154
Beauchemin, K. A. 146
beef cattle (*Bos indicus*; *Bos taurus*) 140, 141
Beerling, D. J. 76
Bergamaschi, P. 84
Betula populifolia 79
BIG/ISTIG (biomass integrated gasifier/intercooled steam-injected gas turbine) technology 227
Bignell, D. E. 63, 66
biocovers, landfills 186
biogas
    anaerobic treatment advantage 152–153
    composition 152
    as energy source 167, 170
    manure 158, 224

production 168–169
wastewater 225
biomass 98, 152, 158
biomass burning 8–9, 97–111
   climate change 102, 103, 110
   combustion 98–101
   emission estimates 110–111
   emission mitigation 227, 232
   emissions calculations 104–106
   geographical distribution 100–104
   global impacts 98–100
   Southeast Asia (case study) 106–110
biomass integrated gasifier/intercooled steam-injected gas turbine (BIG/ISTIG) technology 227
biomass loading 108
black alder (*Alnus glutinosa*) 87–88
Black Sea 51
Blok, K. 214, 215
Boea, K. 169
Bogner, J. 178
Boone, D. R. 21
boreal forests 85, 97, 101–104
Bos indicus (beef cattle) 140, 141
Bos taurus (beef cattle) 140, 141
Bousquet, P. 32, 34
Bowling, D. R. 85
Brassica napus (canola) 80
Brauman, A. 64
Brazil 137, 162, 201
breeding animals, ruminants 147
Breznak, J. A. 63
Britain *see* UK
broadleaf (*Griselina littoralis*) 78
Brown, J. H. 141
Brüggemann, N. 90
Bruhn, D. 79–80, 89
Brune, A. 63
Bryant, M. P. 21
Butenhoff, C. L. 81, 82, 83
butyrate 21, 22
Byfield, S. 227

California, US 190
Canada 140
canola (*Brassica napus*) 80
canopy flux 83–86
Cao, G. 86–87
carbon, biomass 98, 101, 104
carbon dioxide (CO_2)
   aerobic waste treatment 153
   atmospheric concentration 212
   biomass burning 100
   landfills 185–186, 192
   termites 62
carbon dioxide equivalents 3
carbon sequestration 91
cattle 137, 139, 140, 141, 145–146
see also ruminants
CBM (coal bed methane) 203, 219, 220
CDM (Clean Development Mechanism) 189
celery (*Apium graveolens*) 88
chemicals, ruminants 146
China 101, 137, 201, 207, 220
Chlamydomonas reinhardtii (alga) 79
CHP (combined heat and power) generators 167
Christensen, T. R. 32
Cicerone, R. J. 5
Clark, H. 137, 139, 140
clathrates *see* hydrates
Clean Development Mechanism (CDM) 189
climate change
   biomass burning 102, 103, 110
   feedback mechanisms 6, 245
   higher temperatures 4
   IPCC 212
   mitigation 6, 12, 203, 211, 242, 246
   radiative forcing 1, 6, 35, 62, 211
   wetlands 7, 29, 34, 35
Clostridium 20
clouds 100
Clymo, R. S. 28
coal bed methane (CBM) 203, 219, 220
coal beds 49, 203
coal mining 11, 49, 203, 219–220
co-digestion, manure 158, 168
collision model, landfills 190
Colombia 162
combined heat and power (CHP) generators 167
combustion, biomass 98–101
composting 160, 166–167, 226
conifers 85
Conrad, R. 118
continuously stirred tank reactors (CSTR) 154
corn (*Zea mays*) 76, 79
costs, emission mitigation 211, 213–216, 228–238, 239

cover materials, landfills 176
cows *see* dairy cattle
Cox, P. M. 35
Crill, P. M. 32
Crutzen, P. J. 5, 84, 97, 104
CSTR (continuously stirred tank reactors) 154
Cubitermes heghi 64
Czepiel, P. M. 176

daily spreading, manure 223
dairy cattle 145–146
dandelion flowers (*Taraxacum officinale*) 78
deer *see* ruminants
deforestation 227
degasification, mining 204, 220
degassing 42
de Jager, D. 214, 215
Denier van der Gon, H. A. C. 226
Denman, K. L. 31
Denmark 169
developed countries 188, 189
developing countries 176, 189, 192, 215, 232
De Visscher, A. 189
diet manipulation, ruminants 146
digestion, anaerobic 10, 152, 158, 166
Do Carmo, J. B. 84
domestic wastewater 161–162
Donoso, L. 86
dry seeps 44, 46, 50
dry storage, manure 223
Dueck, T. A. 75–76

earth observations, satellite 83–84, 102, 245
Eastern Europe 220
Eaton, P. 97
economic potential, emissions mitigation 238
efficiency improvements 213
EGSB (expanded granular sludge bed) reactors 154
Ehhalt, D. H. 27
electricity 167
EMF21 (Energy Modelling Forum 21) 214, 215
emission estimates
  biomass burning 110–111
  geological methane 52–56
  landfills 188
  rice cultivation 115
  ruminants 137
  termites 63–64, 66, 68
  vegetation 74, 80–87
  wetlands 31–33
emission inventories 127–128, 188–189, 212, 244–245
emission mitigation strategies 11, 211–239
  anaerobic waste treatment 158
  biomass burning 227, 232
  climate change 6, 12, 203, 211, 242, 246
  coal mining 219–220
  costs 211, 213–216, 228–238, 239
  economic potential 238
  efficiency improvements 213
  fossil energy 203–204, 233
  importance of 246
  landfills 176, 190, 192, 225–226, 232
  manure 159–160, 223–224, 233
  natural gas 216–219
  oil 216
  rice cultivation 226–227, 233
  ruminants 144–147, 151–152, 221–222, 233
  scenarios 228–231
  stabilization wedges 212
  technical reduction potential 216–227
  types 214
  ventilation air methane 205, 209
  volume measures 213
  wastewater 224–225, 232
emission reduction *see* emission mitigation strategies
emission trading schemes 144
empirical models, landfills 190
energy generation
  autonomous energy efficiency improvement 212
  cost effectiveness 211
  fossil fuels 11, 201–203, 205
  landfills 175, 176
  VAM 205–209
  wastewater 167, 170
Energy Modelling Forum 21 (EMF21) 214, 215
Engelmann spruce (*Picea engelmannii*)

85
enteric methane 136–144, 151, 221–222
see also ruminants
Envisat (satellite) 245
Ethiopia 137
Etiope, G. 47, 48, 51, 53, 54
Europe 220, 221
everlasting (eternal) fire seeps 46
expanded granular sludge bed (EGSB) reactors 154

faba bean (*Vicia faba*) 80
fats, mineralization 20–21
feedback mechanisms, climate change 6, 245
feeding, ruminants 140–144, 221
fermentation 18–19, 226
Ferretti, D. F. 83
fires 101–103
see also biomass burning
fire seeps 46
first-order kinetic models (FOD) 188, 189
Fittkau, E. J. 62
five finger (*Pseudopanax arboreus*) 78
flaring 167–168, 169, 202, 216, 218
FOD (first-order kinetic models) 188, 189
forests 65–66, 84, 85, 97, 101–104, 107
former Soviet states 218
see also Russia
fossil energy 11, 201–204, 233
fossil methane *see* geological methane
France 180–182
Frankenberg, C. 82, 83–84
Fraxinus mandshurica 88
Frenzel, P. 120
Furyaev, V. V. 97

Gallaher, M. P. 214
gas exchange chambers 86–87
gas hydrates *see* hydrates
gasification, waste 226
gas recovery, landfills 176, 225–226
Gebert, J. 176
Gedney, N. 35
geological methane 7, 42–56
emission estimates 52–56
emission factors 50–52
sources 43–49
term 49

geothermal emissions 43, 48, 51–52
Germany 169, 220
global methane budget 4–11
global warming *see* climate change
global warming potential (GWP), methane 2–3, 125, 175, 209, 242
Glyceria spiculosa 78
goats *see* ruminants
gob-well recovery, coal mining 220
Goldammer, J. G. 97
Gottschal, J. C. 20
grass communities 86
Great Britain *see* UK
greenhouse gas emissions
agriculture 127, 151
assessments 127–128
biomass burning 99, 100
inventories 127–128, 188–189, 212, 244–245
Kyoto Protocol 189
natural sources 242
time horizons 3
UK 1
UNFCCC 137, 139, 188, 212, 242, 243
see also carbon dioxide, nitrous oxide; emission estimates; emission mitigation strategies; methane
grey poplar (*Populus* × *canescens*) 90
Griselina littoralis (broadleaf) 78
Gröngröft, A. 176
Güenther, A. B. 81–83
GWP (global warming potential), methane 2–3, 125, 175, 209, 242

Hackstein, J. H. 63
Halvadakis, C. P. 177
Hammond, K. J. 139
Harmelink, M. G. M. 214
Harriss, R. C. 32
health problems 100
Heimann, M. 35
Helianthus annuus (sunflower) 80
high-rate anaerobic waste treatment 154
Hilger, H. A. 189
Holland *see* Netherlands
Hordeum vulgare (barley) 80
hormones, ruminants 221
hot spots, landfills 180
Houweling, S. 83
Hovland, M. 49

Huang, Y. 124
human population 10
human waste 162
Hunt, T. S. 27
hydrates 7, 42, 216–217
hydrogenotrophic methanogenesis 15–17
hydrophytes 78
hydroxyl (OH) radicals 4, 103

IBP (International Biological Program) 28
ice cores 83
IMAGE integrated assessment model 245
incineration, waste 226
Incisitermes minor 64
India 123, 137, 162
Indonesia 106–110
industrial wastewater 163–165
Innes, J. L. 97
Intergovernmental Panel on Climate Change (IPCC) 3, 127–128, 188–189, 212, 244, 245
intermittent irrigation, rice cultivation 124
International Biological Program (IBP) 28
inventories, emissions 127–128, 188–189, 212, 244–245
IPCC (Intergovernmental Panel on Climate Change) 3, 127–128, 188–189, 212, 244, 245
IPCC-NGGIP (IPCC National Greenhouse Gas Inventories Programme) 188–189, 212, 244, 245
irrigated rice 121, 124
isotopic studies 89–90, 120–121, 185

Johnson, D. E. 140
Judd, A. G. 49
Jugositermes tuberculatus 64

Kalimantan, Indonesia 106, 107, 108–110
Kasischke, E. S. 97, 103
Keppler, F. 74, 75, 82, 83, 88, 89, 91
Khalil, M. A. K. 81, 82, 83
Kim, J. 166
Kirschbaum, M. U. F. 78–79, 81, 83, 88
Kleiber, M. 141

Klinge, H. 62
Klusman, R. W. 47, 54
Kobresia 87
Kobresia humilis 86
Krüger, M. 120
Kyoto Protocol 189, 212, 242, 243, 245

Lacroix, A. V. 54
lakes 32, 34
lambs 142–144
    see also ruminants; sheep
landfills 11, 175–193
    emission estimates 188
    emission mitigation 176, 190, 192, 225–226, 232
    field measurements 178–183
    models 187–190
    sludge treatment 165
Latin America 170
Leadbetter, J. 63
leakage (emissions displacement) 144
leaks, natural gas 202, 218–219
Levine, J. S. 97
Liew, S. C. 107
Limpens, J. 35
liquefaction 218
livestock emissions 151
    see also manure; ruminants
Lobert, J. M. 98
lodgepole pine (*Pinus contorta*) 85
Lophotermes septentrionalis 64
low-rate anaerobic waste systems 154

MacDonald, J. A. 66
macroseeps 44, 46, 50–51, 55
Mahieu, K. 189
maintenance, natural gas 218
Makarieva, A. M. 141
manure 10
    anaerobic waste treatment 156–160, 168–169
    emission mitigation 159–160, 223–224, 233
    see also sludge
Marik, T. 120
marine environment 47–48
marine seepage 43
Martius, C. 66
Matsuo, T. 20
McInerney, M. J. 22
McLeod, A. R. 76–78, 89, 90

Meadows, M. P. 232
meat, demand for 145
Megagnathotermes sunteri 64
MEGAN (Model of Emissions of Gases and Aerosols from Nature) 81–82
Megonigal, J. P. 81–83
MEGTEC Systems 205
Melack, J. M. 33
Messenger, D. J. 90
methane (CH_4)
　anthropogenic sources 6, 8–11, 151
　assessments 127–128
　atmospheric concentrations 1–2, 242
　atmospheric lifetime 3, 103, 175, 203, 209, 242
　global budget 4–11
　global variations 29
　GWP 2–3, 125, 175, 209, 242
　natural sources 6–8
　see also biogas; emission mitigation strategies; methane oxidation; methanogenesis; sinks; *specific sources*
methane oxidation
　landfills 176, 183–186, 190
　rice cultivation 120
　soils 65, 66
　VAM 205
　vegetation 89, 90–91
　wetlands 28
Methanobacteriales 14
Methanococcales 15
methanogenesis 14–24
　archaea 14–18
　fermentation 18–19
　general pathway 15, 16
　isotopic studies 89, 120–121
　landfills 177
　mineralization 20–21
　pH, effect of 156
　rice cultivation 117–121
　significant sources 5–6
　substrates 15–18
　syntrophic degradation 21–22
　temperature 28
　wetlands 28
methanogenic archaea 14–18
methanogens *see* methanogenic archaea
Methanomicrobiales 15
Methanopyrales 15
Methanosaeta 17, 18

Methanosarcina 17, 18
Methanosarcinales 15, 17
Methanosphaera 17
methanotrophs 90–91, 120
methanotrophy 28, 89
　see also methane oxidation
methyl bromide 100
methyl chloride 100
methyl-containing compounds 17
Mexico 137, 162
microseepage 46–47, 50, 53–54, 55
Middle East 216, 218
Milkov, A. V. 53
milk production 145–146
Miller, J. B. 85
mineralization 14, 20–21, 28
mitigation *see* emission mitigation strategies
Model of Emissions of Gases and Aerosols from Nature (MEGAN) 81–82
modelling
　emission mitigation costs 214, 215, 245
　landfills 187–190
　vegetation 81–82
　wetlands 35, 36
modern methane 49
monensin 146
mud volcanoes 44, 50, 51, 53

Nagase, M. 20
Nanninga, H. J. 20
national greenhouse gas inventories 127–128, 188–189, 212, 244, 245
National Greenhouse Gas Inventories Programme (IPCC-NGGIP) 188–189, 212, 244, 245
natural gas 201–202, 216–219
natural sources, methane 6–8
neotectonics 56
Netherlands 158, 168, 169, 170, 188
New Zealand 138, 139, 140, 145–146
Nichol, J. 108
Nicotiana tabacum (tobacco) 76, 78
Nigeria 216
Nisbet, R. E. R. 79, 88
nitric acid 99
nitrogen 98, 120, 192
nitrogen fertilizer 125
nitrous oxide (N_2O) 125, 151, 153,

186–187
'no-regret' measures, mitigation 211
North America *see* US
northern wetlands 32, 33, 34
nucleic acids 20
nutrients, rice cultivation 122–123

Occimum basilicum (basil) 88
oil-related emissions 202–203, 216
onshore mud volcanoes 44
on-site use, oil and gas production sites 218
Oremland, R. S. 5
organic amendments, rice cultivation 121, 122–123
Oryza sativa (rice) 79, 123–124
'other seeps' 44
overloading, anaerobic waste treatment 156
oxidation *see* methane oxidation
ozone 99, 100

Pacala, S. 212
Pakistan 137
Parsons, A. J. 81, 83
particulates 99–100
pea (*Pisum sativum*) 80
pectin 78, 79, 80, 89, 90
petroleum basins 47
petroleum geology 44
phage therapy, ruminants 147
pH, methanogenesis 156
Picea abies 79
Picea engelmannii (Engelmann spruce) 85
Pinus contorta (lodgepole pine) 85
pipeline modernization, natural gas 218
Pisum sativum (pea) 80
plant-mediated emissions 87–88
plant physiology, rice 123–124
plants *see* vegetation
plug flow reactors 154
poplar trees 75, 90
Populus x *canescens* (grey poplar) 90
Portugal 162
Potentilla 86, 87
Potentilla fruticosa 86
process-based models, landfills 190
Procubitermes arboricola 64
production efficiency, ruminants 145, 221

production-enhancing agents, ruminants 221
propionate 21
propionate metabolism 21–22
Pseudopanax arboreus (five finger) 78
Purdy, K. J. 63

Qaderi, M. M. 80, 90

radiative forcing 1, 6, 35, 62, 211
Radojevic, M. 97
rainfed rice 121, 124
Rannaud, D. 189
reactive oxygen species (ROS) 90
Reay, D. 5
recent gas *see* modern methane
recompression, natural gas 218
Reddaway, E. J. F. 28
Reeburgh, W. S. 32
Reid, D. M. 80, 90
Rennenberg, H. 88
residue management, rice cultivation 129
respiratory problems 100
rice (*Oryza sativa*) 79, 123–124
rice cultivation 9–10, 115–129
  emission estimates 115
  emission mitigation 226–227, 233
  isotopic studies 120–121
  nutrients 122–123
  organic matter 121, 122–123
  plant physiology 123–124
  production 116–117
  residue management 129
  water management 124–125, 129
Röckmann, T. 91
ROS (reactive oxygen species) 90
Rouland, C. 64
rumen efficiency 222
ruminants 10, 136–147
  emission estimates 137
  emission mitigation 144–147, 151–152, 221–222, 233
  emissions 137–140
  feed intake 140–144
Ruminococcus albus 18
Rusch, H. 87–88
Russia 102, 137, 202, 220

salt tectonics 56
Sanderson, M. G. 63, 66

Sanhueza, E. 86
satellites 83–84, 102, 245
savannas 84, 97, 102, 227, 232
scenarios 228–231
Scharffe, D. 84
Scheutz, C. 178, 189
Schneising, O. 84
Scirpus yagara 78
Sebacher, D. I. 32
sedimentary seepage 43–44
seepage 42
    abiogenic 49
    marine 43
    microseepage 46–47, 50, 53–54, 55
    sedimentary 43–44
Seiler, W. 104
seismicity 56
septic tanks 154, 169
sewage *see* wastewater
Sharpatyi, V. A. 90
sheep 140
    see also lambs; ruminants
Shindell, D. T. 35
Simpson, I. J. 97
Sinha, V. 85
sinks 4, 65, 66
sludge 165–167
    see also manure
slurry *see* manure
Smil, V. 145
smog 100
Socolow, R. 212
soils 4, 28, 65, 66, 190
South America *see* Latin America
Southeast Asia 106–110
Soviet states, former 218
Spokas, K. 189
stabilization wedges 212
stable isotope analysis 89–90, 185
Steinfeld, H. 137
Stickland reaction 20
Stocks, B. J. 97
stratosphere 4, 103
straw burning, rice cultivation 123
subalpine fir (*Abies lasiocarpa*) 85
submarine emissions 47–48, 51, 52
substrates 15–18, 155–156
sugars 18–19
Sugimoto, A. 63, 66, 67, 120
sulphate 118, 177
sulphate-reducing bacteria 24
sulphur 98
Sumatra, Indonesia 106, 107, 108–110
sunflower (*Helianthus annuus*) 80
Svensson, B. H. 28
Switzerland 168
syntrophic degradation 21–22
Syntrophobacter fumaroxidans 22
Syntrophomonas 21
Syntrophomonas wolfei 22
Syntrophus 21

Tanaka, S. 166
Taraxacum officinale (dandelion flowers) 78
Task Force on National Greenhouse Gas Inventories (TFI) 127–128
temperatures
    anaerobic waste treatment 156
    high latitudes 4
    methanogenesis 28
    wetlands 6, 27, 29
Termes 64
termites 7–8, 62–69
Termitidae 64
Termitinae 64
TFI (Task Force on National Greenhouse Gas Inventories) 127–128
thale cress (*Arabidopsis thaliana*) 79
thaw lakes 34
Thielmann, T. 49
time horizons, greenhouse gas emissions 3
tobacco (*Nicotiana tabacum*) 76, 78
Trachypogon 86
Triticum aestivum (wheat) 76, 80
tropical forests 65–66, 84, 85, 97, 107
tropical wetlands 33, 34
tropics 97
troposphere 102, 103
tundra 32
Tyndall, J. 27

UASB (upflow anaerobic sludge blanket) reactors 154
UK (United Kingdom) 1, 205
uncertainty ranges, emission inventories 244–245
UNEP (United Nations Environmental Programme) 107
UNFCCC (United Nations Framework Convention on Climate Change) 137,

139, 188, 212, 242, 243
United Kingdom *see* UK
United Nations Environmental Programme (UNEP) 107
United Nations Framework Convention on Climate Change (UNFCCC) 137, 139, 188, 212, 242, 243
United States *see* US
upflow anaerobic sludge blanket (UASB) reactors 154
upland rice 121
US (United States)
　coal mining 220
　fires 103
　fossil energy 201, 202, 203
　landfills 176, 180, 182–183, 190
　rice cultivation 124
　ruminants 137, 140, 221
　VAM processing 207

vaccines, ruminants 147
VAM (ventilation air methane) 204–209, 220
van Amstel, A. R. 211, 212, 245
van Wilgen, B. W. 97
vegetation 8, 79–92
　aerobic methane formation 74, 78, 83, 87, 91–92
　emission estimates 74, 80–87
　experimental laboratory studies 75–80
　field studies 86–87
　plant-mediated emissions 87–88
　verification 89–90
ventilation air methane (VAM) 204–209, 220
venting 218
　see also flaring
Vicia faba (faba bean) 80
Vigano, I. 76, 89–90
VOCSIDIZER technology 205–207
volcanoes 43, 48, 51
volume measures 213

Wada, E. 120
Walcroft, A. 78–79

Walter, B. P. 35
Walter, K. M. 32
Wang, S. 87
Wang, Z. P. 78, 90
Wania, R. 35
Ward, G. M. 140
Wassmann, R. 124
waste gas 218
waste management practices 192, 226
　see also landfills
waste treatment *see* anaerobic waste treatment
wastewater 10
　anaerobic waste treatment 160–165, 169–170
　emission mitigation 224–225, 232
water management, rice cultivation 124–125, 129
water seeps 44
wedges, stabilization 212
West, G. B. 141
WestVAMP, Australia 205, 206
West Virginia, US 207
wetlands 6–7, 27–36
　climate change 7, 29, 34, 35
　emission estimates 31–33
　processes 28–31
　seasonal dynamics 33–34
Whale, S. C. 32
wheat (*Triticum aestivum*) 76, 80
Wheeler, G. S. 64
willow trees 75
Wilshusen, J. H. 189
World Health Organization (WHO) 107
World Meteorological Organization (WMO) 107
wormwood (*Artemesia absinthum*) 78

xerophytes 78

Yan, X. Y. 124, 125
Yao, H. 118
yarrow (*Achillea millefolium*) 78

Zea mays (corn) 76, 79
Zimmerman, P. R. 63

For Product Safety Concerns and Information please contact our
EU representative GPSR@taylorandfrancis.com Taylor & Francis
Verlag GmbH, Kaufingerstraße 24, 80331 München, Germany